BLOWN TO
HELL

BLOWN TO
HELL

America's Deadly Betrayal
of the Marshall Islanders

WALTER PINCUS

DIVERSION
BOOKS

For more information, email info@diversionbooks.com

Diversion Books
A division of Diversion Publishing Corp.
www.diversionbooks.com

First Diversion Books edition, November 2021
Hardcover ISBN: 9781635768015
eBook ISBN: 9781635768022

Printed in The United States of America
1 3 5 7 9 10 8 6 4 2
Library of Congress cataloging-in-publication data is available on file

To Ann, who has been the inspiration for whatever good I have done, and to my children and grandchildren in hopes they never see nuclear weapons ever used.

CONTENTS

PROLOGUE

BEFORE DAWN ON March 1, 1954, John Anjain got up from his palm-mat bed and walked out of the plywood, thatched-roof house he shared with his wife and their five children on the Pacific Ocean atoll of Rongelap. At a clearing nearby, he began to lay out rows of coconut pieces; after a day baking in the sun these pieces would become *copra*, Rongelap's chief cash crop.

Anjain, thirty-two, was a quiet, unassuming man, well-liked and respected as magistrate of the atoll. He was slight of stature, brown-skinned, square-faced with his black hair, already touched with gray, pushed back in a slight pompadour.

Only a few could have known that this innocent, humble father and his family, including his then one-year-old son Lekoj, would have their simple lives imperiled by deadly radioactive fallout from a hydrogen bomb test that morning some 120 miles away, detonated by American scientists and military personnel.

Rongelap is one of the twenty-nine atolls and five islands that make up the Marshall Islands, which lie 2,100 miles southwest of Hawaii and 2,400 miles east of Japan.

An atoll is a circular coral reef surrounding a lagoon; the Rongelap Atoll has 61 tiny islets forming a more square than circular chain around a 388-square-mile pristine blue-green lagoon. A smaller atoll, named Ailinginae, eight miles to the west, was generally uninhabited, but the Rongelap people often went there to gather crabs, fish, and turtle eggs.

The islands were originally settled two to three thousand years ago by people who sailed east from Southeast Asia. They first landed in Yap, an island west of the Philippines. Then some moved south to New Guinea, others east to the Solomon Islands, and finally northeast to the Marshalls.

After Spanish explorers in the sixteenth century visited Rongelap and nearby atolls, only whalers and traders passed through that part of the Pacific until missionaries arrived in the nineteenth century. Local chieftains controlled the atolls, with the current Kabua family line "owning" Rongelap going back to the 1860s. In 1885, Germany purchased the Marshall Islands from Spain and established a trading outpost, mainly for copra.

Japan, allied with Great Britain in World War I, won the Marshalls as its price for entering the fight. In the years that followed, Japanese companies developed the copra trade while Japan's military developed Kwajalein Atoll as a military base. In January 1944, after months of fighting, American World War II forces took control of the Marshalls. When the war ended, the US maintained governance under the auspices of the United Nations as part of the Trust Territory of the Pacific Islands with an American high commissioner serving from within the Interior Department's Office of Insular Affairs.

The first American tests of atomic weapons in the Marshall Islands—Operation Crossroads—began in 1946 in Bikini Atoll, 125 miles west of Rongelap. Tests then transferred to the US mainland, at the Nevada Test Site, before being moved back to the Marshalls because the new generation of thermonuclear bombs being developed had yields so large that officials feared radioactive fallout could not be safely contained at any site in the United States.

■   ■   ■

On March 1, 1954, eighty-two men, women, and children were living on Rongelap Island—the largest island in that atoll, although only a half-mile wide and six
miles long.* Their houses were made mostly of palm leaves with coral-pebble
floors, but a few, such as John Anjain's, were built partly of wood. Some houses
had connecting sheds where cooking was done. Windows and doors were covered with burlap or canvas.

The men of the atoll fished the Rongelap Lagoon in wooden outriggers or
along the coral reef several days a week, catching sturgeon and parrot fish. Now
and then they dug arrowroot, a plant with tuberous roots that cook into a dish
resembling tapioca. Whenever they could, they collected fresh coconuts—on
Rongelap, on nearby islands around the lagoon, and on the islands of Ailinginae
Atoll. Coconut meat made up a major part of the Rongelapese diet, and dried
it became copra, the atoll's major export.

Coconut milk supplemented the rainwater they collected and stored in eight
communal cisterns built by the Japanese during their occupation.

The women spent their days gathering food, cooking, and caring for children. They also looked after the cultivation of taro; in a tropical climate taro
plants grow from three to five feet high, with starchy stems that are like mac-
aroni when cooked; the peeled roots, when blanched and sun-dried for a day,
can be milled into flour for bread and doughnuts. Another important food was
pandanus, a stringy fruit the size of a pineapple, which grows on tuberous trees
found all over the Marshall Islands. Pandanus fruit has a sweet juice or thick
nectar that tastes like a mix of sugar cane and mango.

Also on Rongelap Island were 150 chickens, pecking out their food from
tossed-aside pieces of coconut. Their eggs, eaten raw, were considered a delicacy. There were also cats and ducks and ten swine. On special occasions, a
whole pig was cooked for an island feast.

Once every three or four months, a trading ship run by the US Trust Territory government arrived to buy copra, in return for which the Rongelapese
bought rice, flour, and such "luxuries" as canned meats, lemonade powder, and

---

* That March 1, morning, eighteen of these Rongelapese were at neighboring Ailinginae
  Atoll gathering food.

jelly. The children gave the ship's crew and passengers—mostly Marshallese traveling to or from their home atolls—little shells in exchange for gum and candy. The women also sold bags and mats woven from coconut or pandanus leaves, and used the money earned to buy plastic combs and cheap Japanese cloth for dresses. The men traded enormous clam and conch shells from the lagoon for liquor and cigarettes.

While a ship was anchored in the lagoon, all work stopped. But within days of its departure, the new goods were gone, the excitement was forgotten, and the Rongelapese returned to their simple daily struggle to gather food.

■   ■   ■

Eight years earlier, in May 1946, the Americans had temporarily moved the inhabitants of Rongelap to Lae, an island almost three hundred miles away from Bikini Lagoon. The very first atomic tests at Bikini Atoll were to take place that July.

The evacuation was a precaution the Americans followed for all Marshallese living within three hundred miles of Bikini. When Operation Crossroads was concluded in August and the danger of radiation fallout had passed, the Rongelapese had been returned to their home atoll.

Life returned to normal until January 1954, when twenty-eight American military personnel (twenty-five Air Force and three Army) arrived at the Rongerik Atoll, forty-five miles east of Rongelap, to operate a weather station. Meanwhile, a Trust Territory field officer from Hawaii arrived in Rongelap by ship and told John Anjain that the Americans would soon again be testing atomic bombs at Bikini.

Although the tests were of much more powerful nuclear devices than those detonated in 1946, the field officer told Anjain there were no orders to move the inhabitants of Rongelap. For the next six weeks, Anjain and the other Rongelapese went about their daily routines.

■   ■   ■

Just after 3 a.m. on March 1, 1954, some 90 miles west of Bikini, the 23-man crew of the Japanese fishing boat *Lucky Dragon #5* was setting out fishing lines for their final attempt to catch tuna before heading back to their home port of Yaizu.

When the *Lucky Dragon* had left Japan on January 22 for this, its fifth voyage as a tuna boat, her regular skipper, Captain Shimizu, was not aboard. He had to stay in Yaizu for a hemorrhoid operation, so the boat's owner, a businessman named Kakuichi Nishikawa, had selected as his replacement twenty-two-year-old Hisakichi Tsutsui, a newly licensed captain, recently graduated from the Aichi Prefectural Fishery School. The man who really took command, however, was Yoshio Misaki, a twenty-nine-year-old who had the requisite experience to have the title of fishing master. Just before the ship cast off, the boat owner huddled with Misaki and directed him to fish around the Midway Islands, some two thousand miles east, where the seas were rougher than those farther south, but the possibilities for catching large, big-eye tuna were considered better.

The first sign of trouble came just hours after departure, when the crew found they had left behind a spare crank—a key part for the engine. Instead of returning to Yaizu, which would be bad luck, the fishermen went to Ogawa, a nearby port, to buy the part. After loading the extra crank and attempting to depart, the *Lucky Dragon* ran aground on a sandbar, leaving no choice but to wait for the high tide and a final departure four hours later.

For its first five days out at sea, the *Lucky Dragon* was caught in winter storms and several times almost swamped by giant waves. It was not until after the boat had passed Torishima Island, about 370 miles southwest of Yaizu, that Misaki told the captain and the rest of the crew that they would be heading more or less directly east for Midway, amid rougher seas than the calmer southern waters where the *Lucky Dragon* had fished before. The crew was not happy, but they were somewhat cheered when Misaki pointed out that near Midway they had a good chance for landing tuna, popular with Americans. Their pay depended on the size and quality of the catch.

The first fishing day was February 7, when the *Lucky Dragon* was about two hundred miles southwest of Midway. The tricky process of letting out the fishing lines usually began around 3 a.m., when the first twelve-inch floating buoy with a light on top is pushed off the stern with a line attached that drops down vertically into the sea. Stringing out horizontally from it, as the ship moves forward, is the so-called long line to which, every one thousand feet, are attached five branch lines, each with a steel leader and a hook baited with frozen fish; another surface-floating buoy is attached with a light atop every twentieth of them. To release the *Lucky Dragon's* 330 lines generally took about four hours, and the buoys could stretch out to almost fifty-five miles. When that first big chore was finished, the ship halted, and the crew took breakfast.

Four hours later, the even harder job of hauling in the lines began. A power winch pulled in the main line as the boat went from buoy to buoy and the crew brought in the branch lines and the catch. This process could take as long as fourteen hours because the lines must all be carefully coiled with the buoys detached and stored separately.

That first day the crew had caught only one or two big-eye tuna, and they were smaller than the average of nearly three hundred pounds and six feet in length; another eight were even smaller. Because tuna are warm-blooded fish, they must be cut up and put on ice very quickly to keep their red flesh firm. So that first day it was close to midnight before all the work was done. It had not been a good catch, with only several hundred pounds of fish landed—as opposed to the more normal thousand-plus pounds.

The next day the weather turned windy, and it was not until early on February 9 that the lines were again set out, and by afternoon there was engine trouble when the lines were being winched in. Suddenly the men saw that the main line had been severed; perhaps shifting currents had caught some of the hooks on coral reefs. After searching for four days and finding plenty of buoys but only some lines, they worked out that 170 lines, over half the ship's capability, had been lost.

It was too soon to go back to Japan, so the troubled ship went south and then west toward the Marshall Islands where the seas would be less rough. Fishing as they went, they had good and bad days catching various types of tuna, but fuel was running low. By the end of February they were in the Marshalls and fished near Utirik Atoll. They now had some nine tons of fish on ice. They turned west toward Bikini. March 1 would be their last day of fishing before heading home.

When the boat had left Japan, no crew member knew that the US government had alerted the Japanese Maritime Safety Agency of its planned nuclear tests, and that an exclusion zone had been established around Bikini Atoll, extending eastward and very close to where the crew was now setting their lines.

The radioman, Kuboyama Aikichi, was a small, stocky man with a loud voice. At thirty-nine, he was the oldest and most learned of the crew. A seaman most of his life with a wife and three children, he knew that a few years back there had been a restricted area around Bikini for the first American nuclear tests, and he had warned the captain and fishing master to stay away from Bikini.

Still, at 3 a.m. on March 1, the fishing master fixed the *Lucky Dragon's* location at ninety miles outside the old restricted area. Shortly thereafter the crew began to release the lines as the boat headed slowly on its west-southwest course.

· PART I ·

THE FIRST TESTS

1 · IN THE BEGINNING

USING RADIOACTIVE MATERIALS as a weapon of terror or death had been a highly secret part of the US pre–World War II atomic bomb research. In short, part of the initial thinking among American scientists included developing what now is called a "dirty bomb."

The idea had first emerged from a May 17, 1941, National Academy of Sciences report on Atomic Fission by a committee chaired by Nobel Prize–winning physicist Arthur Compton. The men were studying whether the government should devote a major scientific effort toward developing an atomic bomb. At the time, it was believed that Hitler Germany was making progress in such an effort.

The authors of this eight-page, National Academy report wrote that with a vastly increased uranium research effort "it would seem to us unlikely that the use of nuclear fission can become of military importance within less than two years." They went on to list proposed military applications for radioactive materials produced by a uranium fission reaction. They included use as a power source for ships and submarines, or for "violently explosive" bombs.

The first use suggested for the "violently radioactive materials" produced would be "as missiles destructive to life." The paper read, "These might then be carried by airplanes to be scattered as bombs over enemy territory" to cause radiological injury. The report added, "This might be the most promptly applicable military use," although it might not be available within less than twelve months of the first production of a chain reaction—so not earlier than 1943.

In the summer of 1942, Compton believed Nazi Germany was going down this path, writing to James B. Conant, physicist and Harvard University president, "We have become convinced there is real danger of bombardment by the Germans within the next few months using bombs designed to spread radio-active materials in lethal quantities."

In 1943, Manhattan Project research work on the atomic bomb was proceeding at Los Alamos National Laboratory in New Mexico under the direction of Dr. Robert Oppenheimer, who discussed various potential applications with leading scientists, including Dr. Enrico Fermi, head of the team that built the first self-sustaining nuclear chain reaction in Chicago.

Oppenheimer and Fermi talked about using fission materials, particularly strontium, to poison German food supplies. This approach was dropped after Oppenheimer wrote to Fermi that he thought it unfeasible unless "we can poison food sufficient to kill a half a million men."

Such studies continued. An October 30, 1943, report entitled "Use of Radioactive Materials as a Military Weapon," written by Compton, Conant, and chemist Harold Urey, another Nobel Prize-winner, was sent to Lt. Gen. Leslie Groves, director of the Manhattan Project. This report recommended that additional work be authorized on "the use of radioactive materials in order that this country may be ready to use such materials or be ready to defend itself against the use of such materials." One offensive application proposed was the grinding of radioactive particles into microscopic size and, as dust or smoke, using it as "a gas warfare instrument." It could be delivered via a ground-fired projectile or an aerial bomb. "In this form it would be inhaled by personnel," the report said, "The amount necessary to cause death to a person inhaling the material is extremely small. It has been estimated that one millionth of a gram accumulating in a person's body would be fatal. There are no known methods of treatment for such a casualty."

The authors also suggested use of radioactive materials for what they called "terrain contaminant." That meant using a weapon that spread radioactive materials on the ground "in order to deny terrain to either side except at the expense of exposing personnel to harmful radiations." Depending on the quantities of radioactive materials used, "effects on a person would probably not be immediate, but would be delayed for days or perhaps weeks depending upon the amounts of exposure," they wrote. In addition, they said contamination from radioactive material "would be dangerous until the slow natural decay of the material took place, which would take weeks and even months."

Their memo even suggested that radioactive material could be weaponized for ingestion into human bodies by targeting reservoirs or wells. "Four days production [of radioactive materials] could contaminate a million gallons of water to an extent that a quart drunk in one day would probably result in complete incapacitation or death in about a month's time," they wrote.

The authors recommended that "if military authorities feel that the United States should be ready to use radioactive weapons in case the enemy started it first, studies on the subject should be started immediately."

Pursuit of such dirty bombs was eventually set aside because of the successful progress being made in developing the first atomic bombs. But based on their past planning, key atomic scientists were clearly aware of the threat posed by radioactive materials, such as fallout created by detonation of a nuclear weapon, more than seventy years ago.

In the spring of 1945, with preparations for the initial Trinity test of a plutonium-based, atomic device well underway, the main focus of Manhattan Project scientists was getting the maximum destructive effect from an atomic explosion. Since nearly half the destructive energy released from a detonation would come in the form of blast, with another 35 percent coming from heat, those elements drew the most attention.

Radioactive fallout was mostly an afterthought, since radiation would be only 15 percent of the energy the weapon's detonation would release. Focus was on the altitude at which the atomic bomb would be detonated in order for the explosion to create the greatest blast and heat effects over the widest area.

A two-day meeting of what was called the Target Committee was held on May 10 and 11, 1945, in Oppenheimer's office at Site Y, the then-classified name for the Los Alamos National Laboratory in New Mexico. The first topic of discussion was listed as "Height of Detonation," according to a once Top Secret, May 12, 1945, summary of the meetings, declassified in June 1974.

The goal for both the uranium and plutonium atomic bombs was a detonation at a height that would cause overpressure of five psi (pounds-per-square-inch) on the ground. Such a blast overpressure would create a wind speed of 163 miles per hour and collapse most buildings, causing immediate widespread fatalities and universal injuries.

At the Target Committee meeting, expectations were voiced that "Little Boy," the uranium bomb, would have a yield from five to fifteen kilotons (5,000 to 15,000 tons of TNT), and that the detonation fuse should be set to go off at an altitude somewhere between 1,550-to-2,400 feet above ground for the best destructive results.

For "Fat Man," the plutonium-triggered bomb, the expected yield was to be determined after the Trinity test, since that device was of the same design. It was tentatively determined that detonation altitude would be around 1,400 feet.

The declassified memo listed another agenda item as "Psychological Factors in Target Selection." The group agreed there was "great importance" in using the new bomb to get "the greatest psychological effect against Japan" and to make "the initial use sufficiently spectacular for the importance of the weapon to be internationally recognized when publicity on it is released."

To put it bluntly, Oppenheimer and his colleagues wanted to destroy as many buildings as possible and kill and injure so many people that the world would recognize something really terrifying had been added to warfare.

Of course, they didn't want to be seen as going after civilians. Another item on their agenda was "Use Against 'Military' Objectives." But notice they saw fit to put military in quotation marks. They said the initial weapon was to be used on a "small and strictly military objective," but that the target "should be located in a much larger area subject to blast damage in order to avoid undue risks of the weapon being lost due to bad placing of the bomb."

When it came to Hiroshima, the first target, that city was described as "an important Army depot and port of embarkation in the middle of an urban industrial area. It is a good radar target and it is such a size *that a large part of the city could be extensively damaged.*" [emphasis added]

Nagasaki wasn't even on the original target list. It was added after Secretary of War Henry Stimson objected to going after Kyoto, Japan's ancient cultural capital. Even then Nagasaki was the alternate target, the one hit because there was cloud cover over Kokura, the original selection.

The Target Committee's "Radiological Effect" discussion of the bomb was devoted to recommendations associated with safety of the mission's American aircraft and crews rather than the radiation exposure of the Japanese people on the ground. The memo called for no other US aircraft flying closer than 2.5 miles from the detonation site or near the radioactive cloud the bomb created.

On May 31, 1945, another key meeting was held in Secretary of War Henry Stimson's office next door to the White House. The first test of an atomic device was looming in six weeks, and the first use of an atomic bomb was expected shortly thereafter. It was a meeting of what was called the Interim Committee, appointed by President Truman less than a month after he became president, to discuss "recommendations on temporary wartime controls, public announcement, legislation and post-war organization," according to notes from the session.

In attendance were Gen. George C. Marshall, at that time Army chief of staff; James F. Byrnes, then-special assistant to President Truman; William L. Clayton, a successful businessman and assistant secretary of state; Ralph A. Bard, a financier and undersecretary of the Navy; Dr. Vannevar Bush, an engineer, founder of Raytheon Corps and one-time president of MIT, but then effectively Truman's science adviser as director of the Office of Scientific Research and Development; Karl T. Compton, a physicist who helped develop radar and another former MIT president; and Harvard's Conant, who was then also chairman of the National Defense Research Committee.

Also present were Oppenheimer, along with Fermi, Arthur Compton, and Dr. E. O. Lawrence, each of whom had run elements of the Manhattan Project. Finally there was Major General Groves, manager of the project.

Chairing the session was Stimson, a Republican former secretary of state from the Hoover administration whom President Roosevelt had brought into his cabinet in 1940 to show bipartisanship as it became apparent the US would become involved in World War II.

Did that group, almost seventy years ago, get everything right? People have argued that those first bombs should not have been dropped. General Marshall, in fact, at one point during a meeting made that very argument, although he would change his mind.

But listen to what they did say and then do.

At that May 31 session, with Marshall's support, Stimson said the Manhattan Project "should not be considered simply in terms of military weapons, but as a new relationship of man to the universe." He called what was about to happen more important than the discovery of gravity because of "its effect on the lives of men."

For that reason, he said, "It was important to realize that the implications of the project went far beyond the needs of the present war." The atomic bomb, Stimson said, looking toward the future, "must be controlled if possible to make it an assurance of future peace rather than a menace to civilization."

He then laid out the topics for discussion: "Future military weapons; Future international competition; Future research; Future controls; and Future developments, particularly non-military."

On future weapons the speaker was Compton. At this meeting, he, Conant, and Oppenheimer foresaw the second and third generation of what was to become the thermonuclear bomb (the so-called H-bomb). Oppenheimer predicted it "might produce an explosive force equal to 10,000,000-to-100,000,000 tons of TNT."

When it came to future domestic programs, Lawrence pressed for the US "to stay ahead in this field" and insisted that "research had to go on unceasingly."

Oppenheimer "pointed out that one of the difficult problems involved in guiding a future domestic program would be the allocation of materials as between different uses." He said what was being done to build the bomb under war pressure "was simply a process of plucking the fruits of earlier research" and "to exploit more fully the potentialities of this field it was felt that some more leisurely and a more normal research situation should be established."

Bush joined Oppenheimer to emphasize that while concentration had to be on weapons during this wartime setting, in peacetime such focus would be completely wrong. They both agreed "that only a nucleus of the present [Manhattan Project] staff should be retained and that as many as possible should be released for broader and freer inquiry" back in their universities and research laboratories, according to the meeting notes.

The initial nuclear research had been directed at a weapon to shorten the war, but "this development had only opened the way to future discoveries," according to Oppenheimer, who then attempted to illustrate both his practicality and his idealism. First, fundamental knowledge of nuclear fission was "widespread throughout the world," and so "it might be wise for the United States to offer to the world free interchange of information with particular emphasis on the development of peacetime uses."

"The basic goal of all endeavors in the field should be the enlargement of human welfare," Oppenheimer continued, adding a suggestion that never took hold: "If we were to offer to exchange information before the bomb was actually used, our moral position would be greatly strengthened."

Stimson referred to an earlier Vannevar Bush-Conant memorandum that had recommended a future international organization to handle the exchange of nuclear information and promote cooperation. There should be complete scientific freedom, but "the right of inspection should be given to an international control body," he said. Still, he recognized the writing on the wall, asking aloud what kind of inspection would be effective when faced with the difference between "democratic governments as against totalitarian regimes under such a program of international control coupled with scientific freedom."

Although a World War II ally in the fight against Germany and potentially needed to assist a coordinated effort against Japan, the Soviet Union, then under the iron fist of Joseph Stalin, drew special attention.

Bush was first to express doubt that the US could "remain ahead permanently if we were to turn over completely to the Russians the results of our research under free competition with no reciprocal exchange." Oppenheimer joined in expressing doubts "concerning the possibility of knowing what was going on in this field in Russia."

Marshall cautioned against "putting too much faith in the effectiveness of the inspection proposal." He was concerned about the Soviet military's penchant for secrecy and thus questioned whether they would be willing partners. Instead, Marshall said he favored "building up of a combination among like-minded powers, thereby forcing Russia to fall in line by the very force of this coalition."

Byrnes was fearful "that if information were given to the Russians, even in general terms, Stalin would ask to be brought into the partnership." Thus, they would learn program details at a time, he said, when the US program had "to push ahead as fast as possible . . . to make certain that we stay ahead and at the same time make every effort to better our political relations with Russia."

When the discussion turned to Japan, Oppenheimer vividly described how the first atomic bomb would be different from previous conventional, heavy bombing of Japanese cities. "The visual effect would be tremendous . . . accompanied by a brilliant luminescence which would rise to a height of 10,000 to 20,000 feet" while the "neutron effect of the explosion would be dangerous to life for a radius of at least two-thirds of a mile."

There was general agreement, Stimson said, "that we could not give the Japanese any warning; that we could not concentrate on a civilian area; but that we should seek to make a profound psychological impression on as many of the inhabitants as possible."

Following a suggestion of Conant, which reflected the Targeting Committee's view, Stimson agreed that "the most desirable target would be a vital war plant employing a large number of workers and closely surrounded by workers' houses."

In short, the first twistedly logical use of nuclear weapons—and so far their only real proposed use—were as terror weapons, not to fight a war, but to end a war by killing and injuring as many people as possible while pretending the target was only military in nature.

2 · THE FIRST FALLOUT

IN APRIL 1945—JUST three months ahead of the proposed Trinity shot, the world's first nuclear explosion—Hans Bethe, who ran the Manhattan Project theoretical division, had assigned two of his physicists, Joseph Hirschfelder and John Magee, to look at what would be the aftereffects.

The Trinity device was to be detonated from a platform at the top of a one-hundred-foot tall steel tower. The expected explosion, Hirschfelder later wrote, "would pick up dirt, rocks, and assorted debris, some of which [would] pass through the fireball and get plated with radioactive materials."

Hirschfelder and Magee's early predictions—after studying the expected fireball, heat, wind currents, the season, and time of day of the shot—were that enormous amounts of radioactive dust could fall down on surrounding towns, while some of it could drift much farther away. Their initial estimates, according to Trinity historian Ferenc Szasz, represented a danger "so lethal as to make the test virtually impossible."

After studying the effects of a one-hundred-ton, dry-run explosion of TNT that had been mixed with a miniscule amount of fission products, they lowered their estimates but realized that much depended upon the wind. "In spite of all this work, very few people believed us when we predicted radiation fallout," according to Hirschfelder.

The two scientists, supported by Manhattan Project's Health Group, convinced Groves and Oppenheimer to set up a monitoring system for possible radioactive fallout from the Trinity shot, and a map was made of every resident within a 40-mile radius of ground zero. A military detachment of 160 men were made ready to evacuate up to 450 people should fallout occur. Another 25 members of the Army Counterintelligence Corps (CIC) were sent to towns and cities up to 100 miles from the test site. They were to request evacuation troops if needed, or help manage public relations. The CIC members were armed with a cover story that a large ammunition dump with some gas shells had exploded on the test site, which required a clearing of the surrounding area.

Forty-four monitors were stationed to the northeast, which was the direction the wind was expected to blow. The plan was to call them by telephone after the shot and tell them how high, how fast, and which way the radioactive cloud was heading as it crossed New Mexico and drifted into Colorado.

Two hours before the planned 4 a.m. Trinity detonation, there were thunderstorms around the site, but the rain eventually stopped, leaving a still-overcast sky. Detonation was put off for an hour. When the overcast began to break, about 4:45 a.m., the shot was set for 5:30 a.m.

The Trinity device proved more efficient than expected, and the explosion sent the fireball as a reddish-brown stem up ten thousand feet, while above it the mushroom cloud blossomed out at twenty-five thousand feet. Light winds of about ten miles per hour pushed the entire structure in a northeasterly direction. It rose so high that it was ten miles from ground zero before the wind-carried fallout of fission products began dropping from the dissipating cloud.

Hirschfelder and Magee, dressed in white coveralls with dangling gas masks and driving an old sedan, followed the cloud north of the Nevada Test Site, accompanied by some soldiers. At a crossroads store fifteen miles northeast of ground zero, an old man recognized they were from the test site and said, "You boys must have been up to something this morning. The sun came up in the west and went on down again."

The two drove on to a site 19.5 miles northeast of ground zero where Army Searchlight Station L-8's unit had parked, there to light up the cloud in the predawn darkness. The soldiers had breakfast steaks cooking on a grill when

Hirschfelder and Magee arrived—about ninety minutes after detonation—at the same time that a light dusting of radioactive particles was drifting down. Magee's readings of low radiation levels around 8:30 a.m. convinced the troops to bury the meat and leave the area.

Overall, however, the monitors who gathered at the end of the day decided that Trinity's fallout had not caused any serious danger.

That was not the end. It turned out that a "hot spot" had developed in an area twenty miles northeast of the test site, where the road ran through a narrow gorge. The next day monitors discovered a house and a ranch off that road that had been missed by the Army. An elderly couple, the Raitliffs, lived there with some dogs and livestock.

They had been indoors most of the previous day, so they missed most of the fallout—although Mr. Raitliff told a visiting scientist that "the ground immediately after the shot [appeared] covered with light snow" and for several days "the ground and fence posts had the appearance . . . of being 'frosted.'"

Some of Raitliff's animals showed signs of beta burns and within weeks lost hair and had discolored coats. By fall, ranchers in the area sold some seventy-five head of cattle they said had shown burns or discoloration after the Trinity test.

Aircraft equipped with filters had followed the Trinity cloud across Kansas, Iowa, Indiana, and up through upstate New York and New England when it finally went out to sea. Months after the test, the Eastman Kodak Company in Rochester, New York, received complaints of fogging of its x-ray film by tiny, exposed spots, making that film unusable. A company scientist, Dr. John H. Webb, began an investigation and discovered a year later that beta-particle, radioactive fallout from the Trinity explosion had contaminated Wabash River water used by the mill in Ohio that made the cardboard packaging that was supposed to protect the x-ray film.

Webb and the company kept his discovery secret, except for a call made a year later to the Atomic Energy Commission (AEC), which had just been established to manage the development, use, and control of atomic energy.

The AEC's Merril Eisenbud, then-director of its health and safety laboratory, years later remembered hearing of fallout some one thousand miles away from the explosion. "I was amazed, because I got the call from Rochester saying that they had had fallout," he told an interviewer. "And so, when I called around,

I found out that nobody had a monitoring program, which amazed me. So, you know, we got into it in that way."

The Army, from the beginning, was not eager to deal with fallout from atomic testing at the Nevada site. The possibility of endless lawsuits haunted the military, plus the necessity of keeping the whole test program secret.

At the end of July, the War Department authorized Los Alamos medical personnel to "discretely" investigate the health of persons exposed to Trinity fallout "under suitable pretext," according to a 1986 Los Alamos history project report.

After the atomic bombs were dropped on Hiroshima and Nagasaki in early August 1945, the Trinity shot was publicly revealed, but the reasons for the follow-up medical visits to individuals who were exposed to Trinity's fallout "were not disclosed to the residents," the history records.

Dr. Louis Hempelmann, a doctor in the Los Alamos health group, wrote after Trinity, "A few people were probably overexposed, but they couldn't prove it and we couldn't prove it, so we just assumed we got away with it."

"Historical records indicate that pressures to maintain secrecy and avoid legal claims led to decisions that would not likely have been made in later tests," according to the history.

Dr. Stafford Warren, then-chief of the Manhattan Project medical section, in his report on the Trinity test, said no area investigated received "a dangerous amount" of radiation. He did say the radioactive cloud "was potentially a very serious hazard over a band almost 30 miles wide extending almost 90 miles northeast of the site."

One obvious lesson from Trinity was that detonating a nuclear device close to the ground increases fallout, and that doing it at high altitudes, where most of the fireball does not hit ground levels, results in less dispersal of radioactive fallout.

Two days after Trinity, Warren recommended the test site be expanded or a new one found. General Groves agreed. He decided the Trinity test site was too small and a larger site was needed, "preferably with a radius of at least 150 miles without population."

■   ■   ■

Sixteen hours after the United States' August 6, 1945, atomic bombing of Hiroshima, President Truman announced that the nation had produced an explosive power "more than two thousand times the blast power of the British 'Grand Slam' which was the largest bomb ever yet used in the history of warfare."* Truman added a threat: "In their present form these bombs are in production and even more powerful forms are in development."

As a result, the president added, "We are now prepared to obliterate more rapidly and completely every productive enterprise the Japanese have above ground in any city. We shall destroy their docks, their factories, and their communications. Let there be no mistake; we shall completely destroy Japan's power to make war." No details were given, such as the size of the bomb or the plane that had delivered it.

Three days later, a second atomic bomb was dropped on Nagasaki. This time, Gen. Carl A. Spaatz, commander of the US Strategic Air Forces in the Pacific, made the announcement from Guam. It was not the main story on the front page of the August 9, 1945, *New York Times*. That pride of place went to the Soviet Union's declaration of war against Japan.

When news of the Nagasaki bombing reached Tokyo, Japanese Foreign Minister Shigenori Togo proposed acceptance of the Potsdam Declaration, which called for the unconditional surrender of all Japanese armed forces. Japan's Supreme Council for the Direction of the War, meeting in an underground bomb shelter, deadlocked on a decision after debating the matter through the night of August 9.

At 2 a.m. August 10, Prime Minister Admiral Baron Kantaro Suzuki pleaded with forty-one-year-old Emperor Hirohito to decide. Hirohito said, "I do not desire any further destruction of cultures, nor any additional misfortune for the peoples of the world. On this occasion, we have to bear the unbearable."

Surrender announcement was set for August 15. Some Japanese Army officers continued to oppose the emperor's decision and even planned a mini-coup. It failed.

---

* Grand Slam was a twenty-two thousand lb (ten ton) "earthquake" bomb used forty-nine times by the Royal Air Force in 1945 against German targets during the Second World War.

On August 15, Hirohito went on the radio to announce the end of the war, allowing his own people for the first time to hear his actual voice.

It was not until the late 1970s, as a *Washington Post* reporter covering national security affairs, that I gradually learned more details about those initial atomic weapons.

Little Boy, the uranium-based weapon, was 10 feet long, 28 inches in diameter, and weighed 9,700 pounds. The first atomic bomb used in war, it was dropped on August 6, 1945, at 8:15 a.m. from the B-29 named *Enola Gay* piloted by Col. Paul W. Tibbetts Jr.

Little Boy was guided by a parachute and detonated at 2,000 feet over Hiroshima to maximize its yield of 12.5 kilotons (the equivalent of 12,500 tons of TNT). Only later did we hear that it immediately killed some 80,000 of the roughly 350,000 people who lived in Hiroshima, with thousands more having died later from radiation. Five square miles of the city was destroyed.

The United States Strategic Bombing Survey later reported: "The damage and casualties caused at Hiroshima by the one atomic bomb dropped from a single plane would have required 220 B-29s carrying 1,200 tons of incendiary bombs, 400 tons of high-explosive bombs, and 500 tons of anti-personnel fragmentation bombs, if conventional weapons, rather than an atomic bomb, had been used."

Three days later, on August 9, the B-29 named *Bock's Car* piloted by Col. Charles W. Sweeney dropped Fat Man, the plutonium-based weapon at 11:02 a.m. over Nagasaki. It was 10 feet, 8 inches long, and 60-inches in diameter, and weighed 10,800 pounds.

To have maximum immediate impact, the Nagasaki bomb, detonated at an altitude of 1,800 feet, had an explosive power of 22 kilotons (22,000 tons of TNT). It killed an estimated 40,000 of the city's 200,000 population. The hilly nature of the city had "limited" the major destruction to roughly four square miles.

Truman's dramatic picture of what one atomic bomb could do has faded from peoples' minds. Those two atomic bombs were drops in the ocean compared to the power resting in the thousands of thermonuclear bombs and warheads that the US, Russia, and other countries now possess. Nuclear weapons today are thought more of in terms of numbers deployed or stored by a nation, and for use as foreign policy tools or, just as likely, domestic political weapons at home to make those in power appear strong.

3 · WE DID NOT REST

■ **IMMEDIATELY AFTER WORLD WAR II,** the United States felt secure with its monopoly of atomic weapons. Two such bombs had brought the war with Japan to a quick end, but as Oppenheimer and his colleagues had predicted, research and, inevitably, further testing would go on seeking second and third generation atomic weapons leading up to the 1954 Bravo thermonuclear bomb test. The postwar testing between 1945 and 1954 developed little knowledge about radioactive fallout and its long-term effects on people and the environment.

Even before Japan surrendered, American military leaders, particularly in the Navy, were concerned about the future role of its warships in an atomic age and began to press for further testing of nuclear weapons.

One day after Japan's Emperor Hirohito gave a recorded radio address announcing his country's surrender, Lewis Strauss, a Navy officer assisting then-Secretary of the Navy James V. Forrestal, sent a memo to his boss on August 16, 1945, arguing, "If such a test is not made, there will be loose talk to the effect that the fleet is obsolete in the face of this new weapon and this will militate against appropriations to preserve a postwar Navy of the size now planned."

Nine days later, Sen. Brien McMahon (D-CT), chairman of the Senate's new Special Committee on Atomic Energy, publicly called for use of an atomic bomb to see its impact on Navy warships. "I would like to see these Japanese naval ships taken to sea, and an atomic bomb dropped on them. The resulting explosion should prove to us just how effective the atomic bomb is when used against the giant naval ships. I can think of no better use for these Jap [sic] ships."

The proposal was picked up by the top officers of the Army Air Forces, and on September 18, 1945, its commander, General H. H. (Hap) Arnold, presented a proposal that a number of Japanese ships be made available for tests involving atomic bombs and other weapons. The Navy responded a month later with Chief of Naval Operations Admiral Earnest J. King suggesting the Joint Chiefs control the tests with Army and Navy participation—inter-service rivalry has a long history—and that one bomb be dropped by air and another exploded in the water. He also proposed that some older or surplus US warships be included within the targeted area. Joint planning began within a month with an ad hoc subcommittee set up to make a detailed proposal. Then-Major General Curtis E. LeMay—who later became the tough, legendary head of Strategic Air Command—took over as chairman of that steering committee.

Scientists from the Manhattan Project and military officers at earlier meetings had agreed that the tests should gather data on "the nature, range, and duration of radiation danger . . . bomb efficiency, burst location, wave formation, and ship movement," according to a 1984 Defense Nuclear Agency history of what became the Bikini tests. In addition, they sought information "to aid in assessing damage from and designing protection against nuclear weapons" and facts "that would be helpful in learning to detect nuclear detonations."

One member of LeMay's panel was then-Commodore William S. "Deak" Parsons, who months later would become an admiral. Parsons was an unusual Navy officer. Though essentially a military man, he had a scientist's mindset. After the Naval Academy he found himself focused on ordnance. At the beginning of World War II, he had dealt with radar, proximity fuses, and in 1943 had been assigned to Project Y, the code name for the Manhattan Project at Los Alamos, New Mexico. For the next two years he had worked with scientists on ordnance problems associated with building the atomic bombs. He had flown on the *Enola Gay* and, while in flight on the way to Japan, had personally armed the Hiroshima bomb.

It took six weeks of subcommittee meetings to thrash out controversies such as who should command the prospective Joint Task Force and whether target vessels would carry full loads of ammunition, something the Navy opposed since it could heighten the effects of the bomb if fires from the oil or gasoline from one ship ignited other closely anchored ships.

On January 10, 1946, President Truman approved a preliminary plan with Vice Admiral William H. P. Blandy, a deputy chief of naval operations for special weapons, chosen by the Joint Chiefs to run the operation as head of Joint Task Force One. Admiral Parsons, with his Manhattan Project connections, was made deputy task force commander. Because scientists often came crying to him for help, he drew the code name "Wet Nurse."

It was Blandy who gave the plan the name Operation Crossroads, chosen, he said, "because it was apparent that warfare—perhaps civilization itself—had been brought to a turning point in history by this revolutionary weapon."

The initial plan was to conduct three tests to determine, under controlled conditions, atomic bomb effects such as pressure, heat, shock-wave velocity, as well as visual and nuclear radiation. One test, called Able, would duplicate the drop on Nagasaki, which was an air burst, but this time over water. The second, Baker, would imitate an atomic bomb attack on a fleet at anchor, such as Pearl Harbor, but by exploding a preset device at a shallow depth beneath the anchored ships. The third test, Charlie, would also study underwater effects, this time down to two thousand feet with only a small number of vessels above it.

One of the first problems was finding a site for the tests.

An Army and Navy group agreed that a protected body of water at least six miles in diameter was needed to provide anchorage not only for the target fleet but also for the support ships aiding in the tests. The test site had to be uninhabited or have such a small a population that people easily could be evacuated. The site had to be at least three hundred miles from the nearest population center because of possible radioactive fallout—which was still an unmeasured and thus a speculated effect of atomic weaponry. The weather had to be right, guaranteeing no severe cold or violent storms. Another requirement had to be predictable, directionally uniform winds at all altitudes up to sixty thousand feet, so there was no chance of fallout coming back over the task force personnel. Finally, the area had to be under US control with an American B-29 air base within one thousand miles, since the first test bomb would be dropped from such a plane.

After studying areas in the Atlantic and the Pacific, the Navy decided on the Marshall Islands and, in particular, Bikini Atoll. It was a string of 23 coral islands, totaling 3.4 square miles that surrounded a deep, 230 square-mile, blue-green lagoon. Bikini was hundreds of miles from the main east-west shipping lanes of the Pacific. Its weather was temperate, and its winds were considerably more stable than those of other ocean areas. More important, the Marshall Islands were a Trust Territory under US control.

In January 1946, the American delegate to the United Nations Trusteeship Council asked for permission to use part of the Marshall Islands trust area for military purposes and to keep any test sites closed to other nations. (It would not be until April 2, 1947, that the UN Security Council would finally approve the terms of trusteeship, and July 18, 1947, when it entered into force.)

In late January 1946, the US Congress held a series of hearings on a House joint resolution that authorized the use of Navy ships to "Determine the Effect of Atomic Weapons upon Such Vessels." At a Senate Committee on Atomic Energy hearing on January 24, 1946, Blandy said, "The ultimate results of the tests, so far as the Navy is concerned, will be their translation into terms of United States sea power. Secondary purposes are to afford training for Army Air Forces personnel in attack with the atomic bomb against ships, and to determine the effect of the atomic bomb upon military installations and equipment."

There was opposition to the proposed tests, much of it based on false fears that the earth would crack open or that all life would end.

Concern became such that Blandy gave public speeches to try to calm down nervous citizens. During a February 21, 1946, evening speech before a local Philadelphia Red Cross chapter, he said, "The bomb will not kill half the fish in the sea and poison the other half so they will kill all the people who eat fish hereafter. The bomb will not cause an earthquake or push up new mountain ranges . . . The bomb will not start a chain reaction in the water, converting it all to gas and letting all the ships on all the oceans drop down to the bottom. It will not blow out the bottom of the sea and let all the water run down the hole. It will not destroy gravity. I am not an atomic playboy, as one of my critics labeled me, exploding these bombs to satisfy my personal whim."

House Joint Resolution 307 passed on March 12, 1946, by a 313 to 25 vote, but because of concerns about nuclear testing, 94 members, including both Republicans and Democrats, voted present. Among them were Rep. Christian Herter (R-MA), who would later be secretary of state, and then-Rep. Clare Booth Luce (R-CT), a successful playwright, leading conservative politician, and wife of Henry Luce, powerful co-founder and publisher of *Time* and *Life* magazines.

The use of Bikini Atoll required the relocation of its 167 Marshallese inhabitants. With May 15, 1946, as the date for the first test, Blandy wanted the Bikinians off their atoll by March 15.

On Sunday, February 10, 1946, the then-military governor of the Marshalls, US Navy Commodore Ben H. Wyatt, flew by seaplane to Bikini. He assembled the Bikinians, with King Juda, their chief magistrate of the atoll, present. Navy records have reported that Wyatt told the Bikinians "of the bomb that men in America had made, and of the destruction it had wrought upon the enemy (Japan)." Americans, Wyatt said, "are trying to learn how to use it for the good of mankind and to end all world wars." He then asked: "Would Juda and his people be willing to sacrifice their island[s] for the welfare of all men," and for "the good of mankind and to end all world wars?" He asked whether they would temporarily leave their atoll so the United States could begin testing atomic bombs.

The Bikinians were told that they would be allowed to return to their atoll in a matter of months, when the United States no longer needed it for Operation Crossroads.

That same month, the Joint Task Force had selected Rongerik Atoll, 125 miles to the east, as the future temporary home of the Bikinians. It was a previously uninhabited island, but only one-sixth the size of Bikini. No one seemed to care that it had scant resources for long-term survival since the belief was that the move would only be temporary.

Having suffered under Japanese military occupation during World War II and wishing to please their new American guardians, Juda led a meeting of the Bikini Atoll Council to make the decision. He emerged from the subsequent deliberations to announce, "We will go, believing that everything is in the hands of God."

Navy Seabees arrived February 26, 1946, on Rongerik and began constructing cisterns to catch water and twenty-six housing frames for forty families, many of whom would initially live together. These were covered either with pine wood or with pandanus thatch.

Just before the Bikinians were moved from their atoll in early March, US Navy cameramen restaged Wyatt's "consultation" with King Juda, who for the scene was wearing a military uniform. The film was prepared for public relations use in the US, and, according to the George Washington University's National Security Archives, "The performance concluded with the Bikinians having their last church service on the atoll and then meeting at the graveyard" before departing.

On March 7, 1946, with the cisterns initially filled with 25,000 gallons of water from Kwajalein, the 167 Bikinians were loaded on Navy LST 1108, a cavernous landing ship with a crew of more than 200 that normally was used to carry M-4 Sherman battle tanks to an invasion spot. The Bikinians brought with them some outrigger canoes and materials from their church, community building, and individual homes. On the next day, March 8, 1946, they disembarked at Rongerik.

It wasn't long before the Bikinians found that there were very few coconuts on trees in Rongerik. In fact, there were very few fruit trees at all, and the lagoon lacked a big enough supply of edible fish. It was only months after leaving their home atoll that they asked to be moved back to Bikini. However, the Navy and the entire US government temporarily forgot them. America's attention was totally focused instead on the atomic tests that would take place over and in the lagoon that had been their home.

To some degree, comedian Bob Hope summed up the unconcerned American attitude toward Bikini when he joked in 1947, after the completion of Operation Crossroads, "As soon as the war ended, we located the one spot on earth that hadn't been touched by war and blew it to hell."

It has not been a joke for the Bikinians. More than seventy-five years later, and they have yet to return to their atoll.

4 · SERVICE RIVALRY
AT CROSSROADS

- **THE FIRST TEST** of Operation Crossroads was to take place on May 15, 1946. However, that starting date appeared to conflict with Secretary of State James Byrnes's upcoming promotion of the US plan to have international control of nuclear weapons. Financier and government consultant Bernard M. Baruch was scheduled to present the plan in mid-June at the United Nations. Byrnes, in a March cabinet meeting, had said it would be helpful from a diplomatic point of view—particularly in dealing with the Soviet Union—for the Bikini tests to be delayed. As a result, President Truman postponed the first test from May 15 to July 1. The public explanation was that because so many members of Congress would want to attend the Bikini test, much-needed legislation would have to be delayed for more than a month while they were gone. It was better to put off the atomic bomb tests to the summer when Congress was in recess.

On the surface, top Navy and Army officers hoped the tests could provide some answers to basic strategic and tactical questions raised by the new, atomic age. But underlying the tests, and influencing many decisions, was a competition between the services—a problem that has continued through today despite efforts to cloak it over. At the time, syndicated columnists Joseph Alsop and Stewart Alsop referred to the tests as "another grim struggle between the Navy and the Air Force."

This new atomic bomb had ended the war, and it was in the hands of the Army Air Corps, which within a year would become the US Air Force. Rivalry between the services was unabated and particularly so because the War and Navy Departments were also participating in an ongoing Washington debate over what, in 1947, would become a major reorganization of the government's national security framework. At the same time, the Navy was defending its then principal role in America's war plans. No matter what the threat would be in the future, senior US Navy officers believed they needed to prove their warships would survive in the next war, atomic bombs or not.

In the Crossroads tests, they hoped to get some support.

Under attack from atomic bombs, what would be a warship's weakest links; what parts would fail; and how best to build new Navy warships so they could stand up to such an attack? What would such an attack do to the inner compartments of a ship, its boilers, turbines, guns, and fuel supplies? And, of course, what about the ship's crew that work both above and below decks? Navy medical personnel were concerned about radiation and whether a ship's steel shell would provide protection.

Supporting Crossroads were some 42,000 personnel of whom 37,000 were US Navy officers and enlisted men. Most lived and worked aboard some 150 support ships, and also on various Bikini Atoll islands, where they had constructed other facilities, including bunkers, floating dry docks, and steel towers to hold cameras and recording instruments.

The overall targets for the test would be a ninety-ship fleet anchored in the lagoon. The vessels consisted of older US Navy cruisers, destroyers, submarines, as well as a number of surplus auxiliary and amphibious crafts. Military equipment was placed aboard some ships and on Bikini's islands to measure the result of blast, heat, and radiation. There were also three captured ships, including the *Prinz Eugen*, a German cruiser that had been transferred to the US as a war prize; the captured Japanese light cruiser, *Sakawa*; as well as the Japanese battleship, *Negato*. The latter had been the former flagship for Adm. Isoroku Yamamoto, commander in chief for the Combined Japanese Fleet, and where Yamamoto not only plotted the December 7, 1941, attack on Pearl Harbor, but also where he received the first radio report that the strikes had taken place.

The Army Air Force, with its planes based in Kwajalein, conducted pretest, practice bombing runs on the aiming target ship, the *USS Nevada*. The aircraft crews found it difficult from high altitudes to distinguish the battleship from neighboring ships in the Bikini Lagoon, so the task force commanders decided to paint the top deck of the *Nevada* red-orange and install a radio-controlled searchlight for guiding the B-29 on its final bombing run.

Six B-17, World War II bombers and four Navy F6F Hellcat fighters were developed to fly with radio guidance but without crews at Crossroads. For the first time in history these large bombers and smaller operational aircraft were modified to be able to take off, fly, and land without a human aboard. We talk a great deal about drones today, but more than seventy-five years ago the drones used at Operation Crossroads were crewless.

The purpose of these unmanned drones was to fly through the radioactive mushroom cloud after each test shot, taking photos and gathering air samples for radiation levels. The Army Air Corps planes—four B-29s—had their takeoffs and landings controlled by radio transmitters, some mounted on jeeps near the runway. The four Navy aircraft—F6F Hellcat drones—were controlled remotely with radio signals from the flight deck of an aircraft carrier. Once in the air all drones were controlled from so-called "mother ships" flying near them.

Chief of the Radiological Safety Section of the Crossroads Joint Task Force was Dr. Stafford Warren, then a colonel in the Army Medical Corps. Trained in radiation biology, Warren had been a medical officer during World War II, assigned to develop safety around the laboratories and production facilities associated with the Manhattan Project. Eventually he became its chief medical officer and was present at the Trinity test of the first plutonium device in Alamogordo, New Mexico. He later led the team in Japan that assessed medical effects on the ground at Hiroshima and Nagasaki.

Warren's group brought twenty thousand devices to Crossroads to measure nuclear radiation. The most common were film badges. A small piece of unexposed photographic film wrapped to exclude light was the heart of the badge. When gamma radiation struck the badge, it penetrated the wrapping and produced a change in the film's emulsion. When read, after being placed in photographic developing solution, the darkness of the emulsion indicated the amount of radiation to which the badge had been exposed. For the Crossroads tests, thousands of film badges were placed on the target ships, below decks, in compartments, and even where test animals had been situated.

The second most common devices were Geiger counters, which immediately measured gamma radiation, either by motion of a needle on a dial or by producing the famous buzzing aloud or in headphones. A single photon of gamma radiation would upset the electrostatic regime of the Geiger counter's inner workings and set off a discharge of current that moved the dial and/or caused the sound.

The target ships were loaded as closely as possible to battle status to include ammunition, fuel oil, gasoline, and water, while fueled aircraft were placed on decks with other Army equipment.

After Hiroshima and Nagasaki, there were many technical and scientific questions that still needed answers, not the least of which were the effects of the bombs on humans—the kinds of injuries they would suffer whether from flash burns, radiation, or secondary factors such as explosions or fires. Many of the Japanese atomic bomb survivors did not emerge as victims for months or even years after the attacks. Therefore it was apparent to some Joint Task Force members that Operation Crossroads offered controlled conditions that had not been available in Japan. They could permit scientific data to be gathered at the time of the Crossroads explosions and immediately thereafter, which would permit development of accurate information about the biological effects on humans of atomic bombs.

A government report on Crossroads outlined four specific purposes for the biological studies. It would show symptoms produced by the explosion, provide experience in cataloging injuries and detect onset of slow developing injuries, provide experience in treating those injuries, and reveal any new kinds of aftereffects.

A Naval Medical Research Section was created within the Joint Task Force with members drawn initially from the Naval Medical Research Institute at Bethesda, Maryland. There were other individuals involved from the Army and civilian agencies, including the National Cancer Institute, US Public Health Service, as well as university and scientific groups.

The plan was to expose living animals and other biologic materials like bacteria, insects, hormones, vitamins, and even seeds to the bomb effects by putting them aboard twenty-two of the target ships at varying distances from the site of the burst.

The plan for human medical research was to try to gain insights into injuries produced by the bomb's explosion and the resultant radiation on different living organisms. A variety of measuring devices were placed where animals were to be exposed so factors producing injuries could be determined. These included cameras, Geiger counters, gamma-ray detectors, ultraviolet recorders, thermal measuring instruments, and air pressure gauges.

The *USS Burleson*, an attack transport, was converted with animal pens and food bins; laboratories for pathology, hematology, radiobiology, and biochemistry testing; autopsy rooms; and accommodations for seventeen medical officers.

Two hundred pigs, 60 guinea pigs, 204 goats, 5,000 rats, and 200 mice were loaded on the Navy's *USS Burleson* in San Francisco in May 1946, for the trip to Bikini. More than eighty tons of hay and grains were put aboard, along with specially prepared food for the test animals.

Pigs were chosen because their skin and short hair were comparable to those of humans. Uniforms were placed on some pigs to test protection of certain fabrics. Some other animals were coated with antiflash lotions and creams to test their biomedical reaction to thermal radiation. The hair on some animals was cut to human lengths to measure effects.

Goats were selected because they have body weight comparable to humans and body fluids useful for postexposure laboratory analysis. Scientists from Cornell University, who were familiar with goat psychology, recommended that goats be added to the test to see if the explosion would change their psychoneurotic tendencies. The rats were there because of their already known response to radiation. The National Cancer Institute supplied their own different kinds of white mice, chosen because they had a greater or lesser likelihood for developing cancer. They also might help determine if radiation would later produce changes in them or their progeny.

Grain insects were included because their genetic character would permit large numbers of generations to be studied for heredity. Soils samples, such as Caribou loam from Maine, Decatur clay loam from Georgia, and Houston black clay from Texas, were provided with the idea they would be sent back to the states after the tests and planted with seeds to see the results.

By the time it set sail, the *USS Burleson* was a virtual, nuclear Noah's Ark. It arrived at Bikini on June 14, and its human crew prepared for its job of animal distribution to target ships at the appropriate time before each test.

Months before the tests, two Smithsonian Institution scientists were among experts who carried out a biological survey of the plants and animals on and around Bikini and the atolls of Enewetak, Rongelap, and Rongerik.

No dogs were used because civilians had protested their possible employment in the tests. Most surprising, letters were received from some ninety individuals who volunteered to be aboard ships during the nuclear test shots, including one prisoner at San Quentin Penitentiary who was serving a life sentence. No humans were allowed.

5 · THE PRESS AND THE VIPs

■ **MEDIA COVERAGE OF** the Bikini tests would be important. There was immediate worldwide public interest from the moment the tests were announced, and requests to attend from more than a thousand journalists from around the world flooded the Joint Task Force from what appeared to be bona fide media organizations.

For the Crossroads public affairs job, Joint Task Force Commander Blandy had selected Navy Capt. Fitzhugh Lee III, who had some—but limited—past experience with the press. A former fighter pilot aboard the *USS Enterprise* and captain of an escort carrier in the Pacific for the last six months of the war, Lee had been chosen to handle media for Admiral Chester Nimitz, commander of the Pacific Fleet. In that role, Lee was aboard the *USS Missouri* to deal with journalists on September 2, 1945, when the peace treaty with Japan was signed.

Operation Crossroads was to be far different.

There had been a great deal of misinformation about atomic bombs—once America's most secret weapon—and it had increased when the Bikini tests were announced. The most widespread fear was the possible unknown effects that testing a nuclear weapon at sea would have. A government report made public at the time had speculated "enormous tidal waves might sweep across the Pacific and devastate its shores; the very crust of the earth might be parted with unimaginable consequences. The chain reaction in the bomb might spread to the water and the whole ocean might explode. Conjecture had no limitation except man's imagination."

Lee's immediate boss within the Joint Task Force was Brig. Gen. T. J. Betts, the Army's assistant chief of staff for intelligence, whose prime job was to maintain security. At the same time, Betts had been given responsibility for managing the public release of information surrounding Crossroads, the first time an atomic bomb would be exploded with the world's media watching. There would not be any direct censorship, but of course there would be limits on what the media would be told and what they could see. Any official government photographs to be distributed would have to be cleared under security rules set by the Joint Chiefs of Staff.

Lee decided to prepare initially by studying the public relations problems that came out of the February 1921 naval bombing tests generated by Army Air Corps General Billy Mitchell. That episode had been seen as part of the fight between the Army and the Navy, and between those who believed air power—which developed quickly in the post-World War I period—would end Navy war prominence. Those tests had shown how ships at anchor could easily be destroyed, but that those ships maneuvering at sea might escape or not even be found in wide-open ocean waters. Those tests left the public confused. Lee wanted to avoid that type of result.

At a May 13, 1946, press conference. Admiral Blandy announced that the press would be able to see the arrangement of the target vessels at Bikini along with the military equipment aboard them prior to the tests and afterwards, both at sea level and from the air. This was being done, he said, "so that the public will not await the results with the misconception that these tests alone will once and forever establish whether there will be a great or small air force, Navy or ground force."

Among Lee's first problems was the number of media members that would be chosen and how that selection process was to be done. Some senior military officials wanted a limit of just ten, while journalistic groups suggested at least five hundred would be adequate. The compromise was set at around two hundred, but that left the question of who, exactly. Lee tried to get media associations to set up a committee of peers, but because of obvious competitive factors the media backed away from that idea. In the end, Lee and his immediate staff, with guidance from some press associations, did the choosing.

They decided they would select organizations—newspapers, magazines, networks, and radio stations—and allow them to pick the specific individuals who would be sent. The idea of focusing on groups with the most experience in military and scientific matters was dropped in favor of choosing all kinds of news agencies. Most large magazines signed up, but when the tests were postponed for a month and it became clear that their employees would be away for several weeks, a few publications withdrew.

As the dropouts were replaced, Lee noted that several of the magazines and newspapers approved "were not as representative as might have been desired." No individuals or freelancers were to be accepted. However, after the chosen group had set sail across the Pacific, Lee found that the so-called named representative of a newspaper chain turned out to be a public lecturer who had been nominated by a low-level staffer using his publication's stationery.

In the end, there were 187 media attendees. Among them were some major names in journalism, starting with Hanson W. Baldwin, the military correspondent of the *New York Times*, along with the *Times'* Bill Laurence. The latter was back with the newspaper after his leave in order to work on writing reports and press releases for the Manhattan Project.

Others included Bob Considine, a famous World War II war correspondent for the Hearst newspapers, who was accompanied by William Randolph Hearst Jr., then-editor in chief of the chain; Norman Cousins, then-editor of the *Saturday Review* magazine; and several influential radio personalities including Bill Downs of CBS, Raymond Gram Swing of ABC, and Mutual Broadcasting's Quentin Reynolds. *Fortune Magazine* sent an artist, Ralston Crawford. Ten foreign press representatives were selected, including two from Britain, one from the Soviet Union, and the remainder from countries on the United Nations Atomic Energy Commission.

No women reporters were included, despite a protest from the Women's National Press Club. When the Army agreed to fly some women journalists out to the tests, the Navy refused to accept them.

With a group that large, only a few people could go by plane. Most, 117 in all, went aboard the *USS Appalachian*, a three-year-old amphibious flagship designed to carry admirals and/or generals and their staffs. It had served as a troop carrier and delivered US Army elements that reinvaded the Philippines in 1944 and 1945. It took two months at the naval shipyard located near Long Beach, California, to convert the *Appalachian* into a press ship.

A broadcast studio was built on the ship for radio and television reporters, along with installation of high-speed teletype facilities for the print journalists. A setup was created so that broadcasters could speak while standing on the bridge of the ship. Additional sleeping accommodations were prepared for the media and several wardrooms were air conditioned, since the press would be living aboard the *Appalachian* from June 12, when they set sail from San Francisco to Bikini. They would continue living on board while at anchor before and during the tests; and finally, still on board for the trip back to the US in early August.

Lee, faced with the sensitive protocol issue of how to allocate living quarters aboard the *Appalachian*, decided to favor the oldest correspondents with better bunking space along with other nonprofessional privileges. Press packets were prepared to give those with no scientific background some basic data, including an introduction to nuclear physics, the layout of the targeted ships, and the history of Bikini Atoll and the others in the Marshall Islands. Aboard ship, en route to the test site, Maj. John H. Slocum, a radiological safety officer, lectured several times about what to expect when a bomb went off and on the potential hazards from the detonation.

"I had to get these . . . people out there and keep them happy while we were there for some six weeks in the South Pacific in a non-alcoholic, non-airconditioned ship in very overcrowded conditions, and hopefully bring them all back safely without radiation damage and with their stories well told," Lee would later say in a US Naval Institute Oral History.

By January 1946, the Joint Chiefs of Staff had approved invitations for sixty members of Congress. However, the congressional invitations were not sent out until Congress actually approved the tests. Since that did not happen until June 14, only sixteen days before the first shot took place, very few of those invited actually made the trip. Ten House members and four senators traveled to Bikini to view the tests. Two of them, Senators Carl A. Hatch (D-NM) and Leverett Saltonstall (R-MA), as well as Rep. Chet Holifield (D-CA), were members of the President's Evaluation Commission, a group set up by President

Truman to supplement a test evaluation board created by the Joint Chiefs of Staff. Holifield would later become chairman of the House-Senate Joint Atomic Energy Committee.

There were also representatives of eleven countries who were members of the United Nations Atomic Energy Commission—Australia, Brazil, Canada, China, France, Egypt, Great Britain, Mexico, Netherlands, Poland, and the Soviet Union. Those countries chose their own representatives. The twenty-one selected AEC representatives gathered in Washington and traveled by special train to Oakland, California. There they boarded the US Navy amphibious command ship *USS Panamint*, which had been refitted to handle VIP observers for the trip across the Pacific. At Bikini, the *Panamint* served as floating headquarters not only for the UN group but also for congressional and scientific observers.

The *USS Henrico*, a relatively new attack transport that had seen fighting at Okinawa and had brought troops home from the Pacific area, was used to house scientists at Bikini during test preparations. During the day, some scientists collected fish and fauna from the lagoon and surrounding areas. On several evenings, just prior to the initial Able shot, those aboard listened to lectures and twice saw American and Japanese films of the atomic bomb attacks on Japan. One Japanese film showed the human side, while the American film reviewed the physical damage. Another film showed flesh wounds of prisoners of war who had been working three miles from the blast at Nagasaki, according to the log of Lauren Donaldson, a radiobiology scientist from the University of Washington. "The burn wounds and wasting away of the injured [Japanese civilians] leaves one with a sickly feeling," Donaldson wrote in his June 25, 1946, entry.

The British had a separate eight-person military delegation, six of whom were from the Admiralty. William Penney, who had headed Britain's own nuclear program and was part of the Manhattan Project at the Trinity test, led the group. Penney, a mathematician, had calculated damage effects of the bombs and had been part of the Top Secret US Targeting Committee. He had supported hitting Hiroshima and Nagasaki because he believed the hills surrounding both cities would create maximum devastation.

Penney also was among those who advised on the height of the Japanese detonations, so that the fireball did not touch the ground, and thereby avoided contaminating the earth and creating fallout.

When the Bikini tests were announced, the Soviet newspaper *Pravda* charged that the US was aiming to increase its atomic weapons advantage while at the same time claiming it wanted to prevent any further usage. It was no surprise that Moscow took up the US invitation to attend the Bikini tests, sending as its observers Mikhail G. Meshcheriakov, a distinguished physicist from the Radium Institute and head of the physics department at Leningrad University. He was accompanied by Simon P. Aleksandrov, listed as a geologist, but who also worked for Soviet State Security.

The Soviet nuclear weapons program had been accelerated weeks after the US atom bomb was dropped on Hiroshima, when Joseph Stalin established a special committee under the State Defense Committee. On August 20, 1945, Stalin had appointed Lavrentiy P. Beria, the head of Soviet intelligence, to be the committee's chairman.

While the US was preparing its nuclear tests at Bikini, as already noted, Washington had presented a proposal at the United Nations to place nuclear weapons activities under the international organization, leaving peaceful uses to the individual countries. Moscow, at that same time, introduced a proposal that called for an international convention that would ban the production, stock-piling, and use of atomic weapons, with the destruction of any in existence to take place almost immediately.

In short, each country had acted in its own interests. The US wanted to solidify its monopoly; the Soviets, whose proposal contained no authority for inspection, would have relied on other countries to perform self-enforcement.

Among other VIP observers was Stuart Symington, then assistant secretary of War for Air. Symington, forty-five at the time, had been chosen by President Truman, a fellow Missourian, to be the top civilian observer at the first Bikini test. He made attendance at the test shot part of an already planned, round-the-world study of US air bases abroad.

Symington, at the same time, had been picked to be one of the US representatives at the Philippines July 4, Independence Day celebration. Eventually a US senator from Missouri with a continued specialization in nuclear affairs, Symington left Washington on June 26, accompanied by Major General LeMay, then-deputy chief of Air for Research and Development. They first stopped off in Hawaii so that Symington could speak to the local Chamber of Commerce about his support for the Baruch plan to turn nuclear weapons over to international control.

At Kwajalein, Symington stayed with Brig. Gen. Roger Ramey, commander of all the Army aircraft involved with the tests. Ramey at the time was determining which observers among the congressmen and scientists would fly on the several planes that would be in the air as observers when the first test took place. Symington suggested he would fly from Kwajalein on an observer plane with the members of Congress, while the scientists would be in their own plane, which was what later happened.

Another late arrival was then-Navy Secretary James V. Forrestal, who like Symington was also on a round-the-world inspection trip. Forrestal was rumored to be quitting after Truman announced support for the proposed establishment of a unified Department of Defense. Some looked on Forrestal's desire to attend the Bikini test and then go on to visit Navy bases in Japan, the Mediterranean, and the North Sea, as his chance to get out of town and avoid the press and political questioning about the unification issue.

Forrestal arrived at Bikini on June 29 and was quartered on Admiral Blandy's flagship, *USS Mount McKinley*. There he attended the final briefings on the test with the admiral and his staff—a particularly significant event since it was Forrestal's intervention back at the March 22 cabinet meeting that had convinced President Truman to hold the first test on July 1.

6 · TEST ABLE

■ **BEFORE THE FIRST** test of Operation Crossroads, all personnel were evacuated from the target fleet and Bikini Atoll. Military and civilian technicians and staff members who had prepared land-based equipment boarded the support ships, which left Bikini Lagoon and took safe positions at least ten to twenty-five nautical miles east of the atoll.

There had been extended discussion at the Los Alamos Laboratory about which bomb to use for the Bikini tests. The Mark III, Fat Man bomb, the model used at Nagasaki, weighed ten thousand pounds, was ten-feet-long and five-feet in diameter, and thus was awkward to handle. Its fin assembly made bombing accuracy questionable. The scientists had been working on a newer, lighter version, called the Mark IV, which had better aerodynamics. However, Los Alamos scientists did not want to use a new and untried weapon rather than one which had previously worked during combat.

There was also the question of how many usable atomic bombs the US had at the time. In June 1946, there were only nine atomic bombs in the stockpile.

The B-29 chosen for Test Able was named *Dave's Dream*, in honor of Capt. Dave Semple. He had been the bombardier in more than 150 test drops going back to the Manhattan Project. Ironically, Semple had died during a test flight competition on March 7, 1946, at Kirtland Air Force Base outside Albuquerque, New Mexico, to see which plane would carry out this Bikini test mission. Semple had filled in for another bombardier during the practice dropping of a "pumpkin," an unarmed Nagasaki-shaped atomic bomb. Suddenly, that B-29 exploded and fell to the ground from thirty-two thousand feet. Semple and the other nine crew members died. The forty-three-year-old had served in the Army Air Corps for more than twenty years and had won the Silver Star, Legion of Merit, Distinguished Flying Cross, and Air Medal.

Dave's Dream, the so-called Superfortress B-29, had been built by Glenn L. Martin Company * at its Omaha, Nebraska, plant and accepted by the Air Force in April 1945. It was assigned to the 393rd Bomb Squadron, 509th Composite Group, the unit specifically formed to carry out the Hiroshima/Nagasaki atomic bombing missions. During the August 9, 1945, atomic bomb drop on Nagasaki, this B-29, which at that time was called "Big Stink," flew along as a camera aircraft.

After the 509th reached the Marshall Islands, the unit used Kwajalein Air Base as their headquarters. Col. Paul Tibbets Jr., who had piloted the *Enola Gay*, which dropped the bomb on Hiroshima, still was unit commander. He ordered that they continue practice missions, dropping dummy bombs. Unlike the accuracy they recorded back in the states, in the Pacific the practice runs were not as good. The winds appeared to be different.

After midnight on July 1, 1946, technicians at the Kwajalein Air Base began loading a Mark III, Fat Man, into the specially designed bomb bay of *Dave's Dream*. The crew had stenciled the name "Gilda" on the bomb after the actress Rita Hayworth's new film.

Dave's Dream lifted from the runway at 5:55 a.m., piloted by Maj. Woodrow P. Swancutt of Wisconsin Rapids, Wisconsin. It climbed through the clouds and headed 250 miles northwest toward Bikini, where it would arrive nearly two hours later.

On that same early morning, a metronome that had been placed on the deck of the battleship *USS Pennsylvania*, in the bull's eye area of Bikini Lagoon, began ticking. The metronome's ticking sound was broadcast through a microphone in front of it to every Joint Task Force ship located miles away and to radio

---

* In the 1960s, Martin merged with American-Marietta Co. to become Martin-Marietta. In 1995 it merged with Lockheed to become Lockheed-Martin.

networks around the world. The idea was that after the explosion caused the ticking to stop, other microphones on target ships further away could pick up the sound of the explosion.

Dave's Dream, after taking one practice run over the target area to check the wind and gauge the approach, carried out a second pass, this time simulating a drop at 8:31 a.m. Visibility was excellent, and a flashing light on the targeted, orange-red painted *USS Nevada* was clearly in sight of the bombardier.

While the B-29 practiced its bombing runs, some seventy-eight other aircraft joined it aloft in the area, gathering data. One of them, a Navy F6F drone, went out of control and crashed.

The final run of *Dave's Dream* began at 8:50 a.m., about fifty miles away from the target ships. When the voice of B-29 bombardier Major Harold H. Wood announced, "Release minus two minutes," observer ships ordered those aboard to put goggles on and those without goggles to look away from Bikini.

At 9 a.m., from twenty-eight thousand feet, bombardier Wood called out "bomb away, bomb away," and Gilda was dropped. It was fourteen seconds before 9 a.m. in Bikini. In California, it was 2 p.m., Sunday afternoon, June 30.

As planned, 48 seconds later, the bomb had dropped to an altitude of 520 feet above the Bikini Lagoon, and its fuse caused detonation.

NBC Radio's Elmer Peterson came on the network from San Francisco, interrupting regular programming to bring what he called "a history-making broadcast."

He announced, "Today, in just a few moments from now, many questions about this new power will be answered, in part or in their entirety—the radius of destructive activity, the effect on differing types of vessels, the future role of naval surface vessels. There are many questions of immediate significance—will the atom bomb, exploding over water, form radioactive cloud formations? How will it react on the waters of the Pacific? Will it create atmospheric disturbances that will originate a typhoon? In such matters the very safety of the men making the test is involved."

On the US East Coast it was just after 5 p.m., that Sunday evening, when suddenly the metronome stopped, and the world waited for the sound of the atomic bomb blowing up. John Crosby, the *New York Herald Tribune* radio-TV critic, wrote in the next day's paper, "The metronome came through clearly despite static, then stopped, started again haltingly like [boxer] Billy Conn trying to rise in the eighth round of his last fight and then stopped for good."

Despite all the practicing, the bomb fell way off target, 980 feet short and 1,870 feet left of the brightly-painted *USS Nevada*. There was only one ship within 1,000 feet under the point where the bomb exploded, the *USS Gilliam*, an attack transport, which sank along with four other vessels. The target miss, four times the expected probable bombing error, resulted later in a government investigation of the flight crew, which found the fault was supposedly a collapsed tail fin on the bomb.

In the first two seconds after detonation, light from the fireball, which contained a good proportion of the released energy, was the first indication to task force ships, ten to twenty-five miles away, that the explosion had taken place. On the nearest observer ships, the light was several times brighter than the sun at noon. With a surface temperature of one hundred thousand degrees Fahrenheit, and an even hotter core, the fireball caused flash burns on the target ships and animals aboard them.

Ten to twenty seconds after initial explosion, the fireball dissolved into the now familiar mushroom cloud. First, there was the dazzling, circular white condensation cloud. But within five seconds it had enlarged to a horizontal ring two miles in diameter, enclosing for moments the fireball. Within another few seconds, from its center, as the condensation cloud dissolved, came the mixture of radioactive fission products, soot, vapor, and smoke, wriggling upward so fast that within twenty seconds it was one mile high. Seven minutes later it was at seven miles high, one mile higher than Mt. Everest.

As it quickly rose, the stem expanded to two miles across, and at the top a cap of ice crystals appeared.

For thirty seconds it had been quiet, and then CBS correspondent Bill Downs was the first to speak from an observer plane over Bikini saying, "There was no tidal wave. The airplane did not receive any shock wave. I can't see any damage below."

The *USS Appalachian* had been safely anchored eighteen miles from Bikini. Journalists had lined the starboard side of the ship to watch the bomb go off. They had put on goggles and covered their eyes with their forearms, as had been suggested. They had expected a tremendous explosion followed by a blast, and a great heat flash. As Navy Captain Lee would later write, "Those things didn't happen. Many didn't recognize the heat flash when it came, but it was there. I felt it myself. All of this anticipation ended up with a sense of let-down, and no immediate news, so there was a tremendous rush of writing about everything and anything, right or wrong."

It was not the first time the media missed the story when it came to nuclear weapons, and certainly not the last.

Complicating the initial radio reporting was the lack of experience with local atmospheric conditions in the Bikini area. There was static and thus poor reception in the United States for the "pooled" broadcast immediately following the Able bomb blast, leading to complaints from American broadcast companies.

From the *USS Appalachian*, the *Washington Post*'s Gerald G. Gross cabled to his newsroom, "Many observers seemed disappointed by the reports after the terrific anticipation and suspense which had been built up within us. Some had expected terrific destruction: perhaps they expected too much." Others reported the *USS Nevada* was still afloat, and on Bikini Island palm trees continued to wave. The *New York Times* initial story was headlined BLAST FORCE SEEMS LESS THAN EXPECTED. It recorded that while the initial explosion appeared "ten times brighter than the sun" only two ships immediately sank, or capsized, and eighteen were reported damaged. Associated Press reporter Don Whitehead, also aboard the *USS Appalachian*, wrote that along with the sound of the blast "came the shock wave. I noted a sudden sharp pain in my ears, and felt the rush of the wind. But it was only a small, sharp, shock—far from the wind anticipated by those who had heard descriptions of the New Mexico blast which swept men from their feet at 20 miles."

Simon P. Aleksandrov, the Soviet geologist/KGB agent, pointed to the mushroom cloud and muttered, "Not so much," and a Brazilian observer told a reporter he felt "so so" about the blast. Gen. Joseph "Vinegar Joe" Stillwell, a hero of World War II in the Pacific and a member of the Joint Chiefs Evaluation Board, was aboard an observer plane. "The damned Air Corps has missed the target again," was his reaction when he saw the *USS Nevada* still afloat after the detonation.

Vice Admiral John H. Hoover, assistant chief of naval operations for materiel and also a member of the Joint Chiefs Evaluation Board, also voiced belief that the bomb had not gone off as planned. Admiral Parsons, the weaponeer on the Hiroshima bombing mission, felt that the Able bomb was less powerful than either the Hiroshima or Nagasaki atomic bombs.

Symington, aboard a C-54 observer plane, would write in a memo that his was the closest aircraft some seventeen miles away from Bikini Lagoon. "The test itself was awesome," he wrote. "We were given dark glasses to protect our eyes and warned not to be concerned at the shock wave, which would hit our plane a few seconds after the explosion." Looking down at the scene, "I noticed a large ship turn over in the lagoon as a child would turn a toy boat over in a bathtub," Symington wrote. He was referring to the USS *Gilliam*, which in November 1944 had carried one thousand troops of the US 11th Airborne Division to Leyte during the reinvasion of the Philippine Islands. The *Gilliam* had been less than fifty yards away from zero point, above which the bomb had exploded, and thus the ship closest to the blast.

An hour and a half after the bomb was exploded, Admiral Blandy broadcast an announcement from the USS *Mount McKinley* to other ships in the task force: "The bomb was dropped with very good accuracy. I must say here that I am very well pleased indeed with the excellent performance of our task force in the operation. It could not have been better." Of course, the bomb had missed the target by one-third of a mile.

Blandy added, "It might well be at least a month before we are able to determine accurately the efficiency of this bomb, but we should have an approximate idea of it in a day or two. Our necessarily complex safety plans have been functioning perfectly and continue to do so. There are no known deaths or injuries to our men, nor do we expect any. The radiological safety parties are already functioning and the drone boats and planes are now returning the information we need."

Just after Blandy's announcement, NBC interrupted its network broadcast with the following: "A few hours ago the atomic bomb was dropped on the target in Bikini. Aboard this vessel [the USS *Panamint*], which carries the scientific observers and United Nations representatives, there is, without a doubt, a keen sense of disappointment. Disappointment with what we witnessed this morning. It was a successful experiment for the Army and Navy, but from twenty miles away it was a pretty poor spectacle."

7 · ABLE'S AFTERMATH

■ **WHEN THE ABLE** bomb detonated, its fireball did not touch the earth's surface, in this case the water of the Bikini Lagoon. As the fireball rose and then cooled, small, smoke-like particles formed and were carried up with the cloud to what was called a stabilizing altitude, some twenty thousand to sixty thousand feet in the air. The spread of this fallout material depended on the winds and weather. Gradually it returned to earth and created some localized, hazardous areas.

Soft and fluffy as it seemed from miles away, the interior of the mushroom cloud was deadly. The fission products, predominantly the vaporized bomb elements, were radioactive, the equivalent of tons of radium. Any human who had inhaled its vapors would have died. And what made it more deadly was that, after an hour or so, its unique identity from the normal atmospheric clouds vanished, while the unseen deadly fission products remained.

The hazard from radioactive materials tends to decrease with time depending on the so-called half-life decay rate of that material plus its dispersion and dilution while traveling in the air.

As part of the protective coverage from fallout, the Joint Task Force had established a team of five "downwind" US Navy destroyers. After detonation, they sailed up to sixty miles in patterns downwind from Bikini. Following the cloud path, they measured for any possible fallout. The destroyers continued their patrolling until the distant limits of detectable radioactivity was reached. In addition, two teams of B-29 and F-13 aircraft with photographers aboard were designated as Cloud-Tracking Aircraft Units. They flew parallel on either side of the cloud mass and reported its position as it went downwind.

Eight minutes after detonation, a B-17 drone entered the mushroom cloud at 24,000 feet, followed a few minutes later by three other B-17 drones at 30,000, 18,000, and 13,000 feet. The top turret of the B-17 drones had been replaced by a large filter box. In each bomb bay was a large rubber bag. When the drone aircraft entered the radioactive cloud, on command of its controller on the ground, the air filter opened for twenty seconds to allow ninety cubic feet of air into the rubber bag, which then shut tight. The drones were then guided back to Enewetak where the air was examined.

The Navy, meanwhile, sent in three F6F drones, assigned to the carrier *USS Shangri-La*, at altitudes of 20,000 feet, 15,000 feet, and 10,000 feet.

The B-29 drones had Geiger counters and tracked the radioactive clouds over deserted Pacific Ocean areas where, after several hours, the fission products lost much of their dangerous nature. Other drones collected air samples while their automatic cameras took photographs before returning to base.

In his post-shot broadcast, Joint Task Force Commander Admiral Blandy reported, "The radioactive cloud is drifting as we had estimated. It will not endanger personnel, ships or other Pacific Islands." The Navy's 1947 official report on Crossroads said, "Only deserted ocean areas received the continuous and invisible fallout of radioactive materials. Fortunately the contaminated areas healed themselves relatively promptly."

The radioactive cloud had lingered above Bikini Lagoon for almost an hour after the explosion, causing the PBM Martin Mariner aerial observer aircraft, which provided no protection from radiation for its crew and passengers, to circle before approaching its outer limits. After the clouds had cleared at 10 a.m., an observer aboard PBM Charlie, the lead reconnaissance aircraft, reported that the *USS Gilliam*, which was closest to the actual explosion point, had sunk along with the *USS Carlisle*, another attack transport ship. The *USS Anderson*, a 338-foot destroyer that had taken part in many of the Pacific sea battles and was within a half mile of the explosion, had also already sunk. Pictures from a PBM

Martin Mariner photo seaplane taken through the smoke cloud one minute after detonation showed the *Anderson* burning amid ship and half the upper part of its superstructure gone. The *USS Lamson*, a 344-foot destroyer, was observed by an aircraft 45 minutes after detonation lying on its starboard side with its bridge structure under water. Sometime later, the boat rolled over, floated awhile with its bottom above the surface, and then sank. The Japanese light cruiser *Sakawa*, about a half-mile from the detonation, was heavily damaged with major breeches in its hull, fires on the stern, and severe superstructure damage.

At 10:15 a.m., the *USS Begor*, control ship for drone boats, moved to a position ten miles outside Bikini Atoll and released the drone boats, which were guided into the lagoon to carry out sampling of radiation. Transmitters aboard the *USS Begor* guided the boats, and a cylinder emitting a yellow-green smoke allowed aircraft monitoring the drones from above to follow their courses. When a prescribed Geiger counter reading was transmitted back to the controllers on the *Begor*, a water sample was taken. As a boat completed its mission, it returned to the *Begor*, was washed down by fire hoses, and boarded by a safety officer. When the boat was declared safe, a radiation chemist boarded and gathered the collected water samples.

The drones were followed by radiation safety monitors who traveled aboard six motor patrol gunboats and twenty landing craft. Their measurements of radiation intensities were sent by radio to a control center where they were mixed with information that had come from the aircraft measurements already collected.

By noon, the drones had covered most of the lagoon. As their data was collected, military and civilian personnel kept a chart with a red line indicating where the water was clear. That second wave of boats and landing craft helped determine areas declared safe from radiation and they were indicated by a blue line. By 2:30 p.m., the lagoon was declared safe, and the first of the initial ships with boarding teams sailed in to start the inspection of the vessels in the outer part of the target array. At 10:30 p.m. that evening, eighteen target ships on the outer perimeter had been boarded and declared safe.

One of the radiological safety officers assigned at Crossroads to monitor the aftermath of the test shots was Dr. Robert A. Conard, then a thirty-three-year-old Navy commander. Little did Conard—a brown-haired, handsome, soft-voiced, southerner—know that almost the rest of his career, in the Navy and out, would be spent dealing with the aftermath of American nuclear testing.

Born in 1913, Conard graduated with a BS from the University of South Carolina in 1936 and received his MD in 1941 from the Medical College of South Carolina. That same year he enlisted in the US Navy Medical Corps and spent more than two years in the Pacific as the medical officer on the light cruiser *USS Montpelier*. After the Solomon Islands campaign, and when the *Montpelier* was on its way to the Mariana Islands for landings at Saipan, Guam, and Tinian, Conard had his first look at the Marshall Islands. At that time, he saw Kwajalein and Enewetak, both of which showed ruins from the fighting that had ejected the Japanese.

In November 1944, Conard was assigned temporarily to the National Naval Medical Center in Bethesda, Maryland, with the intention of going on to the Navy hospital in Charleston, South Carolina, to specialize in internal medicine. However, as Conard would put it in a 1993 oral history interview, "Something happened that changed my whole career—and that was the atomic bomb. I got a call from Navy officials in Washington. They told me they were looking for doctors that would specialize in radiological safety and become associated with an atomic bomb test which was to take place. They wondered if I might be interested. I was floundering at that point, and I said it sounded very interesting."

In early 1945, he was picked by the Navy's Bureau of Medicine and Surgery to be one of ten medical doctors from different services to be trained in what was to be a new field—atomic medicine. They were to study radiation effects, radiation measurements, and radiological safety. As part of their education, the group spent four to five months visiting various Atomic Energy Commission (AEC) laboratories, learning how to use instruments to measure radiation and study its effects on humans.

Conard would later say, "We turned out then to be the corps of medical experts to join in the Bikini operation—Operation Crossroads. . . . At that time the effects of atomic weapons were practically unknown, and so this was a new field we were getting into."

Conard had watched the Able shot from one of the support ships which then sailed closer to Bikini. On the afternoon of July 1, when the clearance was given, he found himself on one of the first patrol boats with eight other men.

As Conard later wrote, "We steamed into the lagoon, and with some trepidation, because we really didn't know what we were going to face." Because it was an airburst, he said, there was "no fallout and there was little radiation involvement. We were able to clear the ships for boarding soon."

Seven hours after the detonation, they began retrieving test animals from the twenty-two ships on which they had been placed, taking them to the *USS Burleson* for examination and medical care for those with injuries. About 10 percent had died instantly from the impact of blast—mainly pigs, goats, and guinea pigs—either from lung hemorrhages or contusions. Reactions were obviously dependent on distance from the detonation point.

The fireball in the first half-second, although it had not reached the lagoon's surface, still had produced serious flash burns on the animals, particularly those in direct line of sight from the detonation. Fur had protected some, while skin creams and clothing on others had proved to have provided some protection.

Radiation had greater impacts.

Gamma radiation, in animals and in humans, slows production of red and white corpuscles, which in turn causes excessive bleeding and, over time, lack of protection against infection. Heavy gamma ray exposure leads to hyper-irritability, muscular weakness, diarrhea, reduced breathing, and loss of appetite. Lesser gamma exposure causes later effects, such as hair loss or blood rupture patches or purple or dark red blotches on the skin.

Some animals when found were nearly dead, could not stand, or had blood in their diarrhea. Goat 53, penned in the open on the deck of the original target battleship *USS Nevada*, which had been some 625 yards from the actual detonation, died two days later from the fireball's radiation. Goat 119 had been tethered on the *Nevada* inside a five-inch gun turret and shielded by armor plates. Nonetheless it had been exposed to gamma rays and died four days later of radiation sickness. An examination showed enlarged blood vessels in Goat 119's brain.

Some goats proved to be unmoved by the experience. A camera had been trained on one goat aboard the transport *USS Niagara*, which was quite distant from the detonation. It showed the animal had kept chewing on hay as the shock wave hit and debris fell around it.

A fifty-pound, six-month-old China Poland sow was found swimming in the lagoon on July 2. As Pig 311, it would become famous for having escaped from the Japanese officer's toilet on the main deck of the cruiser *Sakawa*, just after that vessel had taken on water. Later examined aboard the *USS Burleson*, the pig showed signs of radiation sickness and had a low blood count. Within weeks, Pig 311 recovered and eventually was sent back to the Naval Medical Research Institute in Bethesda, Maryland, for observation. In April 1949, Pig 311, then six hundred pounds, was given to the Smithsonian's National Zoological Park in Washington, DC, where she became a favorite of visitors. Attempts to breed her failed, and she died on July 8, 1950, four years and seven days after her exposure. The cause of death was not disclosed.

White mice were flown back to the National Cancer Institute as soon as possible. They were mated, and in late September a few litters were born that apparently were normal.

On August 2, 1946, the Presidential Evaluation Board made a preliminary report on the animal studies after the Able test air burst and related them to what would have happened to the ships' crews. It found "initial flash of principal radiation, which are gamma rays and neutrons, would have killed about all personnel normally on station aboard the ships centered around the air burst and many others at greater distances. Personnel protected by steel and other dense materials would have been relatively safe on the outlying target vessels. The effects of radiation exposure would not have incapacitated all victims immediately, even some of the most severely affected might have remained at their stations for several hours. Thus it is possible that initial efforts at damage control might have kept ships operating, but it is clear that vessels within a half mile of an atomic bomb air burst would eventually become inoperative due to crew casualties." Three years after Crossroads, only twenty-eight of 5,500 animals that took part in the tests were still alive.

On July 2, as more areas of the lagoon were inspected and ships found clear of radiation, further on-board inspections began. Blandy took Navy Secretary James Forrestal on a tour in a picket boat with a Los Alamos scientist holding a Geiger counter to determine radiation levels. As they approached the submarine *USS Skate*, one of the ships on the surface closest to the detonation point, the Geiger counter measuring radiation levels went "off the scale" and Blandy was reported saying, "Let's get the hell out of here." Still, they inspected the submarine and noted the sub's superstructure and weather deck were demolished. The bridge structure was opened up and bent upon itself in front of the conning tower, and nearly half the forward superstructure had been wrecked.

It was later determined that the submarine's pressure hull was undamaged and operable. However, while all the ship's control and propulsion machinery were undamaged and operable, all electronic equipment failed to work.

Blandy explained to Forrestal that submarines at that time were built with hulls thicker than surface craft because they had to withstand terrific pressures when they submerge. Later that same day the *Skate* was towed to Bikini Island and beached to prevent her from sinking.

Late effects from the test shot were still taking place. The heavily damaged, Japanese cruiser *Sakawa* had taken on water. While the boat carrying Forrestal and Blandy maneuvered nearby, the cruiser sank as it was being towed by a salvage group to be beached on Bikini Island.

The Blandy-Forrestal boat circled other ships looking at damage and observing fires. Finally, they boarded the thirty-two-year-old battleship *USS New York*, which had been anchored more than one thousand yards from the detonation point. It had been declared safe earlier that morning and had an "affirmative" blue and white barred banner flying to show it was free of damaging radioactivity. Aboard ship, however, there were indicators of the blast. The initial hot flash had left a burn shadow against a gun turret; it was in the shape of a folding chair that had been leaning against it. Fires started by the heat had blackened the battleship's smokestack; the blast had shattered glass from searchlights and crushed cans of Army test materials that had been placed on the deck. Blandy noted that "the heavy parts of the ship itself were undamaged."

The admiral and Forrestal then posed in front of a blue-chalked inscription, "Old Sailors Never Die." It had been written by a sailor as he had departed before the test shot. Blandy reminded Forrestal of a promise he had made to then-New York Governor Thomas E. Dewey, that the state would eventually receive the battleship as a museum piece, if it survived the atomic tests.

Speaking to the press after his lagoon tour, Forrestal described the first Crossroads test as "tremendous," and that "heavily built and heavily armored ships are difficult to sink, unless they sustain underwater damage." Even in the atomic age, he continued, "the American Navy will continue to be the most efficient, the most modern and the most powerful in the world."

■   ■   ■

Another task force team approached the carrier *USS Independence*, which had been less than a half mile from the detonation spot. They tried to extinguish fires aboard because some were near ammunition stores. The flight deck was torn its entire length, and the superstructure was gone. Holes were in her sides, and radiological readings showed levels above those permitted for boarding. Like the *USS Skate*, the *Independence* was towed out of the target area. It would take two more days before teams on July 4 boarded the *Independence* for decontamination and cleanup work. It was another week before a full crew returned.

■   ■   ■

Stuart Symington, who had been on Guam listening to the radio broadcast of the post-Able activities, decided to skip the Philippine independence celebration that had been an original stop for his Pacific trip. He returned to Kwajalein to arrange for his own survey of the Able bomb's effects.

A supporter of air power, Symington would later write that he "was disturbed at the way in which various commentators, encouraged by the Navy, were downplaying the results of the tests and the importance of nuclear power in warfare." He arranged to inspect the target ships with Blandy and wrote, "The force shown us was unbelievable. On one ship, steel doors were blown off and in, then blown around and out of another entrance. Clearly this was a great new force with which the world would have to contend."

On the other hand, a Navy veteran would later write that he went back the next day aboard the carrier *USS Saratoga*, which had been more than a mile from the blast site, and found the only damage was a bent smokestack and small fires on the flight deck. Along with the rest of the crew, they returned to their quarters aboard ship and began preparing for the next shot, Baker, which was to come in three weeks.

A group of scientists and journalists visited Bikini Island, which was four miles from the explosion point, and, as one reporter recorded that monitoring towers set up for the shot were intact and, in his words, "Palm trees were not singed."

8 · THE MEDIA STORY I

■ **WITHIN TWENTY-FOUR HOURS** of the Able test, journalists had filed some 250,000 words, the greatest amount of immediate coverage that would follow any nuclear test event. However, in succeeding days, the media continued to convey a mixed message to the public at large.

The Navy controlled most of the reporting and downplayed the destruction. Two days after the test, Captain Lee shepherded a group of journalists around the lagoon for an inspection of the ships. Lee, a loyal Navy officer, underplayed the damage, saying, for example, that the *USS Independence* could in short order be put underway by a good crew, ignoring its hull and superstructure damage. Passing the *USS Skate*, he claimed the submarine was "good as ever," although its topside was a mass of twisted steel.

Blandy would tell the reporters that if they had been inside the *Skate* at the time the bomb exploded, "you would never know there had been an atomic blast." As for the aircraft carrier *Independence*, he said he had seen more damage done to ships by Japanese kamikaze attacks.

As expected, Army Air Corps officers saw things differently. Brigadier General Ramey had gone on the radio hours after the shot and said Bikini's lagoon was "still reacting from the effects of the blast and radiation of the world's mightiest bomb." Four days later he told a press conference that, had it been real combat, the entire fleet of ships would have been wiped out.

A week after the test, on July 7, 1946, the *New York Times'* Hanson Baldwin found himself reporting not only on the Able shot but also a newly released United States Strategic Bombing Survey of the atomic bomb effects on Hiroshima and Nagasaki. His article was entitled, ATOM BOMB IS PROVED MOST TERRIBLE WEAPON: SURVEYS IN JAPAN AND BIKINI TEST ARE ENOUGH TO CHANGE CONCEPTS OF WAR.

In the piece, Baldwin seemed to play down the damage from Able, writing that "the results at Bikini must . . . be qualified" because of the close stationing of the target ships, and that much of the damage "could have been avoided had there been fire-fighting crews and damage control parties aboard."

The main message of his article, based primarily on the survey in Japan, "emphasize[d] that the atom bomb [was] a terribly potent weapon of destruction." He pointed out that the prime targets in Hiroshima and Nagasaki were human beings, as part of the effort to end the war. In that sense the Bikini tests could not gauge that type of threat, since the Able shot was only aimed at ships, deserted by military and civilian personnel who were safely miles away.

Baldwin concluded, "After the let down from Crossroads and the fourth atomic bomb [including Trinity], the general public has not assessed properly the new knowledge now available. The atom bomb is primarily a weapon against cities, and the Bikini tests can give no answer to that."

On July 11, the Joint Chiefs Evaluation Board released its preliminary statement on Test Able. Members had observed the shot, surveyed the ships, and interviewed specialists, although all the test data had not yet been compiled.

The board's report said that within a half mile of the detonation point, which was some 1,500 to 2,000 feet west of the original target, a destroyer and two troop transport ships sank immediately. Another destroyer capsized and sank the next day, as did the Japanese cruiser *Sakawa*. The *USS Skate*'s heavily damaged superstructure made it impossible for it to submerge. The carrier *Independence* was badly wrecked, gutted by fire and damaged by internal explosions, some from torpedoes stored below decks. The other fifteen badly damaged but afloat ships that were within one thousand feet of the explosion point took up to twelve days to patch up well enough so they could depart under their own power for Navy bases where they could get needed major repairs.

Three major combatant ships had been within a half-mile of the explosion, including the *USS Nevada*, the original target ship. While their hulls suffered little damage, their superstructures were "badly wrecked," and thus "these ships would unquestionably be put out of action." They, and other ships within three-quarters of a mile of the detonation, would have required major repairs at a principal Navy base.

More important were the board's estimates of the potential effects on personnel. It stated that flash burns, produced by instant radiation from the explosion, "indicate that the casualties would have been high among exposed personnel." Those below or even sheltered on deck "would not have been immediately incapacitated by burns alone." It reported that "lethal doses of radiological effects" would reach personnel within ships that received extensive blast damage to their superstructures, and that casualties from blast effects alone would be high to exposed personnel within a half mile, although more medical data was needed.

The board said one thing was clear: "The atomic bomb dropped at Bikini damaged more ships than have ever before been damaged by a single explosion."

Assessing "combat readiness" in the wake of Test Able, the Navy's own Bureau of Ships group found many of the "surviving" vessels would be virtually dead in the water, their boilers, radar, radio, and equipment out of commission, and their crews dead or dying from radiation.

9 · BAKER TEST PREPARATIONS

■ **A FEW HOURS** after the Able shot took place, Joint Task Force Commander Adm. William Blandy told his staff he wanted the Baker underwater test to take place as soon as possible. Rear Adm. Thorvald A. Solberg, the Joint Task Force director of Ship Material, commanded some ten thousand personnel at Baker and was responsible for all the Navy ships involved. He balked, saying he would need six weeks to get the target ships in order. But Blandy gave him three weeks, the time scientists had told him they needed to have the tests take place or they would have trouble holding their staffs together.

The Baker test was set for July 25, 1946, and it would turn out to provide more important lessons for nuclear weapons than Able. Baker's results, not appreciated at the time, have become an important lesson in the study of radioactive fallout.

Back in October 1945, then-Navy Commodore Deak Parsons, fresh from his experiences at Los Alamos and the atomic bombings of Japan, was working at Navy headquarters in Washington, DC. He had set up what was called the Navy Atomic Bomb Group of eight people, six other officers and two civilian scientists. His group had the assignment of putting together a plan for possible tests of atomic bombs to be exploded above and/or below Navy ships. One of the group's four recommendations was for "a shallow-water explosion within a harbor, within an assemblage of moored ships," according to a letter Parsons sent that month to Norris Bradbury, the Los Alamos physicist who Oppenheimer had just chosen to replace him as director of the national laboratory.

The other three suggested tests called for exploding an atomic bomb suspended from a blimp tethered above target ships; detonating one from within a bathysphere a half-mile below the ships; and dropping one from a B-29 that burst above the ships.

In December 1945, Parsons was named senior Navy member on General LeMay's joint Army-Navy subcommittee that was to plan the tests. In January, Parsons became a rear admiral and was appointed one of Blandy's two deputy Joint Task Force commanders for Operation Crossroads. It was no surprise that the Parsons group's suggestion for a shallow-water explosion became Crossroads Baker test shot.

Almost immediately after Blandy announced it at his January 24, 1946, press conference, the impact of an underwater atomic bomb explosion became a subject of controversy. Senator McMahon, from his position as chairman of the Senate Special Committee on Atomic Energy, speculated that a subsurface nuclear explosion might set off a chain reaction. A Navy spokesman expressed some scientists feared it might set off a tidal wave. There also was concern about the potential spread of radioactivity.

In public appearances, Blandy would often repeat lines from his February 21 speech, such as, "The bomb will not start a chain-reaction in the water converting it all to gas and letting the ships on all oceans drop down to the bottom. It will not blow out the bottom of the sea and let all the water run down the hole. It will not destroy gravity."

Others spoke out against the public anxiety, including renowned physicists Edward Teller and Hans Bethe, who had taken part in the Manhattan Project. They said the test explosions, including the one scheduled to be underwater, would not cause much more than local disturbances and certainly not result in the "explosion of the globe."

At Los Alamos, however, there was recognition that no atomic weapon had ever been detonated with water as the surrounding medium. As one paper from the laboratory put it, "While no serious doubt existed on this matter, it was nevertheless not completely certain that an atomic weapon would transfer energy to a more dense medium [meaning the water] in exactly the same manner as it would transfer energy to air."

A Defense Nuclear Agency (DNA) report released before Crossroads had speculated that the initial radiation caused by an underwater atomic detonation would be "absorbed by the water surrounding the device" with the intense heat vaporizing the water near the burst. This would form "a bubble beneath the surface of the water" that would expand as the energy released in the explosion worked against the mass of water. This expansion would continue until the energy is expended, at which point the bubble would begin to collapse as it rose toward the surface.

As planned in Baker, if the burst were close enough to the lagoon's bottom, an underwater crater may be formed, and the material excavated from it would be radioactive and contribute to the residual radiation from the explosion, according to this DNA projection.

To prepare for this first underwater nuclear detonation, Blandy's task force group and independent scientists carried out different simulations to help predict not just wave reaction but also the spread of radioactivity. One involved the detonation of one thousand pounds of TNT in shallow water with the results extrapolated upward to the 20,000-TNT-level expected from the test bomb. In another case, scientists examined how high the column of radioactive water would rise.

They determined a column going up ten thousand feet would cause the greatest hazard because most of the radioactive fallout would come down on target ships within one thousand feet of the detonation, making reboarding them dangerous for weeks. Although predictions were that the radioactive steam from the shot could go up anywhere from ten thousand feet to sixty thousand feet, the most likely guess was thirty thousand feet. Nevertheless the pre-Baker plan put the maximum at fifteen thousand feet.

The final estimate was that radioactivity from the water would cause target ships within 1,500 yards of the detonation to be seriously contaminated. It would be dangerous to reboard those within 1,000 yards of the detonation for weeks because of radioactive water contamination, according to one study. Yet, in a July 10 conference aboard the flagship, *USS Mount McKinley*, Dr. Stafford Warren, the task force radiology chief, suggested access to the ships might be possible within five days.

Ironically, while there was recognition that radioactivity in fallout would play a bigger role in the Baker test, the total number of civilian and military officers in the Radiological Safety Section of the Joint Task Force dropped from 303 for the Able test to 258 for Baker. Not only that, but eight members of Congress departed along with more than forty of the journalists who had come out for the Able shot.

While Baker preparations were underway, the *USS Appalachian* sailed back to Pearl Harbor with some of the departing news personnel. While there, it picked up other reporters who would cover the second test. At the same time, some remaining journalists along with those observers who remained were taken on a cruise to other South Pacific sights such as Ponape, Truk, Majuro, and Guam to kill time.

During that same period, on July 16, there was a ceremony on Rongerik Atoll, where the Bikini Marshallese Islanders had been moved. A US official read a message from President Truman that offered his, thanks to the 167 Bikinians who had left their homes. Sen. Carl Hatch (D-NM), one of the congressional observers for the Able shot, addressed King Juda, who was dressed in a Navy fatigue uniform, with the Bikinians seated behind him in a semicircle. Hatch said that the president in Washington "knows the sacrifice you have made and he is deeply grateful to you for that. You have made a true contribution to the progress of mankind all over the world." He added, "The President of the United States extends to you, King Juda, his thanks for all that you have done."

Juda was then given as gifts from the grateful US "a pipe, a cigarette holder, matches, a carton of cigarettes and a complete set of photographs of the atomic cloud over Bikini," according to a news report. In addition, he was also offered an invitation to witness the upcoming Baker test shot.

Meanwhile, repair work had to be done on several target ships damaged by the Able shot. The submarine *Skate*'s superstructure was repaired and a temporary bridge built. The aircraft carrier *Independence* needed substantial work to make certain it was watertight. There were substitutions for ships sunk. Those damaged by Able were moved to less exposed positions in an exchange with those hardly hit by that first test—thus putting in place a better means of testing gradation of damage from the second shot.

The Japanese battleship, *Nagato*, along with the *USS Independence* were moved closer together, and within 1,500 yards of the planned detonation point, as were several other larger ships. In a pattern that simulated an amphibious landing, 24 landing craft were beached on the lagoon side of Bikini Island. Altogether there would be 68 ships and submarines as targets.

Four days before the planned shot, six unmanned submarines were hung with weights in order to keep them in a submerged position for the test. When two of the subs popped up, they had to be resubmerged in a process that lasted until early morning on test day. The damage to the *Skate*'s topside prevented it from being submerged, but since its pressure hull was sound, it and one other submarine were moored on the surface along with two anchored seaplanes.

For Baker, a limited number of animals were available after the effects of the Able shot. Also limited were the anticipated types of injuries expected from a shallow underwater nuclear explosion, beyond radiation exposure from fallout. Some twenty pigs and two hundred rats were placed aboard four attack transport ships, exposed so they would provide data about injuries from any blast-plus-radiation.

The day before the shot, the animals were all photographed in their locations and supplied with an adequate supply of food and water since there was an expectation that there would be delay in recovering them after the shot because of irradiated fallout.

The vessel central to the Baker test was *LSM-60*, a two-hundred-foot, World War II amphibious landing craft that had been built in 1944. It had delivered troops and equipment during the 1945 invasion of Iwo Jima, for which it received a battle star. It had been converted so it could hold in place a heavy, watertight, steel container, suspended into the water under the ship's hull at a fixed point 90 feet below and halfway to the bottom of the 180-feet-deep Bikini Lagoon.

Inside the steel container was the nuclear device, nicknamed "Helen of Bikini," whose explosive power was to be twenty-three kilotons.

Two air-conditioned laboratory spaces had been created aboard *LSM-60*, where scientists worked preparing the Nagasaki-sized "Fat Man" implosion device. The ship also had been equipped with electronic receivers and transmitters, plus a high antenna on its forward deck that would allow it to receive the coded signal to trigger the device. Within the container, "Helen of Bikini" had a coaxial cable running from it up into one of the newly-constructed laboratories in the *LSM*.

■　　■　　■

King Juda's arrival at Bikini surprised the task force leadership, including Blandy, who apparently had forgotten the invitation for the visit, which still awaited Joint Chiefs of Staff approval. Newsreel cameras were already rolling when in a small boat, Juda, at 2:35 p.m. July 23, approached the task force flagship. Since Blandy was elsewhere, his chief of staff welcomed the Bikini chief aboard and settled him temporarily in a cabin while they conferred on what next to do with him.

The black-haired, stocky, muscular, Bikini chief, wearing a white tank top and shorts, was taken on a tour of his home island with an interpreter and officers. He saw there was no damage from the Able nuclear test. He remarked that the older of his people wanted to return to the atoll, if Bikini remained intact after all testing, but the younger ones did not care where they lived. That night, Juda slept in a Seabee barracks on Bikini Island.

On the morning of July 24, 1946, Blandy confirmed that the Baker shot would take place the next day at 8:35 a.m. Juda, now in khaki shorts and shirt, wearing black Navy shoes with one laced with white string, attended a briefing on the upcoming test and was told the device would blow a great deal of water high into the air. He asked to stay aboard one of the ships to witness the test, but Blandy had to admit he still had not received approval from Washington.

Late on July 24, just twelve hours before the test, the Joint Chiefs finally sent a radiogram to Blandy saying he was to "use his [own] judgment" with regard to Juda, so the Bikini chief was added to the distinguished observers on the admiral's flagship.

It began raining heavily late on July 24, but more favorable weather was predicted for the next day, although it would make less difference because the shot was to be underwater.

Evacuation of support ships from the lagoon and people from Bikini had started that July 24 morning. Personnel and ships not needed in the immediate aftermath of the shot were sent to Rongelap Atoll some 120 miles away. Because of fear of potential impact of the detonation, five C-54 transports were placed on Enewetak, 190 miles west of Bikini, to evacuate personnel there if it became necessary. Also on Enewetak were two C-54s that would fly the next day through the test cloud gathering samples and then go directly to Kwajalein so the samples could be analyzed.

By late afternoon July 24, all those on Bikini Island had been picked up, and at 5:35 p.m., sunset, all but 13 support ships plus the 68 large and small target ships were left in the lagoon.

Deak Parsons and several colleagues remained aboard the *LSM-60* that night, making sure the vessel was moored correctly in position and that there were no problems with the device itself or with the communications that would trigger it. Parsons was the person who had flown on the B-29 *Enola Gay* less than a year earlier. He had been aboard the *Enola Gay* to avoid having to arm the weapon on the ground before takeoff, thereby preventing an atomic disaster should it have crashed while taking off.

At 5:30 a.m. on July 25, 1946, Parsons and his team made their final checks and opened the transmission path for the coded signal, the equivalent of arming the device so it could accept the trigger order. Some thirty minutes later, they left the *LSM-60* in a small boat for the *USS Cumberland Sound*, a seaplane tender that for the tests had been converted into a laboratory ship. It was anchored thirteen miles away at the entrance to the lagoon.

Meanwhile, an unusual mission took place. During that early morning, the *USS Conserver*, a rescue and salvage ship, was among those doing last-minute work on the target ships in the lagoon. At 5:30 a.m., bunting was seen flying from the yardarm of a target ship, the *USS Gasconade*. It was the approved signal for evacuation, and turned out to have been raised by three sailors that had been overlooked the previous day when their colleagues had been taken off that ship. The three were picked up by the *Conserver* and by 6:20 a.m., Bikini Lagoon was officially declared evacuated.

As the *USS Cumberland Sound* sailed away from Bikini to a safe position out of range of the shot, Marshall Holloway, a thirty-four-year-old Los Alamos physicist, went down that ship's stairs toward the transmission room. The head of Los Alamos's group at Bikini, Holloway also served as Parsons's technical deputy within the task force and had been among the overnight group on *LSM-60*. He quickly settled himself in the transmitting room before a double console that contained dozens of buttons that controlled the firing mechanism. He had described the setup earlier to reporters as a "glorified alarm clock."

Under the preplanned schedule, when Holloway periodically closed a switch or two it would set off one sequence in the firing mechanism. Seated next to him with a microphone was British physicist Ernest W. Titterton, who had participated in the Manhattan Project and had joined Operation Crossroads at the request of the American group. Titterton, in a heavy British accent, was to broadcast the countdown over radio to loudspeakers on the other ships in the task force and to a worldwide audience

It was Titterton who slowly counted down the last thirty seconds.

At 59.7 seconds after 8:34 a.m., Bikini time, Holloway pushed the final button that sent the radio signal that detonated the fifth US nuclear device at exactly 8:35 a.m.

10 · THE BAKER SHOT

▪ **BECAUSE BAKER WAS** an underwater shot, the media and observers were aboard ships positioned ten miles closer to the target area than they had been for Able, and personnel did not have to wear either dark glasses or turn away at detonation.

"Things happened so fast in the next five seconds that few eyewitnesses could afterwards recall the full scope and sequence of the phenomena," wrote Adm. William A. Shurcliff, the Joint Task Force's official historian. Deak Parsons would later write, "Even from 15 miles, Baker was a spectacle." More dramatic was the *New York Times'* William Laurence, who reported, "For a time it looked as though a giant mountain had risen from the sea, as though we were watching the formation of a continent . . . and then it took the shape of a giant chain of mountains, covered with snow, glistening in the sun."

It was the same amount of energy that had been released by the Able shot, some twenty-three kilotons or the equivalent of twenty-three thousand tons of TNT, but now detonated below the surface of the lagoon.

In a fraction of the first second, thermal radiation created an underwater fireball or gas bubble which rose as a giant dome under and around the 750-ton, *LSM-60* and surrounding target ships. Within that first second, as the bubble broke the surface, the fireball shot upwards as part of a frothing column of water.

The water column hit one mile up in fifteen seconds, and a mile-and-one-half by a minute. At that point, the water column had expanded to two thousand feet wide.

That crown spread with the appearance of a fluffy, six-square-mile-wide, cauliflower cloud with a hollow center that itself was about a mile-and-a-half in diameter. At its height, the column contained some two million tons of water. After another few seconds, this highly radioactive cloud of fission products and water began falling back to the surface. It then created what was later called the "base surge," in effect a deadly ring of highly radioactive mist which then moved outward at some 200 feet per second—equal to 135 miles an hour—over the entire target area.

Admiral Shurcliff described it as appearing like "a steadily expanding and fattening doughnut," rising from three hundred feet above the lagoon's surface to an ultimate two thousand feet.

An Army radiological monitor looked down from a Navy patrol aircraft circling the site and reported, "The flash seemed to spring from all parts of the target fleet at once. A gigantic flash—then it was gone. . . . And where it had been now stood a white chimney of water reaching up. Then, a huge hemispheric mushroom of vapor appeared like a parachute suddenly opening. . . . By this time the great geyser had climbed several thousand feet. It stood there as if solidifying for many seconds, its head enshrouded in a tumult of steam. Then slowly the pillar began to fall and break up. At its base a tidal wave of spray and steam rose to smother the fleet and move on towards the islands. All this took only a few seconds, but the phenomenon was so astounding as to seem to last much longer."

From another plane, an observer saw a ship in the initial column of water "on its nose before it sank." It was the battleship *USS Arkansas*, which had been moored just 250 yards from the detonation spot. The 26,000-ton, 560-foot-long warship had been lifted up in a second by the initial underwater fireball, thrust into the water column, and then driven down by the descending cascade to the bottom of the Bikini Lagoon, where it still sits today.

Then-Col. Kenneth Nichols, a Manhattan Project veteran, who watched from an observer ship, described the shot aftermath this way: "Niagara Falls in reverse shot up over an area fully 2,200 feet in diameter; millions of tons of water rose about 5,000 feet and finally vapor and steam came out on top.

As the tons of water came tumbling back into the lagoon, what appeared like a tremendous breaking wave broke out of the mass of water and advanced towards the next circle of target ships. Momentarily I thought, 'My God we have miscalculated the height of the wave, the alarmist may be right.'" It turned out to have not been solid water, merely steam and spray.

Two drone B-17 sampler aircraft had flown directly over *LSM-60* at the time of detonation. The one at six thousand feet had its bomb bay doors blown in and a tail gunner's hatch blown into the aircraft. The B-17 drone above it at sixteen thousand feet had jumped some three hundred feet from the blast but was not damaged.

The Baker shot's dramatic show above the surface water gave no indication of what had gone on below.

The Baker detonation dug a 2,000-foot-wide crater that went 32 feet deeper than what had been the 180-foot-deep Bikini Lagoon. The original sedimentary bottom had been primarily coarse-grained organic organisms mixed with a much smaller amount of sand and mud. After the shot, the bottom under the burst point had become primarily layers of mud, some 10-feet-thick.

The shock wave produced by the detonation was "probably the most severe shock wave ever produced on earth," Admiral Shurcliff wrote. Since water can hardly be compressed, it was a great medium for transmitting shock waves.

Underwater gauges that had been set prior to the test showed that near the explosion pressures were greater than 10,000 pounds per-square-inch; at up to one-half-mile they ranged down from the thousands of pounds-per-square-inch to the hundreds. The first surface wave 11 seconds after detonation and 330 yards away was 94 feet high; 12 seconds later at 660 yards it was still at 47 feet high, lifting the stern of the aircraft carrier *Saratoga* 43 feet into the air; 25 seconds later, at a mile from detonation spot, the wave was 24 feet high.

Baker had created an awesome sight to see, but the invisible, radioactive contamination caused by the underwater shot had initiated fallout problems that were unlike anything dealt with before.

Much of the initial nuclear radiation had been absorbed by the sea water. However, a significant fraction of neutron radiation had been taken in by sodium in the sea water, which created radioactive salt crystals in water droplets. Gamma radiation, too, had been absorbed by the sea water. What became more threatening were the radioactive, contaminated solid particles from the lagoon bottom, which had been pulled up into the water column.

As the base surge had expanded over the target ships, it bathed 90 percent of them in this radioactive contaminated water, sand, coral, metal fragments, and dust. The deadly, gamma-radiation emitting mist covered decks, along with exposed metal, wood, canvas, and most certainly the pigs and rats left out in the open.

Ninety-foot waves hit the closest target ships, but it was the radioactive base surge that covered the entire fleet, which Admiral Shurcliff later described as "a kiss of death on a majority of the target vessels."

One hour after the detonation, some target ships emitted radiation levels that were three times the lethal dose to humans. An individual exposed to those circumstances would have reached the allowable exposure dose in three seconds, which for Operation Crossroads was 0.1 roentgen per 24 hours.* That was, at the time, an amount of radiation that a human could tolerate without harmful effects on health. It was a rate that had been established in 1934 by the National Bureau of Standards, but in 1946 reaffirmed by the Joint Task Force Radiological Safety Section, which said that dosage was verified by two years of experiments with dogs and mice and a workforce of some eight thousand people.

Time passed after detonation. The sun emerged and dried off areas once wet or moist, but the alpha-, beta-, and gamma-radiating materials remained at levels that had not been foreseen.

Such fission products are submicroscopic particles and therefore not visible to the eye. Their presence can only be determined by using monitoring devices such as Geiger counters. The radioactive particles turned out to be extremely difficult to remove, imbedded as they were in paintwork, metallic structures, wooden decks, and absorbent materials such as ropes, clothing, and canvas.

There had been little experience in handling radioactivity like this outside controlled areas in the nation's nuclear laboratories. It would become a totally new experience for Navy and other personnel on the scene, where those involved in the decontamination operations could themselves become exposed.

---

* An old unit of ionizing radiation derived from exposure to x-rays or gamma radiation.

11 · BAKER'S AFTERMATH

FORTY MINUTES AFTER detonation of the Baker nuclear test, two drone boats had been launched from the *USS Begor* to enter the lagoon for water collection near the target ships. Directed by aircraft orbiting above, they gathered ten samples in five-gallon containers and by 10:30 a.m. were back at the *Begor*, still anchored miles outside the lagoon.

The drones, however, were so radioactive that they could not be boarded to retrieve the sample water containers. Instead, the small boats were directed by radio to another ship, the *USS Albemarle*, a seaplane tender which had been reoutfitted in order to carry the two atomic test bombs to Bikini. The *Albemarle* had air-conditioned laboratory spaces where the near final assembly of the devices took place, so it was equipped to deal with radioactive materials. Still, it would be four hours before those first water samples were placed onboard.

As preparations began for reentry of salvage ships and others into the lagoon, the drone boats were directed toward the target center and then halted when they ran into high surface areas of radioactivity. In some cases, the water around some outer rim target ships was pronounced clear, but then deemed unsafe because of drifting radioactivity.

After two hours, the radioactive cloud began moving to the north-northwest and task force ships anchored miles away in that direction were ordered to move westward.

At just past 11 a.m., ships of the salvage unit entered the lagoon. Only four ships on the outer edge of the target area were declared safe. None had been touched by the contaminated downpour from the base surge.

A landing party was sent to Bikini Island to pick up photographic and other technical equipment before that material became overexposed. The team had to leave quickly when their originally cleared entry channel showed ever higher radioactivity as they exited out of the lagoon.

A second manned patrol boat attempted to get to the carrier *USS Saratoga*, which was apparently sinking. That team also had to retreat as the radioactivity level grew dangerous. On its exit from the lagoon, the patrol boat's measured contamination was so great that it had to be moored and its crew evacuated for the night.

Five-and-a-half hours after detonation, eight initial boarding teams aboard patrol boats and salvage ships with radiological monitors could only watch as the *USS Saratoga* continued to sink. Blandy had ordered tugs to attempt to attach lines to the carrier in order to tow her for beaching, but approaching her or other ships that were clearly damaged could not be done because of high radioactivity.

At 5:45 p.m. the nineteen-year-old *USS Saratoga*, built originally as a battle cruiser but converted to be one of America's first aircraft carriers and the oldest one still afloat, slowly settled, stern first, to the bottom of Bikini Lagoon. Someone said, "Hail and Farewell" over the task force loudspeaker as sailors aboard other ships watched. In the *New York Times* the next day, Hanson Baldwin wrote, "Perhaps she might have been saved, had there been a crew aboard. But she died a lonely death, with no man upon the decks once teeming with life, with pumps idle and boilers dead."

Just before midnight, the task force reported to the Joint Chiefs that along with *LSM-60*, the ships immediately known to have sunk were the battleship *USS Arkansas*; the aircraft carrier *Saratoga*, which had been 350 yards south of the detonation; a landing craft; and an oil barge. The destroyer *USS Hughes* and the attack transport *USS Fallon* were both reported listing, as was the Japanese battleship *Negato*. All three had been about five hundred yards from the explosion in different directions and still too radioactive to permit being approached by salvage personnel.

By the end of the first day, only seven ships of the original sixty-eight had been cleared for boarding. Among the crews of patrol boats that entered the lagoon that first day, thirty-six returned with film badges that showed mild radiation overdoses.

On July 26, it was decided that radiological conditions within the lagoon had remained high, and while manned patrol boats would sweep areas around target ships over the next few days, only drone boats would be sent inside until contamination levels went down to permit reboarding of ships.

Dr. Stafford Warren wrote his wife that day saying he had taken Blandy and some press through the target area on a thirty-foot patrol boat and within thirty minutes instruments showed they had "got just a tolerance dose," with their Geiger counter still clicking away. Blandy later told reporters that passengers on that boat faced being "goners—if not immediately, at least later on," had they remained in that area because of their exposure to radiation.

That afternoon, the *USS Hughes* was towed to Enyu Island and beached to prevent its sinking, though it was left sitting four feet below the water. An attempt to tow *USS Fallon* for a similar beaching was delayed for a day because of high levels of contamination surrounding that ship. Study of a radioactive oil slick within the target area indicated one or more of the submerged submarines had been breached and had gone to the bottom.

On July 27, animals were removed from three ships still considered "hot" with radiation. Removing animals from two others was delayed because of high contamination. The last animals were collected five days after the test. Longer exposure to radiation had meant higher mortality rates. Most animals were on ships within five hundred yards of the explosion, and many of them were in interior rooms. Nevertheless, six pigs were found dead from radiation aboard target ships. None of the remaining pigs survived more than a month; nor did one-third of the rats. All the remaining rats died within four months from gamma radiation.

Based on what had happened to the animals, Admiral Shurcliff wrote, "Had the target array been manned, casualties and both physical and psychological injuries would have been very great. Rescue and attention to casualties would have been difficult and dangerous." Within just over a mile from the explosion, he said, ships would have been inoperable for weeks.

The Joint Task Force leadership then realized, according to a later Pentagon technical study, that "no plans had been prepared for organized decontamination measures," because no such radioactive fallout had ever been encountered before.

The Radiological Safety Section convened a meeting on July 27 to discuss what methods could be developed to help rapid removal of radioactive materials from the target ships. Porous materials, such as wood, paint on metal, and rough and rusted metal surfaces found all over ships were a major source of radioactivity, but it was also found on a variety of other organic materials.

A Radiological Safety Group was formed to carry out experiments on contaminated equipment taken from target ships. Electric lanterns, copper pipe, brass junction boxes and even a plastic coffee maker were blasted with gritty materials that were on hand, such as ground coconut shells, rice, barley, and sand. Reagents such as soap powder, lye, and naphtha were tried, but were unsuccessful. A long washing in a 5 percent solution of acetic acid did work, but it was not really applicable for mass usage on ships.

The next day, July 28, while radiation levels remained high, the order went out to resurface submerged submarines. By late afternoon, one was released to the surface while four others were reported to have sunk with ruptured tanks and leaks. It was no surprise that on the third day after the test, Blandy issued an order that warned: "Dangerous radioactivity persists in the vicinity of all target ships. Except for boats carrying [radiation] monitors, all boating shall be confined to the immediate vicinity of prescribed anchorages of ships authorized to reenter the lagoon."

It was another two days before all task force ships were able to take their pretest berths within the lagoon.

12 · CLEANUP ATTEMPTS

■ **INITIALLY, TREATMENT OF** the fallout from the July 25, 1946, Baker test involved washing down decks and then entire ships, first with water and then with foamite in order to remove fissionable material. That was followed by scrubbing down and scraping open surfaces and ships' hulls that had been in contact with algae, which had become contaminated from radioactive materials in the water.

In most cases this ship-cleaning process reduced levels of radioactivity, but only to allow safe reboarding of some vessels for very short times. Additional methods were needed.

Over the next two days experimental treatment of target ships continued, primarily using foamite and high-pressure saltwater wash downs. The Japanese battleship *Negato*, which had been taking on water from breeches caused by the detonation, was too radioactive for salvage work and sank the night of July 29.

On July 31, Admiral Solberg issued a memorandum entitled "Decontamination Procedures on Target Vessels," which recognized "most target vessels are contaminated to a greater or less degree with fission products and therefore present varying degrees of radiological hazards which at the present time prevent re-boarding." It said the current decontamination procedures were expected to reduce the radiological hazards on the topside of the target ships "to a point where it will be possible for personnel to be aboard for a period of at least four (4) hours at one time." But it also noted that so far there was no telling what radiation problems could exist below decks.

His memo then described the difficulties created by fallout of radioactive materials, which even today people would have difficulty contemplating.

"Fission products are sub-microscopic particles and therefore not visible to the eye and their presence can only be determined by the use of monitoring instruments. These products are extremely difficult to remove in as much as they are quite firmly imbedded in the paintwork, metallic structures, wooden decks and particularly in such absorbent materials as [rope] lines, clothing, bunting, etc. Great care must also be taken in all decontamination procedures to prevent personnel engaged in these operations from becoming contaminated themselves."

The continuing radioactivity produced by the Baker test fallout consisted of gamma rays, neutrons, and beta and alpha particles.

All cause some damage to human cells that compose living tissue. Human cells can recover depending on the radiation dosage, degree of initial damage, and the health conditions of the individuals involved. The degree of injury differs based on the depth to which radiation penetrates the body and the amount of energy absorbed. Gamma rays have a high penetrating power. Alpha particles have a very high effect but a very small range, and thus normally are not harmful unless taken into the body. Beta particles are fast electrons but lose energy as they pass through cells.

On August 3, Dr. Warren reported that some key personnel and enlisted men had received more than recommended radiation doses, and while they were not seriously harmed, more care had to be taken to avoid greater overexposure. As they continued to deal with the problem, he expected exposure to get worse. "There is very little leeway in this since approximately 10 percent of the daily permitted dose is already taken up by the exposure to which all the Task Force are subjected every day while living in the lagoon," Warren wrote in an August 7 report.

One example he used was in the case of the hands and faces of those sent to clean up targeted ships: "It is almost impossible to enforce the wearing of gloves continuously on badly contaminated ships during the clean up stages under present circumstances where large numbers of men are involved. Nor is it feasible to expect them to take the proper care of their contaminated clothes."

Adding to the problem was that there were not enough film badges, some other devices gave off erratic readings, and many of them failed altogether.

On August 4, Admiral Solberg, as director of Ship Material, issued instructions for soon-to-begin decontamination on the lesser-contaminated target ships.

The Joint Task Force had settled on the slow wholesale washing process that, each time done, drove down radiation levels about 50 percent so that personnel could spend at least two hours onboard target ships. Working in relays timed to limit their exposures, the sailors used a method of scrubbing with abrasive and paint removal materials to further reduce contamination. The process required a safety officer and a radiological monitor to accompany working crews to make certain they did not get overexposed.

Some individuals did receive more than the permissible level, but in those cases the persons were given time off for one or two days. Because radiological monitoring equipment broke down or became limited, it became necessary to establish a special group just to repair and maintain those instruments.

Personnel in cleanup crews had to wear rubber boots and gloves, since cloth or leather were quickly contaminated with fission products and were difficult to clean. All personnel had to be fully clothed at all times despite the hot, humid weather. They had to take showers after each work session and change clothes. Special change facilities and showers were arranged on salvage ships for work groups.

In addition, a system had to be set up to examine and wash clothing worn by the cleanup personnel to prevent the spread of radioactive matter picked up during work periods. Separate laundry facilities were established to wash contaminated clothing to avoid mixing it with general laundry.

Dangerous radiation areas found after initial cleanup attempts aboard ships were marked and roped off. New detergent mixtures were applied using long-handled scrubbers and each time used were followed by a hosing down with saltwater to wash all contaminated materials from the ship. It was a process repeated many times in areas of various target ships to drive radiation levels down to tolerance limits.

Meanwhile, a new problem arose.

The support ships that had entered Bikini Lagoon and anchored outside the immediate target area began reporting that their hulls and saltwater lines that led to evaporators—which turned sea water into fresh water—had themselves begun to accumulate radioactive material.

Navy Commander Conard, as a radiological officer, had the job initially after the Baker shot of recovering technological instruments from target ships. He also had to make sure the personnel involved were properly monitored and took showers after performing their jobs so they washed off any radiation contamination picked up on their bodies.

During a later oral history, Conard recalled, "We didn't know too much about radiation at that point, and so we were trying to be very careful and bent over backwards to be sure that people didn't get too much radiation." He described sailors with little protection being initially sent to hose down decks to reduce radiation. "People didn't understand, the concept was so new, the whole idea of radioactivity," Conard said. While the cleanup was going on and sailors returned to their own ships, Conard remembered they found that "algae had taken up radioactivity from the water and the sides of the ship became radioactive, so we even had to move people for sleeping in more toward the center of the ship, so they wouldn't get irradiated."

Although that water in Bikini Lagoon, where the ships were anchored, showed minute amounts of radiation, radioactive elements had collected on algae, barnacles, or rusted areas of some ship hulls and had begun causing readings immediately inside some vessels of 0.1 roentgens per 24 hours, the established limit for exposure. Some of those ships were sent out into the open ocean for a day and, with an accompanying washing, were able to reduce contamination levels. But for a few, officers attempted decontamination of radioactive areas on their hulls by scraping with a chain, wire, or rope.

For the water evaporators, orders were issued to the support ships that no one was to open this equipment without specific authorization of the Radiological Safety Section, and then only with a monitor present. To meet the problem, testing showed that operating evaporators at low rates limited radioactive products being carried over in the process of turning the lagoon's saltwater into distilled water.

At this point a disagreement took place between Dr. Warren, whose concern was radiological safety of personnel, and Admiral Solberg, who was responsible for decontaminating the Navy ships.

By August 3, Dr. Warren had told Admiral Blandy that except for target ships with relatively little radioactivity, the rest "should be declared hopelessly contaminated and be towed to shallow water and beached and time allowed for radioactive decay to take place." Solberg, on the other hand, had told Blandy that he believed the cleanup was going on satisfactorily and with time would be successful, a position Blandy initially supported.

By August 8, however, Blandy recognized some ships could not be saved. He sent a telegram to Admiral Chester Nimitz, chief of naval operations, which discussed decommissioning thirty-nine of the target ships. "They cannot all be made absolutely safe to board in the near future for sufficiently long periods to either prepare them for movement to Pearl [Harbor] or to assess fully in all cases the damage sustained," Blandy wrote.

On August 9, an inspection of decontamination progress took place below decks on the German cruiser *Prinz Eugen*. The senior Radiological Safety officer and Target Group commander viewed samples taken from the wardroom that showed Geiger counter radiation levels that could cause unacceptable exposure to personnel working there for extended periods. In addition, other samples showed the presence of alpha emitters from plutonium, which also would cause biological effects in humans, but were not readily detected with instruments available within the Joint Task Force.

Although alpha emitters can be blocked easily by clothing, or even a piece of paper, if that radiation were inhaled or absorbed by the body, it could be deadly.

Further examination showed that alpha emitting fallout was probably widespread in the target ships, and that its "presence was considered a serious and indeterminate menace to personnel exposed for indefinite periods of time on contaminated target vessels," according to a later Navy study. Only trained detection specialists could deal with alpha emitters.

On August 10, Blandy called a conference to discuss this new discovery. Writing later to his wife, Warren said he told Solberg that he "was only fooling himself & risking a lot of men." Warren had enlisted Los Alamos scientists to review the alpha emissions and had gotten the results just before the meeting. He later told his wife that the Los Alamos analysis came "in the nick of time to nail it down & when Parsons saw it, he said, well this stops us cold alright."

At the meeting Warren showed Blandy and others that scales from skin of an exposed Bikini Lagoon fish put out enough alpha emissions to create an x-ray picture. It was determined, as one report later said, "Months would pass before natural decay would lower intensities to the point where crews could occupy and operate the ships."

Warren later wrote his wife that at the end of the meeting, Blandy said, "If that is it, then we call it all to a halt."

All decontamination procedures then underway were ordered to end, and further work on target ships was limited to recovery of instruments, surveys, and preparations to tow vessels from the area. Blandy cabled the Navy Department, "The tendency of radioactive matter to concentrate and accumulate in ships, especially in evaporators and in marine growth on the hulls, makes it mandatory to remove the ships of the Task Force from this atoll with its small and decreasing but nevertheless cumulative hazard."

On August 19, the Navy began towing target ships remaining afloat to Kwajalein, some 217 miles away. Support ships set sail to Hawaii or the US mainland where they faced inspection and further attempts at decontamination. It was September 5 before the last target ship was towed out of Bikini Lagoon, and it was September 26 when the final support ship departed.

The chief of naval operations later declared Bikini Lagoon a "defensive sea area," in a Notice To Mariners, and thereafter shipping and personnel were restricted from entering without authorized proper authority.

In his book *Operation Crossroads* published in 1994, Washington, DC, attorney, Jonathan Weisgall writes: "For all its thousands of pages of detailed plans, the US Navy managed to expose tens of thousands of men and more than 200 ships to radioactive contamination more than 2,000 miles from decent port facilities without ever having attempted experimentally to irradiate a ship or parts of one to determine how—or whether—a ship could be decontaminated."

There should have been a deeper lesson for people back then, and one passed on for those of us today—realization of what the long-term effects could be from radioactive fallout from a single nuclear weapon should one ever again be exploded underwater or, whether by error or design, detonated in a way that has the fireball hitting the ground.

13 · THE MEDIA STORY II

▪ **THE BIKINI TESTS** made the front page in newspapers across the globe the day after the Baker test, but not as the major story. The *New York Times* had it in the middle of page one on July 26, 1946, under a three-column headline, 11 SHIPS SUNK, 6 DAMAGED IN MAELSTROM OF ATOM TEST.

On July 28, the *Times*' Baker test story, written by Hanson Baldwin and headlined RADIOACTIVITY BAR TO BIKINI SURVEY, was back on page thirty-one. Baldwin and other reporters had boarded a Kwajalein-based C-54 and been flown back to Bikini where they had circled over the target ships in Bikini Lagoon for about thirty minutes. They were at altitudes "between 1,500 and 1,000 feet, but no lower because of radioactivity," Baldwin wrote.

After the C-54 returned to Kwajalein, half the correspondents from the *USS Appalachian* remained to fly back to the US, while the rest returned to the press ship for the trip back to Bikini, during which Baldwin completed his article.

That same day—July 28—an Associated Press reporter, in a short dispatch written from the *USS Mount McKinley*, wrote, "Bikini Lagoon was so dangerously radioactive today, three full days after the underwater atomic bomb explosion, that no one was permitted to make close-range inspection of the damaged, target fleet."

Earlier, the day after the Baker test, an editorial writer for the *New York Times* had written, "Radioactive air is terrible enough; radioactive water is worse because it takes time to evaporate. Radioactive air plus radioactive water will kill a large percentage of the crew." The writer went on to see a further danger from fallout, writing that "rain can also be contaminated by the cloud tossed up when an atomic bomb explodes under water so that the crews and passengers of remote, neutral merchant ships are exposed to new risks."

While most media coverage of the Able shot had seemed to downplay the power of the bomb, the eruption of water caused by Baker—followed by the giant, misty, three-mile-wide vaporous cloud—impressed some of the journalists. Also impressive was the nearly immediate sinking of several larger ships.

A University of Michigan Survey Research Center poll taken after the tests found most people answered "yes" when asked whether discovery of the atomic bomb made it easier to keep the peace. The *Arizona Republic* supported the idea of keeping secrets about the bomb because it was America's "ace in the hole."

Nevertheless, shortly after the Baker shot it seemed that the tests had not lived up to earlier fears about the deadly, destructive power of the atomic bombs being detonated. Captain Lee, who handled the press for the Joint Task Force, would later write, "The public was led to believe that something much more cataclysmic was going to happen than actually did." Nine days after Baker, the *New York Times*' Laurence wrote, "Before Bikini, the world stood in awe of this new cosmic force. . . . Since Bikini this feeling . . . has largely evaporated and has been supplanted by a sense of relief unrelated to the grim reality of the situation."

The average citizen, Laurence continued, "had expected one bomb to sink the entire Bikini fleet, kill all the animals . . . make a hole in the bottom of the ocean and create tidal waves. He had even been told that everyone participating in the test would die." None of that happened, apparently leaving the feeling among some Americans that powerful as it may seem, the atomic bomb was just another weapon.

Laurence had written his article after having returned home with other reporters and before Joint Task Force leaders like Blandy, Parsons, Solberg, Warren, and others began seriously to deal with the unexpected, high levels of lingering radiation from Baker.

The military men associated with Operation Crossroads had their own views of the press. Captain Lee later wrote, "If I were doing this job over again, I would investigate a little more, and would check on the background of some representatives." Having just 25 trained journalists rather than 187 would have been better and would have been easier, he wrote.

Smaller numbers would also have limited the backup of stories being sent. "I would say that a woman scorned is the acme of serenity compared to a newspaper man who has written his story and has dropped it into your hands, and who finds that it hasn't kept moving, through negligence or failure somewhere along the line," Lee wrote. "Rapid communication is the life blood of the press."

Army Colonel Betts, whose concern was with military intelligence and the potential release of classified information, agreed with Lee that the large press group had been "unwieldy." Along with the official observers, they had created "a very complex transportation problem and a resultant morale problem." Betts recommended that for future tests, the press and observers should be limited "to an overall total of one hundred or less."

The larger than expected turnout of journalists had turned out to be a problem. Their demands overwhelmed both the Joint Task Force's transmission capabilities and spokesman Lee's ability to answer questions or provide experts who could. Beyond that, many officials were unhappy they could not control the stories in the way they wanted.

For example, Los Alamos scientists had a different view of the purpose of media exposure. In a chapter in the history of the Manhattan Project dealing with Crossroads they wrote, "The main object of submitting the spectacular phases of the operation to the glare of publicity despite the security risks which this policy was involved was obvious—more obvious at the time, perhaps, than now. . . . It is desirable that the whole world should know and appreciate the devastating power of the atomic bomb in order that peoples of the world might properly appraise the organized efforts to prevent atomic bombs from ever being used in warfare again."

Commodore J. A. Snackenberg, the Joint Task Force chief of staff, was more direct. His recommendation was, "Future tests with military atomic weapons should be conducted without provision for the attendance of observers for whom special and extraordinary security arrangements must be made." The commodore reflected the view of many senior military officers who in those days were irritated by having to divert themselves or subordinates from "serious work" in order to take care of reporters and/or radio soundmen or camera crews, knowing that any complaints from media members could get them in trouble. In short, keep out the press.

14 · THE MARSHALLESE

■ **FOR THE ABLE** and Baker tests in July 1946, the Navy had sent an LST landing ship, normally used to carry tank units, to Rongerik to temporarily put the Bikinians at sea in case there was fallout on the atoll's islands. Neither detonation, some 125 miles away, were deemed to have any effect on Rongerik, and within hours of each detonation the former Bikini residents were returned to the atoll.

King Juda's report back to his subjects after observing the Baker shot made the Bikinians long to go home. After all, there was no visible damage to their atoll, and they were unhappy with their new home. They felt worse when they learned that the inhabitants of Rongelap, just forty-three miles away, as well as Marshallese from Enewetak and Wotho, who also had been briefly moved from their atolls during the tests, had been allowed to return to their homes when the Crossroads tests were concluded.

A scientific staff team that had accompanied King Juda after the tests studied Rongerik's uncontaminated ecology and gathered data to be used to compare with what was to be found on Bikini. There was great concern that radioactivity from the tests could produce genetic changes in the environment of Bikini Atoll.

A few months after Baker, the Navy reported in a technical study that about 50 percent of the total radioactive material produced by the explosion remained in the waters of Bikini Lagoon. Initial readings had shown the largest part of the radioactive material remained on the surface, and very little was present near the bottom, but that turned out to be untrue.

Pretest surveys had found that Bikini Atoll and Lagoon had more than one thousand species of organisms. However, the Navy Seabees had used the insecticide DDT in the areas on Bikini's islands where they built facilities. In the 1960s, because of its toxicity to animals and humans, it began to be prohibited for use. *Science News Letter* later noted, "Use of the DDT predictably clouded reasons for insect and plant deaths on Bikini," after the nuclear tests had concluded.

Nonetheless, the Navy examined tens of thousands of specimens that had been exposed to radioactivity in the area on and around Bikini, many of them having reproduced at least once since testing began. They "failed to show definite evidence of aberrant forms," according to a 1946 Navy report. That was expected since the view up to that time was that "mutations produced by radiation almost invariably do not survive," the report said.

Those same surveys did find large amounts of radioactive material remained at the lagoon bottom, particularly near the sunken target ships. That radioactivity had made its way through the food chain as, for example, sea cucumbers and worms took up radioactive mud. Plants on the lagoon bottom also were found to have some radioactivity and, when eaten by small fish, passed it on to larger fish and crustaceans.

As a result, investigators found some low levels of radioactive fission products in fish, clams, snails, oysters, corals, sponges, octopods, crabs, sea urchins, spiny lobsters, and shrimp. As already noted, the algae had also soaked up radioactive matter.

None of this information was passed on to the Bikinians on Rongerik. However, there already was a Marshall Islands legend, which the Bikinians believed at the time they were sent to Rongerik, that the atoll had once been the home of an evil spirit named Litobora. When she died, according to the legend, poisons from her body seeped into the ground and flowed into the nearby waters around Rongerik. Fish in those waters became dangerous to eat, fruit tasted oddly, and coconut trees became stunted.

In May 1947, having had food shortages the previous six months, the Bikinians suffered a major fire on Rongerik which damaged many of the coconut trees. The Bikinians had already wanted to move, and this time to an atoll where there were no other people, and one where they would not be under the authority of a hereditary chief or iroij, other than King Juda.

A Naval Board of Investigation was established, and a medical officer who visited Rongerik reported back that the Bikinians were "visibly suffering from malnutrition." The decision was made to find a new place for their resettlement. The choices came down to Kili, one of the smallest islands in the Marshalls, and Ujelang, an atoll three hundred miles southwest of Bikini. Ujelang was thought most likely after the Navy, in July 1947, took some Bikinians to inspect the uninhabited atoll.

At about the same time, the governmental status of the Marshall Islands changed. On July 18, 1947, the United Nations Security Council by resolution brought the Marshalls, along with the Caroline and Mariana Islands, into its Trusteeship System with the eventual expectation that they would become independent. League of Nations mandates to the Japanese after World War I, the Marshalls, Caroline, and Mariana Islands had been liberated in 1944 by US forces and since had been under control of an American military governor. Now, as the United Nations Trust Territory of the Pacific, they were to be administered for the UN by the US government in Washington.

The UN had already designated the Trust Territory as a "strategic area," and the US had been given authority "to establish naval, military and air bases and to erect fortifications . . . [and] to station and employ armed forces in the Territory." As part of the arrangement, the US also committed to develop political institutions in the Territory and "promote the development of the inhabitants . . . toward self-government or independence as may be appropriate."

In addition, the US was to "protect the health of the inhabitants" as well as "promote the economic advancement and self-sufficiency of the inhabitants, and to this end . . . protect the inhabitants against the loss of their lands and resources."

The United States of America, at the time the world's richest, most scientif-ically advanced, and most powerful democracy, pledged to care of and educate toward self-government the far less advanced Marshallese. At the same time, US officials were withholding from these islanders information about the possible radioactivity threat from their environment, while closely keeping track of any signs that their health was being affected.

Pawns or guinea pigs, the Marshallese were being used by their American overseers.

15 · MORE CLEANUP

■ IN LATE AUGUST 1946, the Navy decided that the radiological contamination of most Operation Crossroads target ships, increased by fission material picked up from the Bikini Lagoon, required decommissioning of all vessels that could not be sufficiently decontaminated and thus manned. Sixty-three target ships were towed to Kwajalein, including the battleships *Nevada* and *New York*. Another twelve target ships that were lightly contaminated took on crews and were sailed back to the US where they were to be further inspected.

Decontamination work continued from the fall of 1946 into 1947. In the end, forty-one of the target ships brought to Kwajalein remained there until they were sunk.

The decontamination process followed by the Navy provided a preview of what would have to be done after future nuclear detonations, particularly should that nuclear warhead, bomb, or artillery shell have its fireball hit the earth's surface—ground or water—and create radioactive fallout.

At Kwajalein, a 1,500-person Kwajalein Maintenance Force (KMF) was established to work on the still-contaminated Crossroads target ships. A radiation safety unit (radsafe) was formed for the KMF, which had its own laboratory flagship—the hospital ship *USS Haven*. The attack transport *USS Geneva*, one of the few target ships to be decontaminated after Baker, became sleeping quarters for the radsafe unit. The Navy also set aside *APL-27*, a non-self-propelled, so-called barracks craft as the "change" ship, where those working on the contaminated vessels would put on protective clothing, and where after work concluded they would shower and put on new clothing.

One of the first tasks for the Maintenance Force was to offload ammunition that was still aboard contaminated target ships and carry out its disposal. The battleship *Nevada* had more than 1,100 tons of ammunition aboard. Originally part of the military materiel tested to see its response to an atomic bomb, most ammunition was considered highly stable, but some was experimental, and others had been obtained from foreign navies. The initial concern was that some ammunition had become unstable from the heat and/or the emitted radiation.

A first step was to check radiation levels aboard ship, to make sure they were low enough to permit work crews enough time to accomplish tasks. That required an initial boarding team to inspect for hazardous areas using radsafe monitors. They also had to make certain flooded areas were pumped out enough so that wearing knee-level rubber boots would be enough protection.

Working parties prepared each day by going to the "change" ship, entering from the "clean" side where they were given laundered fatigues, canvas or rubber gloves, and rubber boots or field boots with removable canvas covers. They were also issued a film badge and a breathing apparatus to prevent inhalation of radioactive matter. The work crews left from the "contaminated" side of the "change" ship and boarded a landing craft to reach one of the contaminated ships.

The ammunition removal was exhausting and dangerous. Most work was below decks where often a sailor could last little more than thirty minutes before the heat generated by his clothing and breathing mask combined with Kwajalein's humid climate forced him to go up on deck for a break in the fresh air.

Since the target ships' own equipment and electric facilities were unavailable, portable lighting was used below decks, and portable pneumatic hoists lifted the ammunition topside. From there it was initially transferred to a freight barge, which took it out to sea to be dumped. Later in the process, the decision was made to simply leave the ammunition on deck and let it fall into the sea when the target ship itself was sunk.

When crews finished their workday, they returned to the change ship's contaminated side. Each sailor or officer turned in his clothing for laundering; each breathing apparatus was checked for contamination and sterilized; canvas gloves and boot covers were thrown overboard; while rubber boots and gloves were washed. Each boarding party member showered twice and was checked with a Geiger counter to make certain all contamination was gone. He then changed into his regular clothing and left by the clean side of the change ship.

As time went on, the Navy's chief of the Bureau of Medicine added safety regulations that recognized the radiation hazard for persons boarding target ships. One of the new rules: All persons who were to board target ships and encounter radiation had to have a preduty physical, which had to be followed up with monthly physicals as long as that work continued.

Work parties had to be accompanied by a monitor, and while aboard a contaminated ship, no one was to eat, drink, chew gum or tobacco, or lounge around. By June 1947, US units working on Kwajalein and at all shipyards where target ships were located had to provide monthly reports of personnel film badge exposures. But orders weren't always followed. Things got sloppy. Future Navy reports would show that, as the number of monitors and radsafe officers declined, there were violations of a variety of procedures, with some groups even avoiding the change ship altogether before going aboard some target ships.

Then-Lieutenant Commander Conard recalled transferring to Kwajalein and being one of the monitors checking incoming target ships and the ammunition that was removed. "We had our men wear respirators since we were afraid that plutonium and other fission products might be present on the ships," he later recalled. "That proved to be a very tricky operation, with the difficulties of going down in a hot ship and bringing back all the ammunition."

When the *Prinz Eugen* was towed into Kwajalein, Conard went on board the still-contaminated ship and admired the beauty of the German cruiser. "Inside was all the silverware, all the fine furniture, everything was left intact and it was in pretty good shape we thought." However, because of the radioactivity, one of the small seams of the ship that had been loosened by the Baker shot had not been repaired. By late December 1946, the cruiser was listing, and an attempt was made to beach her. On December 22, the *Prinz Eugen* went mostly under in the lagoon, her stern still seen above water level.

Conard was later transferred to Pearl Harbor. When former target ships were towed there, he said he was the officer "forced to order the sinking of many beautiful small boats such as captain gigs [motorboats used to ferry commanding officers from ship to shore] because radioactive contamination was considered too high."

While work continued on the target ships, Navy officials began to question radioactivity that might have been picked up by support ships that had entered Bikini Lagoon after the Baker shot and had remained there for ten days or more thereafter.

The expectation had been that decay and steaming in the open sea on the way home would be all that was necessary to dilute any radiation hazard. But on August 29, 1946, Navy headquarters told the commanding officers of those ships to avoid dry-docking until further notice. A subsequent order dictated that each support ship be examined at Pearl Harbor or San Francisco to determine its radioactive status.

A medical office was established in late August at the San Francisco Naval Shipyard, and radsafe monitors were brought on to handle decontamination experiments on the non-target ships that had been returning to the United States. Most trained Navy radsafe monitors up to that time had been working at Kwajalein.

The Navy then realized it had not yet established service-wide standard decontamination procedures, or even set tolerance levels for radioactive emissions. It had left the decision up to each command or facility. More important, the Navy had not designated a person or persons to be responsible for the decontamination program or even how such a program would be developed.

In early September 1946, the destroyer USS Laffey became one of the first support ships subjected to inspection for radioactivity at the San Francisco Naval Shipyard. At Crossroads, the Laffey had been one of five destroyers that served in the Surface Patrol Group, which at times had kept track of ships entering and leaving Bikini Lagoon and also carried out posttest detection, measuring radioactive levels for the radsafe teams.

On September 5, two radiologists watched as the Laffey entered a floating dry dock at the San Francisco shipyard and had the water slowly pumped out of its hull so it could be scraped off for samples of rust, paint, barnacles, and algae by workers in protective clothing and breathing apparatus. Using handheld Geiger counters, the radiologists detected radiation levels that were safely below accepted tolerance levels, persuading officers to permit overhaul work on the rest of the ship. Thereafter, samples of the ship's pipes and tubing, which had been previously exposed to saltwater in the Bikini Lagoon, as well as sections of the evaporator were cut off and sent to laboratories for examination.

Those tests showed higher radioactivity than found on the hull, but still safe enough for workers to continue under monitored conditions. Ironically, brushes and some canvas used in collection of samples showed enough radioactivity that it was decided to encase them in cement and have them later dropped into the ocean. Another precaution taken: The *USS Laffey*'s hull—now fully exposed—was washed down four times a day to prevent any radioactive dust particles from gathering on it.

Over the next days, workers tried a variety of techniques to determine the best method for decontaminating the evaporators, other saltwater-exposed valves and pipes, and areas where inspection had found radioactivity still existing on the ship. Mixtures of acids with ammonium hydroxide were used as dips or sprayed through pipes at equipment to seek their effects on reducing contamination.

Navy officials closely watched these decontamination experiments. After Admiral Solberg conducted an inspection of the *Laffey* and other ships on September 16, a Navy report of the visit noted the Bureau of Ships was establishing a "Decontamination Section" in Washington that would follow all experimental work at the San Francisco shipyard.

The report also stated that "the subject matter of how to rid a ship free of radioactivity should be classified as 'Top Secret.'"

Sandblasting, chipping, and scraping the bottom and propellers were methods used to decontaminate the *Laffey*'s deck and hull over a period of several days. The process also included washing down the area, collecting the quarter-inch of wet sand that accumulated beneath the ship at the bottom of the drydock, and measuring it for radioactivity. It showed very small amounts on a Geiger counter held a half-inch from the source, so the sand, itself, was then washed down with a fire hose for two-and-one-half minutes, thus reducing levels further by one half.

It was concluded that such a decontamination process would work well on surfaces of steel, but not as well on wood, which—of course—would be the surface material of most homes there in San Francisco or elsewhere in America or around the world where, God forbid, radioactive fallout could be carried by the wind for twenty or thirty miles from where a future nuclear weapon hits the ground.

16 · LESSONS FROM CROSSROADS

■ **IN HIS FINAL** classified report as commander of Joint Task Force One, Adm. William Blandy considered the results of Crossroads and wrote, "It is absolutely vital to the safety of the United States that no country obtain an ascendancy over us in the production or use of this bomb . . . [since its] destructive capacity dwarfs anything ever before produced." He added that the "radioactive contamination produced by . . . Test Baker appears to constitute a blow which is even more paralyzing than the actual blast and heat effects."

"Accordingly," he went on, "unless and until atomic weapon development is prohibited by international agreement, I recommend that development of the military applications of this weapon be pushed with all possible vigor and that underwater uses receive special consideration."

Blandy peered into the future and, seeing an inevitable nuclear arms race, concluded, "Even with a satisfactory peacetime control, I trust that the lessons of Bikini—in design, radiological safety, tactics and strategy, for all the armed forces—will not be forgotten, for in war, peacetime controls do not function, and peacetime promises are often forgotten. Our only course is then to assume, as we did with poison chemicals in World War II, that the enemy if technically capable, may develop and use atomic weapons at some stage of a long war, and prepare ourselves as quickly as possible both to defend and to retaliate against them."

While those were Blandy's official views, passed on in a once classified report, in writing for the Navy publication *All Hands* in August 1946, the admiral described what he said were his personal feelings: "One fact, however, stands starkly clear in my mind. The only certain defense against the atomic bomb would be the knowledge and assurance it would never be used again in warfare, a knowledge backed by unbreakable, permanent international guarantees and checks. It is my greatest hope that the atomic bomb can be abolished as a means of waging war. This must await the test of time."

The President's Evaluation Committee released its own statement after the Baker shot in August 1946, which echoed Blandy's concerns: "As was demonstrated by the terrible havoc wrought at Hiroshima and Nagasaki, the Bikini tests strongly indicate that future wars employing atomic bombs may well destroy nations and change present standards of civilization. To us who have witnessed the devastating effects of these tests, it is evident that if there is to be any security or safety in the world, war must be eliminated as a means of settling differences among nations."

For mankind in this twenty-first century, it bears emphasizing that these were words from those "who have witnessed the devastating effects" of atomic bombs—which almost no one today, as of this writing, has done.

Of course, not all who watched the tests took that view.

Army Air Force Major General W. W Kepner, the deputy task force commander for aviation at Bikini, wrote most presciently in his section of the final Joint Task Force report. He called for all military services to immediately review organizing, training, equipping, and deployment of units "to use the atomic bombs" because "obviously there can be no lengthy period for reorganization and training after the outbreak of hostilities."

The ability to use atomic weapons "must keep pace with its development in the scientific laboratory," Kepner wrote, calling for "further advances [to be] initiated to exploit the use of atomic weapons for both offensive and defensive purposes." He even proposed what since has become dual-purpose weapons, recommending that "thought be given to the feasibility of developing a bomb case suitable for the housing of either high explosive or atomic charges."

Kepner also foresaw a move toward a first-strike strategy, having written, "A future war may certainly include the use of atomic bombs by both adversaries, in which case victory may well go to the nation which most quickly and completely exploits the capabilities of this new weapon."

In the aftermath of the Baker test, the two senior Joint Task Force officers most concerned with the impact of radiation, Admiral Parsons and Dr. Warren, concluded at different times what should have been one of the most important matters learned from the Crossroads tests.

While Blandy and Kepner's contributions to the Joint Task Force final report concerned preparing for nuclear war, Admiral Parsons wrote instead, "The one lesson from Test Baker which has served to punctuate the deadliness and awesomeness of the Atomic Bomb" was "the radiological danger." He continued that the radiological hazards from the test "were vividly impressed on the consciousness of all the members of the Task Force and it is hoped that the impact of this tremendous problem will be fully realized throughout the services, the scientific world, and the nation at large."

However, in a public speech on December 27, 1946, before the American Association for the Advancement of Science in Boston, Parsons took a slightly different approach. He confessed, "We hardly realized all the ramifications and implications of this persistent radioactivity," adding, "The paralyzing radioactivity of the second bomb had to be experienced to be believed . . . [and] without it we would have had only incomplete and unconvincing theory to guide us in our preparations and plans for the future."

Yet, Parsons concluded his speech on a more positive note, saying, "Not one man was injured by the blast or radioactivity from the bomb experiment . . . Tens of thousands of people have seen the two atomic bombs set off under controlled conditions and have thereby gone part of the way toward substituting a rational fear of the known for the irrational fear of the unknown in this Atomic Age."

Stafford Warren, lower ranking in the military chain and having returned to civilian life, took a different tack in a thoughtful article he wrote in the August 11, 1947, issue of *Life Magazine*. Looking back over the year since the Baker test occurred, Warren wrote that during the time it took for the initial radioactivity of fission products to lessen "a more insidious hazard was discovered." He then described slight contamination spreading outside the target area and the process in which lagoon algae absorbed radioactive particles, which were transferred to the small fish that ate them and then to the larger fish which swallowed them. Finally, that radioactivity returned to the algae when the large fish died, and the radioactive material cycle would continue because fission products continued to produce radioactivity in the next life cycle.

Warren's broader conclusion is worth repeating in full: "Some of us have reflected, however, what might happen if Bikini had been a populous harbor with a wind blowing in from the sea. Most of its people would have inevitably died. A smooth-working evacuation system might have saved a few of them, but no defense would have been effective."

He then echoed thoughts Blandy had privately voiced a year earlier, "The only defense against an atomic bomb still lies outside the scope of science. It is the prevention of atomic war."

■   ■   ■

Major General Leslie Groves and other senior Los Alamos officials opposed a third Crossroads test. They had several concerns.

Too many of their staffs were involved with Crossroads, and they were uneasy about using another atomic bomb. The bombs dropped on Hiroshima and Nagasaki had to be assembled by scientists, as were those used in the Crossroads tests. Meanwhile, at Los Alamos, scientists were developing a new version of the plutonium-triggered Fat Man, the Mark III, but key components were in short supply.

After the Baker shallow-water explosion, Groves had questioned what more would really be discovered from the third planned test, Charlie. And since most harbors or offshore areas where such a weapon would be used were shallow, questions arose as to what military purpose would emerge from testing at the planned greater depth—two thousand feet down. In an August 7 memo to the Joint Chiefs, Groves wrote that Los Alamos had to accelerate work on its new weapon and that it was "imperative that nothing interfere with our concentration of effort on the atomic weapons stockpile which constitutes such an important element in our present national defense."

Groves had kept secret the number of atomic bombs available, both before and after the first two were used over Japan. Four days after the Nagasaki bomb dropped, General Marshall asked his war planner, General John E. Hull, to find out from Groves' assistant, Colonel L. E. Seeman, when the next bomb would be available. Seeman replied that one was ready for shipment and another "should be ready the 19th," some six days later. After that, Seeman said, the Army could expect about three a month. Hull noted that usually it would take six days after a bomb was received before it could be prepared for use.

Seeman wrote that production was "still in the midst of development" and that "we are changing amounts and proportions of the active [nuclear] material so that certain ones may, or will, have lesser power. They may be more nearly equivalent to the one at Hiroshima [12.5 kilotons] than the one at Nagasaki [23 kilotons]." In fact, from August 1945 to September 1946, the US atomic stockpile had gained just nine deliverable atomic bombs—and two had been used at Crossroads.

While Blandy argued that the Charlie test was still needed, Army Secretary Robert P. Patterson and members of the President's Evaluation Commission agreed with Groves. It was all but settled when Gen. Dwight David Eisenhower, then-Army chief of staff, met privately with Truman and informed him about the small size of the atomic stockpile. On September 6, 1946, Truman told his cabinet he had decided to postpone Test Charlie. The White House statement the next day said that based on the information from the first two tests, "the Joint Chiefs of Staff have concluded [Test Charlie] should not be conducted in the near future."

While the Crossroads tests had been taking place, there had been a hard-fought battle on Capitol Hill over civilian control of what had been, under Groves, a military-directed, wartime Manhattan Project. While the original bill proposed in Congress would have kept the military in charge of nuclear weapons, effective lobbying by a group of scientists who had been part of the program got that measure changed.

Along with the future role of the military, among the other issues hammered out by legislators was peacetime control over atomic weapons, along with direction of design and testing of potential weapons. A key issue was agreement that almost total secrecy would be placed on information surrounding atomic programs.

On August 1, 1946, just days after the Baker shot, President Truman signed legislation creating the Atomic Energy Commission (AEC), which shifted control of the atomic bomb and development of atomic energy from the military to a civilian-controlled board. The AEC's five civilian commissioners, appointed by the president, were given complete control over production plants, laboratories, equipment, personnel, and testing. Truman chose David E. Lilienthal, a Harvard Law-educated liberal, who had been chairman of the Tennessee Valley Authority, as chairman. One of the other commissioners appointed by Truman was Lewis Strauss, the conservative Republican who had promoted the Crossroads tests and would later become AEC chairman.

A Military Liaison Committee and a civilian General Advisory Committee were established within the AEC to coordinate both military and scientific applications of atomic energy. On Capitol Hill, a Joint House-Senate Committee on Atomic Energy was created, and for decades it was the only congressional venue where nuclear weapons could be discussed.

While the legislation became law in August 1946, the AEC did not take over until January 1, 1947, leaving four months during which time the commission members were approved by Congress and the staff put together. Groves supervised the transition, although he had lost the legislative fight for the military to retain control of the manufacture of atomic weapons, including their assembly, storage, and protection. He had supported ending the Army's administration of atomic energy civil programs.

"Everyone knew that I was in a caretaker's position, and they had no assurance that my views would be those of the Commission. After the commissioners were finally appointed, it was quite evident that my views would not be accepted without a long, drawn-out delay," Groves later wrote.

All weapons and weapons facilities temporarily had remained with the military, while civilian scientists at Los Alamos and the other national laboratories would retain responsibility for developing new types of bombs and other atomic weapons. Guidance would continue to come from the military. Another part of Groves's plan involved his deputy, Col. Kenneth Nichols, who was nominated to become the AEC's director of military application, who under the law had to be a member of the armed forces.

On December 30, a final agreement was reached. The military transferred to the AEC all atomic weapons, parts, and fissile materials, but the president was to decide by March 1, 1947, what would be retransferred back to the military and under what circumstances. Nichols did not get the job. Instead, Groves appointed him to be the Pentagon's military liaison officer to the AEC.

Groves, himself, had been removed from a direct role with the AEC, and in January 1947 became chief of the newly-created Armed Forces Special Weapons Project, which was to be an inter-service organization responsible for all aspects of atomic warfare for the Army and Navy.

Six months later, in July 1947, Groves's continued intervention in all atomic weapons activities led to his Armed Forces Special Weapons Project organization being limited to training military personnel in the handling and assembly of atomic weapons. Unhappy at this role, Groves retired in February 1948, at which time he was promoted to the rank of lieutenant general.

Ironically, five years later in 1953, President Eisenhower asked Nichols to retire from the Army in order to become general manager of the AEC, with one task being the creation of better relations between the commission and the military.

17 · NEW TESTS, NEW DEVICES

■ **WHEN THE FIRST** atomic bombs were being dropped on Japan, scientists at Los Alamos had already been working on modifications to the Nagasaki bomb model.

Central to their work was changing the makeup of the "pit" or central, nuclear, trigger of the weapon, which initiated the wider nuclear explosion. It was a hollowed-out sphere of highly enriched uranium, slightly larger than 3.5 inches in diameter, which was surrounded, all the way around, by high explosives.

Within the uranium sphere was a round, solid plutonium center.

When the explosives were fired, the resultant shock waves from all parts of the outer sphere compressed the uranium/plutonium core, creating a fission—or atomic—reaction that set off the final explosion.

In April 1947, the AEC's General Advisory Commission proposed that elements of a new, more efficient Los Alamos core design and two other design changes be proof tested—effectively setting the US down the road toward a second-generation atomic bomb. The aim was to sharply increase the power of the nuclear explosion.

The new Los Alamos approach called for changing what had been a solid plutonium core as trigger for the bomb to what was called a "levitated core," which had a space between the hollowed-out uranium sphere and inner pluto-nium center. The new "pit" became more efficient thanks not only to the space change but also to the mix of fission materials used.

On June 27, 1947, after an Oval Office presentation by AEC Chairman David Lilienthal, President Truman approved what was called the Operation Sandstone test program to begin in April 1948. Sandstone would be three proof tests of design changes for a device with an even more powerful explosion than Crossroads, where the tests were to measure actual performance of a deliverable bomb.

By the time of Truman's approval, the US had thirteen atomic bombs of the Nagasaki variety stockpiled. But they had been handmade. Another reason for the new designs was that they would allow atomic weapons to be produced in an assembly-line process and thus permit a more rapid growth of the nuclear stockpile.

Within three months, the Joint Chiefs had formed Joint Task Force Seven, with Army Lt. Gen. John E. Hull named commander. He chose as his depu-ties two Crossroads veterans, Admiral Parsons and Air Force Major General Kepner.

One of the first decisions by the AEC and the new Joint Task Force was that secrecy would be maintained about timing of the tests and their purposes, although the fact that tests were being held would be announced. However, unlike Crossroads, there would be no on-scene press coverage, and the number of observers would be limited. Forrestal, as defense secretary, wrote he would limit "non-participating observers" to those whose work requires "a first-hand knowledge of, or direct action with respect to, the results of the tests."

The next decision to be made by the AEC was where to hold the tests, since all agreed that Bikini Atoll was too small for what were to be larger explosions than Crossroads. In addition, because these new tests involved devices and not deliverable bombs, the detonations were to be done from three, 200-foot towers. The atoll had to be one with widely separated islands.

Enewetak Atoll is located in the northwestern part of the Marshall Islands. It is an elliptically-shaped atoll with some forty separate islets that enclose a large lagoon roughly seventeen miles at its widest point and twenty-three miles at its longest. The total land area of Enewetak is just 2.75 square miles.

In February 1944, a US Marine regiment had needed two weeks to defeat a light but determined Japanese force in a series of battles on Enewetak's Islands. Some thirty-seven American servicemen were killed or went missing in the operation before the atoll was secured. Soon a US airstrip, 6,800 feet long, had been constructed on Enewetak Island to house a bomber squadron used in the Pacific War. Piers were built on other atoll islands, plus a fighter strip on Enjebi, where the Japanese had had a fighter base. When Enewetak was selected for nuclear tests, one of the first things done was to take from the atoll the remains of US servicemen buried there and return them to their next of kin.

To the north and west of Enewetak there was open ocean, the directions the steady trade winds flow, and thus the predominant path that any fallout would travel from the test shots.

Bikini, unoccupied since Crossroads, was 190 miles to the east. Some 110 miles further east of Bikini was Rongelap, which would be the closest then-inhabited atoll to any test shots at Enewetak. Just west of it was Rongerik, where the unhappy Bikinians had been living for almost two years, since they left their own atoll for the Crossroads atomic tests.

The Bikinians, however, were preparing to move to Ujelang Atoll, their recently chosen new home. In late November 1947, twenty Navy Seabees had taken a group of ten Bikini men to Ujelang where they picked out a community area and began construction of housing.

What the Navy did not know when it started construction on Ujelang for the Bikinians was that the AEC, which was running site selection for Sandstone, had surveyed Enewetak in September and by mid-October had selected it for the next nuclear test series. The AEC decision had not yet been approved by the White House, but the commission staff had already started looking for a new home for the Enewetak inhabitants—with the eyes on Ujelang.

On December 2, 1947, President Truman announced his approval of Enewetak for the next atomic tests. On that same day, the US notified the UN Security Council that the atoll and surrounding areas were to be a safety exclusion zone to ships due to American experiments relating to nuclear fission.

At the same time, the AEC put out a press release that said Enewetak was chosen because it had "the fewest inhabitants to be cared for, approximately 145, and . . . from a radiological standpoint, it is isolated and there are hundreds of miles of open seas in the direction in which winds might carry radioactive particles."

The southern islands of Enewetak Atoll were to be used to house members of Joint Task Force Seven, the northern ones were to be sites for the testing devices. The twenty-mile separation between the two sets of islands was believed to create a safety barrier for people and facilities from where the explosions took place.

As for the 145 Enewetak Marshallese, their new homes, the AEC press release read, "will be selected by them . . . The inhabitants concerned will be reimbursed for lands utilized and will be given every assistance and care in their move to, and re-establishment at, their new location. Measures will be taken to insure that none of the inhabitants of the area are subject to danger; also that those few inhabitants who will move will undergo the minimum of inconvenience."

Attached to Truman's order to the Defense Department to move the Enewetak people was a memo that stated that the displaced Marshallese should be "dealt with as wards of the United States for whom this country has special responsibilities."

Not mentioned at the time, but brought out years later in 1999 before the Marshall Islands Nuclear Claims Tribunal, was the Enewetak people's attachment to their land on their atoll. "For Marshall Islanders in general, and Enewetak people in particular, land is a part of one's person and one's entire identity," according to Montana State Professor Laurence Carucci, a consultant to the Enewetak Claims Commission would later tell the Claims Tribunal. "Not only is land hyper-valued because it is scarce, land is extremely highly valued because it represents the collective labor of generations of people who have worked the land, transforming it from bush into habitable space," he said.

The day after Truman's announcement, December 3, 1947, US Navy Captain John P. W. Vest, the military governor of the Marshall Islands, flew to Enewetak and proposed to the chiefs that they move to Ujelang Atoll, despite the fact it was in the process of being prepared as a relocation site for the Bikini people.

"I told them they would be able to return to Enewetak fairly soon after the tests were completed; perhaps in three to five years," Vest would later say, adding, "It certainly was not in my mind that it would be longer than that, or that the taking of Enewetak for the testing program was permanent."

Days later, it was announced that the Enewetak people, and not the Bikinians, would be moved to Ujelang. On December 21, 1947, temporary living quarters, some built earlier by Seabees aided by Bikini workers, were available when the Enewetak people arrived. Permanent housing and other facilities were built in the spring of 1948 by Seabees and Marshallese from Enewetak.

■   ■   ■

Meanwhile, the Bikinians still on Rongerik were suffering from lack of food and the need to ration water, according to a January 1948 report to the UN Trust Territory high commissioner. That news came from Dr. Leonard Mason, a University of Hawaii anthropologist, whom the UN had sent to Rongerik. Mason recommended that food supplies and a medical officer be flown to Rongerik. In February, the resident medical officer on the atoll reported about starvation conditions among the Bikinians. The situation grew so bad that on March 14, 1948, the Bikinians were evacuated to a tent encampment on Kwajalein that had been hastily set up near the island's airport.

They had spent just over two, lonely years on Rongerik, and suddenly they were on Kwajalein where, as Bikini elder Kilon Baumo put it, "We were frightened by all the airplanes that landed close to our homes. We were also frustrated by the small amount of space in which we were permitted to move around. We had to depend on the US military for everything. We were afraid of this alien environment and almost from the day we got there we began thinking about other places to live."

In June 1948, the Bikinians chose Kili Island, located 400 miles south of Bikini, as their new living site. Whereas Bikini was an atoll with 23 islands surrounding a 243-square mile lagoon, Kili was an island one-and-a-half-miles long and only three-quarters of a mile wide at its widest point. It had no lagoon and a total land area of 230 acres, one-sixth the size of all Bikini's islands.

The reefs that encircled Kili and the surrounding high seas made it difficult to get canoes or small boats in and out from the beach. In the nineteenth and early twentieth century, German planters had developed a business from copra. Coconut oil, extracted from copra, became a moneymaking, exported commodity for the islands, although that trade had ended when the Japanese took over after World War I.

The Bikinian leaders has chosen Kili because it was uninhabited and had no chief or iroij who claimed authority over it. Another reason was their belief that they could renew the growing of copra and thus have a source of outside income. On November 2, 1948, what were now 184 Bikinians boarded a Navy vessel and were taken to Kili, where they occupied buildings that had been under construction since September by Seabees and fellow Marshallese.

On November 24, 1948, The *Honolulu Star-Bulletin* reported that if the Bikinians "do not adjust to Kili, the United States will have a new headache." The newspaper quoted a Navy official saying, "The Navy is running out of deserted islands on which to settle these unwitting, and perhaps unwilling, nomads of the atomic age."

Without sheltered fishing grounds, the Bikinians could not practice their natural skills. Kili's sandy soil produced little in the way of edible food, and its coconut trees were not as abundant as expected. For three months a year, no landings on the island could take place because of rough seas. In early 1949, the Navy sent a special demolition crew to Kili to blow an opening in the reef that created a channel so that landing craft could deliver food and receive copra.

It was apparent from the restart of American thermonuclear tests on their atoll that the Bikinians would continue to face hard times. As of 2019, a majority of them with their children and relatives, numbering six hundred, still live on Kili. But the remainder have moved to other Marshall Islands, Hawaii, and even to mainland United States—nuclear refugees from their home islands.

18 · MORE NEW TESTS

▪ **OPERATION SANDSTONE WAS** no Crossroads. It was designed to proof-test future, second-generation, Los Alamos-designed atomic weapon elements, rather than actual atomic bombs. Also, fewer personnel would be involved in its three tests: only some ten thousand individuals of whom less than five hundred were civilians.

And there would a limited number of observers and no journalists present.

The Sandstone tests were held as the Cold War worsened. In February 1948, the Soviets managed a Communist coup in Czechoslovakia. On March 20, 1948, the four-power Allied Control Council in Berlin ended, and on March 31, the Soviets ordered inspection of all allied military trains moving from West Germany into Berlin. In response, Gen. Lucius D. Clay, commander of the American zone in Germany, on April 2, ordered all US military train shipments to Berlin halted and required supplies from then on to be shipped by aircraft. It began what grew into the round-the-clock Berlin Airlift.

As a result of the Soviet-initiated crisis, there was talk of postponing the first Sandstone detonation, scheduled for April 15, to save the nuclear materials involved. They might be needed if war broke out. AEC Commissioner Lewis Strauss even warned Defense Secretary Forrestal to limit the number of atomic weapon assembly experts sent to Enewetak, in case there were a Pearl

Harbor-type attack on that atoll, which could cripple US ability to assemble additional atomic bombs from the parts and nuclear material it possessed.

Despite these concerns, preparations for the tests pushed forward, but with precautions.

US Navy destroyers escorted the ships carrying nuclear elements and equipment from Pearl Harbor to Enewetak. The task force traveled using wartime-style maneuvers, zigzagging off a direct course, with ships' crews standing around-the-clock watches. When the tests got underway, eight destroyers carried out surveillance on the ocean areas surrounding the atoll. One of those destroyers was assigned anti-submarine warfare patrols outside the Enewetak Lagoon in between test shots in the event that Soviet submarines were prowling the area.

Even publicity was muted in this new Cold War environment. The AEC wanted to make a public announcement as the April 15, 1948, date for the first Sandstone detonation approached. The Defense Department objected. The commissioners feared a leak, while the military worried that the Soviets would send ships to collect airborne samples to analyze any radioactive cloud. A compromise was reached that an announcement would be made after a long enough delay so that any fallout would have dissipated in the atmosphere.

■　　■　　■

The Sandstone tests were designed to lead to a new type of core trigger for nuclear weapons—one that would produce higher yields but use less costly nuclear materials.

The yield of the first test, X-Ray, detonated from a two-hundred-foot tower on Enewetak's Enjebi Island, the most northern of the atoll, was thirty-seven kilotons. It was, up to that time, the most powerful explosion ever recorded, and more than 50 percent greater than the Nagasaki bombs used at Crossroads.

Three days after detonation, when the radioactivity had died down, scientists returned to Enjebi to retrieve the measuring instruments they had placed there. Data collected was then fed into the next test, Yoke, which took place on May 1, on Aomon Island. It went off at forty-nine kilotons. On May 15, Zebra, the final test, took place on Runit Island with a yield of only eighteen kilotons.

For security reasons, the Enewetak cleanup included destroying any physical evidence related to test results. In addition, the task force ships kept the waters around the atoll closed and a fifty-person garrison was established to guard the land area.

Overall, the tests verified the Los Alamos scientists' new design principles and opened the way for more powerful future bombs that could be manufactured on an assembly line. The composite core, using both uranium-235 and plutonium, reduced the amount of scarce and costly plutonium needed; and the levitating of the core increased its efficiency and the power of the chain reaction. Less than two weeks later, Norris Bradbury, Oppenheimer's successor at Los Alamos, told a mixed military-AEC group that the Sandstone tests had made most current atomic bomb elements obsolete, and significant changes had to be made for the next generation of atomic weapons.

In September 1948, as tensions over Berlin increased, Forrestal again brought up the question of whether or not the US would quickly use the atomic bomb should war break out. At a September 13, White House meeting with Truman, after a briefing by the Air Force, Forrestal recorded in his diary, "The President said he prayed that he would never have to make such a decision, but that if it became necessary, no one need have a misgiving but what he would do so."

The next night, September 14, Forrestal, along with Secretary of State Marshall, attended a dinner for twenty newspaper publishers and editors at the home of Philip L. Graham, who then ran the *Washington Post*. The purpose was to discuss the Berlin crisis, but Forrestal raised the issue of the atomic bomb. There was "unanimous agreement," Forrestal wrote, "that in the event of war the American people would not only have no question as to the propriety of the use of the atomic bomb, but would in fact expect it to be used."

On April 4, 1949, the North Atlantic Treaty was signed in Washington, and the NATO alliance was formed. A month later, the Soviet blockade of Berlin ended, thanks to the airlift. At a June 23, 1949, press conference, Secretary of State Dean G. Acheson said that "the position of the West had greatly grown in strength, and that the position of the Soviet Union in regard to the struggle for the soul of Europe has changed from the offensive to the defensive."

But American smugness, based in good part on its nuclear weapons monopoly, was about to change. In the summer of 1949, the Soviet Union prepared to test its own nuclear device at a remote site in Kazakhstan.

 The site had originally been chosen by Soviet Secret Police Chief Beria in 1947 in his role as head of the country's nuclear weapons effort. The steppes region he picked was almost uninhabited, some eighty-four miles northwest of the city of Semipalatinsk. Military units and workers were brought in from gulags to build a complex called Semipalatinsk-21. It was made up of residential and office buildings, laboratories, plus a small hotel for tests observers. The town was later named Kurchatov, to honor Igor Kurchatov, the nuclear project director.

A one-hundred-foot-high tower was constructed some forty-two miles south of the complex, with a workplace nearby where the final assembly would take place. Parts for the test device were shipped by train and aircraft from Arzamas-16 (now Sarov), the Soviet's Los Alamos, located 250 miles south of Moscow and 2,000 miles from the test site. When put together, the Soviet test device would look like the US Nagasaki atomic bomb, Fat Man, since much of its detailed design had come from Klaus Fuchs, the German-born, British scientist who worked as a spy for the Soviets while at Los Alamos in 1945.

The Soviet physicist, Mikhail Meshcheriakov, who had attended both shots of Operation Crossroads as a foreign observer, had written a 110-page report to Beria's Special Committee for Nuclear Affairs on what he saw at Bikini, including target placements and the aftermath of the blasts. His data and the film he shot had been used in helping prepare the Semipalatinsk test facilities.

It was not surprising that within the target area were elements similar to those at US Nevada test areas—segments of highways, railroad bridges, military vehicles, trucks, automobiles, a railroad car, and locomotive. The Soviet test preparers even had built a subway tunnel segment, fifty-feet to one-hundred-feet in depth, within one-quarter-of-a-mile from the detonation point. Of course, there were tanks and artillery pieces; two bombers; a one-story, wooden house; a four-story, brick home; and water towers.

Beria arrived at the site in late August along with Kurchatov.

At 7:00 a.m., on August 29, 1949, the Soviet Union conducted its first nuclear test, codenamed RDS-1, or First Lightning, at the Semipalatinsk test site. The device had a yield of twenty-two kilotons, about the same as the bombs used at Operation Crossroads.

Two days later, a US Air Force WB-29 weather reconnaissance plane was on a routine patrol near the Soviet Union Siberian coast as it flew from Japan to Alaska. Such flights had begun in 1947 and had been designed to detect signs of atomic testing. This flight picked up signs of radioactivity in the air. The Air Force then sent other aircraft in the Pacific area to collect samples, and they, too, found signs of abnormal radioactivity, which when analyzed was traced back to between the 29th to the 27th of August, and in an area that included Kazakhstan. Meanwhile, a Navy Research Laboratory project that had begun in April 1949 as Project Rain Barrel, discovered fallout from rain in Kodiak, Alaska, on September 9 and days thereafter.

On September 23, 1949, President Truman announced, "We have evidence that within recent weeks an atomic explosion occurred in the USSR." Not said was that US scientists had concluded it was a plutonium bomb with uranium surrounding the pit and similar to the American Fat Man, Trinity test design.

Within the American scientific community, a long-raging dispute over the future of nuclear weapons development burst into public view. The main question was whether scientists should create a weapon even more powerful than the atomic bomb—using technology developed in constructing the A-bomb—and employing not just uranium and plutonium but tritium and deuterium gas, which are heavy isotopes of hydrogen.

Hydrogen fusion causes a speeding up of the fission reaction and can more than double the weapon's release of energy.

Oppenheimer led one faction that believed atomic bombs were destructive enough and wanted laboratory and experimental work aimed at designing cleaner, more usable atomic weapons. Edward Teller, another brilliant physicist, led a second faction that wanted to develop the hydrogen or thermonuclear bomb.

Both sides presented their cases to the AEC and to President Truman late in 1949. Oppenheimer stressed among other things the dangers from radioactive fallout that would accompany the explosion of a powerful H-bomb. Up to that point, little was known about dangerous radioactive fallout, other than the experience with the Baker shot at Operation Crossroads.

In January 1950, President Truman sided with Teller and directed the AEC to begin development of the H-bomb.

At the Pentagon, officials had been concerned about the costs of carrying out more tests in the South Pacific and had already been looking a new test site within the US, one that would ease their preparation, prevent danger to civilians outside the test area, and limit knowledge of what was taking place. AEC Chairman Lilienthal warned about "policy and psychological considerations" involved with the potential of radioactive fallout on American populated areas.

The Armed Forces Special Weapons Project, at the direction of the Joint Chiefs, had undertaken a secret study, called Project Nutmeg, that assessed the physical feasibility for conducting large nuclear weapons tests within the continental United States. Navy Captain Howard B. Hutchinson, a meteorologist who had been at Sandstone, was chosen to determine where tests could be undertaken without radioactive fallout causing harm to American people, places, or the US economy.

This meant Hutchinson had to find sites where wind, weather, and atmospheric stability would favorably disperse any radioactive sand, soil, or water created by an atomic detonation before it contaminated any populated areas. And at the site, he would specify that the device be detonated from a tower high enough so that any radioactive ground material—sand, soil, or water—would be lifted into the column of hot gases and eventually dispersed in the atmosphere.

Hutchinson chose five sites to study in Utah, New Mexico, Nevada, and North Carolina, settling on two as most promising—the arid southwest parts of Nevada and a stretch of the North Carolina coast on the Atlantic Ocean between Cape Fear and Cape Hatteras.

In March 1949, the AEC set aside Hutchinson's report. They deemed a continental site undesirable except in the case of a national emergency.

The Soviet First Lightning test and escalating military engagements in Korea changed all that. Planning was hastened for a new test series to determine the feasibility of the H-bomb, and by necessity, because of high yields, they would have to go back to the Marshalls, instead of using areas in the continental United States.

By spring 1950, a four-shot thermonuclear test series, called Greenhouse, was planned for Enewetak. The series was to take place in the spring of 1951. One test would demonstrate the notion that small amounts of the hydrogen isotope deuterium could increase the yield of fission explosions; a second would use tritium; while two others would prove that a fission trigger could cause a thermonuclear reaction. As with Crossroads, a group of military and civilian experts was put together to organize the tests, and Joint Task Force Three, under Air Force Lt. Gen. Elwood "Pete" Quesada, was formed. For a year it operated out of temporary buildings on the Mall in Washington before moving out to Enewetak.

Its initial planning was threatened when, on June 25, 1950, North Korea attacked across the 38th parallel, and President Truman quickly ordered air and naval units to South Korea. On July 5, 1950, the first US combat troops arrived in Korea.

The fighting raised questions about whether military units would be available for the planned nuclear testing in the South Pacific. At the same time, the war created added pressure on the US to regain its nuclear supremacy, so the hydrogen bomb project was again speeded up.

While one of the Greenhouse tests would have too large a yield for a continental site and had to be tested in the Pacific, others with smaller yields were candidates for a US test site, if one were approved.

Project Nutmeg was revived, and the military and AEC quickly agreed on the Nevada site.

An expert panel that included Manhattan Project veterans Edward Teller and Enrico Fermi met on August 1, 1950, to discuss the Nevada site to be located at Frenchman's Flat. They concluded a "tower-burst bomb having a yield of 25 kilotons could be detonated without exceeding the allowed emergency tolerance dose . . . outside a 180-degree test area sector 100 miles in radius," according to an Energy Department history.

Fermi even suggested that inhabitants potentially subjected to the upper emergency tolerance dose should be warned to stay indoors and take showers. The expert panel determined there was "not a probability that anyone will be killed, or even hurt . . . but . . . the probability that people will receive perhaps a little more radiation than medical authorities say is absolutely safe," according to the history.

A week later, President Truman delayed approving the Nevada domestic test site when Defense Secretary Louis Johnson asked for a quick decision. By now, the AEC was convinced the domestic test site was needed, given growing international tensions, and so while waiting for Truman's approval, the commissioners got the military to do preparatory work for a Nevada facility that could house 1,500 people.

On October 25, 1950, the new AEC chairman, Gordon Dean, talked to Truman in the White House again about the need for the Nevada site, because the first Greenhouse test at Enewetak had been pushed back to December 1951. Again Truman refused, but asked his National Security Council (NSC) to look at the issues involved.

The NSC went back to Dean and others for advice. Los Alamos scientists came out strongly in favor of the Nevada site and in a November 1950 paper wrote that there would be no radiation hazard from detonations "possibly as high as 50 KT [kilotons] and certainly none for a 25 KT [kilotons] detonation." A major reason for the nonhazard determination: Scientists said that only 4,100 people lived downwind within a 125-mile radius of the chosen Nevada site.

On December 18, 1950, President Truman finally approved the site, and on January 11, 1951, the first test program, Operation Ranger, was given the go-ahead by the White House. That same day, the AEC announced the tests and distributed handbills in the area surrounding the Nevada site. They said test work would be undertaken on the part of the Las Vegas Bombing and Gunnery Range for "work necessary to the atomic weapons development program."

The handbill warned that "NO PUBLIC ANNOUNCEMENT OF THE TIME OF ANY TEST WILL BE MADE," but that health and safety authorities had determined that no danger from—or as a result of—AEC activities "may be expected" outside the closed-off testing area. It also read that "radiological surveys and patrolling of the surrounding territory will be undertaken to ensure that safety conditions are maintained."

19 · A RACE TO THE H-BOMB

• **THE 1951 GREENHOUSE** series of tests were to verify principles that would lead to a thermonuclear bomb. But the AEC was in a rush to develop the H-bomb, and before Greenhouse had taken place, the commission had already begun plans for tests in 1952, codenamed Ivy, of two, full-scale thermonuclear devices.

Greenhouse's George test at Enewetak on May 9, 1951, used heavy-hydrogen deuterium in the core trigger, which boosted the yield of the surrounding uranium. Teller's colleague Stanislaw Ulam had the idea of directing a mechanical shock of a nuclear explosion to compress the deuterium fusion fuel; Teller had come up with the idea of placing a plutonium rod through the center of that fusion fuel to "sparkplug" the explosion. The Greenhouse experiment was essentially developed by Teller, based on a device patented back in 1946 by Johann Von Neumann and Klaus Fuchs. When detonated, it created the largest nuclear explosion to date, 225 kilotons.

Test planners recognized the US was not only in a nuclear arms race with the Soviet Union but also that war was being waged in Korea.

The US commander in chief of the Pacific, Adm. Arthur W. Radford, had ordered a Navy force to provide surveillance of the Enewetak area during both Greenhouse and following Ivy tests. The force included an escort aircraft carrier, four destroyers, an underwater detection unit, six F4U fighter aircraft, four torpedo bombers, and a squadron of P2V maritime patrol aircraft.

Joint Task Force Commander General Quesada had recommended a maximum of twenty official observers for each Ivy detonation, which had been agreed to by the AEC. Of the total number of spaces allocated, the Department of Defense and AEC each received ten spaces for each shot.

The National Security Council planned for limited public disclosure of the Ivy tests, with just one brief statement to be made in early September 1952 that new tests would be taking place in coming months. After that, there would only be an announcement after each test had taken place.

Security was tight. There was concern that before and after the planned Mike shot, which was to prove an H-bomb was possible, personnel participating in that test would return to the US, increasing the danger that additional test information would be made public through leaks to the press. In addition, officials worried that there could be pressure on task force personnel to talk before the second shot took place.

A study was done during preparations for Ivy to determine whether the task force could censor personal mail sent from personnel at Enewetak back to family or friends in the US or elsewhere. Both practical and legal issues arose: Individuals would be leaving Enewetak, both before and after tests, and once gone they would no longer be under task force supervision, leaving open the question how their mail could be censored.

In addition, under US law, peacetime censorship of personal mail by service members could only be established by the president or secretary of defense. Civilians, on the other hand, could not come under some censorship without special legislation from Congress, which no one wanted to seek.

In the end, the task force decided to have security indoctrination of personnel, coupled with an appeal for self-censorship when it came to the nature of the mission.

Evacuation plans were made since no one knew for sure how big the detonation of the first thermonuclear device would be. Plans were made to move American personnel at Bikini Atoll, for example, 190 miles west of Enewetak. Those individuals were on Bikini to prepare the next tests, based on Mike succeeding. A plan had even been prepared for evacuation of Kwajalein, which was 389 miles away.

Exercising caution, the AEC decided that the Enewetak Marshallese now living on Ujelang, along with their animals, should be evacuated from that atoll for the duration of Ivy, despite being nearly 150 miles away from the planned points of detonation.

On October 27, 1952, a Navy LST landing ship picked up 157 Enewetak Marshallese, aged three months to over eighty. Instead of relocating them to a new atoll, the Navy kept them aboard ship where, as passengers, they occupied ad hoc quarters as the vessel moved to what was considered a "safe" distance from the test site. There was rough weather and torrential rains the first days at sea, and many Marshallese became seasick. They and others were unable to eat, while a few remained ill most of the time they were aboard. When the Mike shot took place, despite their distance from the site, the Enewetak Marshallese saw the explosion and named it the "Big Drop."

Mike was truly an experimental vehicle as it stood on Elugelab Island, in the northern rim of Enewetak Atoll. It was a twenty-two-foot-tall, cylindrical steel device nicknamed the "sausage," installed inside a giant shed. Its casing was manufactured by American Machine and Foundry Co. at its facility in Buffalo, New York. Attached to the casing was an equally large cryogenic cooling device to prevent the canisters of liquid deuterium and tritium from evaporating. At the top of Mike sat the uranium/plutonium trigger (itself an atomic bomb), which would cause the fusion explosion.

On Enjebi Island, three miles to the east of Elugelab, a four-story reinforced concrete building had been constructed before 1951, for an earlier test. For the Mike shot, it was to measure the strength of the hydrogen bomb.

To assure that the thermonuclear reaction took place with the desired force, Mike was essentially overdesigned. Scientists had no idea beforehand what its power would be, with estimates running from two megatons (equivalent to two million tons of TNT) to over ten megatons (ten million tons of TNT).

Up to that time, it was expected to be the most powerful explosion in history created by human beings.

The shot was planned for 7:15 a.m., November 1, 1952, Enewetak time.

That was three days before the US presidential election, where General Eisenhower, the Republican candidate, was favored to defeat the Democrat's candidate, Illinois Gov. Adlai Stevenson. Truman had been made aware that some in the AEC were interested in delaying the test, fearing it could have an effect on the election. President Truman understood, but did not agree to a delay, indicating instead that he hoped some technical glitch might cause postponement.

When that did not occur, less than a week before the planned shot, the Defense Department sent an official to Enewetak to see if a delay still could be arranged. He reported back that by then it would be extremely difficult to have a postponement.

The last attempt to delay the test came on October 29, a day in which Truman had been campaigning in Iowa and Illinois for Stevenson. At 8 p.m. that evening, just before the president was to make a major speech at a Chicago hotel, he was again offered a chance to put off the test, and again demurred.

At about the same time in the Pacific, a British cargo ship, the *SS Hartismere*, was detected sailing into what had been expected to be the path of any fallout, and a Navy patrol plane was sent to get it to change course. The P2V patrol aircraft had engine trouble, and before reaching the British vessel, made an emergency landing on the by-then-abandoned Enewetak air strip. A helicopter had to pick up the crew that night and fly them to ships outside the danger zone.

Meanwhile, the *Hartismere* sailed on until predawn on November 1, when it was located by another Navy patrol plane and warned to leave the area. When Mike went off as scheduled, at 7:15 a.m. on November 1, the *Hartismere* was two hundred miles away.

Mike's roaring nuclear fireball burst upwards like a rising sun on the horizon to observers thirty miles away. In seconds, it had spread over three miles in diameter, followed first by a roiling white mushroom cloud that then turned red-brown as it soared more than forty thousand feet into the sky in one-and-a-half minutes. In those ninety seconds, a thirty-miles-in-diameter stem emerged. Some three minutes after detonation, an upper cloud, some sixty miles in diameter, had grown out of the stem.

It kept growing and by six minutes was topped by an upper cloud that rose over one hundred thousand feet and spread fingers across a circular area one hundred miles in diameter. After fifty-six minutes, the entire monumental structure appeared to stabilize, although winds at different levels, blowing at different speeds, at times in different directions, affected drift of the cloud.

However, as late as sunset, "distant and high portions of the cloud could still be observed," according to a Joint Task Force report.

The first estimate from shipboard observers put the explosive power at 12 megatons, but later analysis set it at 10.4 megatons, more than 700 times the power of the Hiroshima atomic bomb. Where the Mike fireball was three-and-one-quarter miles across, Hiroshima's had been about one-tenth of a mile.

Where the Mike device had once stood, there was a crater covered with water one mile across and 164 feet deep. The Air Force described the crater as going down the height of a seventeen-story building and spreading wide enough to "comfortably" hold fourteen Pentagon Buildings.

A month later, Dr. Lauren R. Donaldson wrote in a report to the task force commander, "Following Mike shot, the radiation level increased many fold, especially along the northern and western portions of the atoll. . . . The amount of radiation found on and in the specimens was sufficient to destroy or damage these forms over a very wide area."

At Enjebi, three miles from the blast, the cement building had cracked but was still standing. The island itself had been swept by a radioactive tidal wave. Two days later, no living animals were to be found, and only stumps of vegetation remained.

Fish found off Enjebi had been burned on one side—as if, according to the field notes of one scientist, they had been dropped into a hot pan on one side only. Trees fourteen miles south of the blast had been scorched and wilted. Birds at that distance had been burned and were sick.

Teller, as one of the device's key designers, had been invited by Los Alamos Director Norris Bradbury to attend the Mike test, but he declined. He was only eight weeks into setting up his new Lawrence Livermore Laboratory in Livermore, California, and was preparing for that laboratory's first tests, and so did not go to Enewetak.

He did, however, make arrangements that permitted him to monitor the explosion on a seismograph located on the University of California campus at Berkeley.

Although it was an above-ground test, the intensity of the shock waves had been expected to reach Berkeley. The wave took some twenty minutes, traveling under the Pacific Ocean, but as Teller described it in his memoir, a luminous spot and a seismograph screen "did a little dance" at the correct time.

When he developed the film, Teller wrote that the result was "clear, big and unmistakable." He decided to send a telegram to Los Alamos, which he wrote in a way that would not break security. He addressed it to Elizabeth Graves, herself a physicist who had worked with her husband on developing Mike.

"It's a boy," Teller wrote, and all who saw it realized the meaning, and that Teller was claiming paternity. He would thereafter be known as the father of the hydrogen bomb.

When later reporting the Mike results to President-elect Eisenhower, AEC Chairman Dean said, "The island of Elugelab is missing!"

Manned F-84G jet fighters were used to collect air samples because they were less costly to operate than drones. However, to make the flight from their base in Kwajalein, they had to be refueled in flight. The aircraft's speed limited the exposure of pilots to radioactivity, but their highest altitude was forty-five thousand feet, so there was no sampling of the upper cloud. Although future tests of their clothing showed there was low-energy gamma radiation present in the Mike cloud, the exposure was "in all cases well within the prescribed limits," according to the Joint Task Force's final report.

One of the jet fighter pilots became the sole immediate casualty of the test. When the pilot failed to get refueled and attempted a landing at the vacant Enewetak air strip, his plane crashed into the water some three miles from the runway, flipped over, and sank. Air rescue helicopters later searched the crash site, but the pilot was never found.

In the aftermath of the test, some fifteen rafts and twelve buoys deployed to gather fallout were recovered. One recovered buoy taken aboard the destroyer USS O'Bannon actually was radioactive enough to slightly contaminate twelve crew members. Years later, in 2005, Los Alamos scientist Benjamin C. Diven, who worked on developing the Bravo bomb, told an Energy Department Oral History interviewer, "They didn't find any significant fallout on Mike. There was fallout but not lots. They couldn't account for most of it, so it was generally thought that, still, it had just carried all of this great amount of dirt and other things up clear above the troposphere, the lower atmosphere and into the stratosphere, and that it was such fine particles that it just hung there and gradually spread around the world and didn't put much down anywhere."

Diven, with hindsight, concluded, "Well, that's not what happened. Their model of what Mike produced [in fallout] was simply wrong. They hadn't sampled the ocean all around the Mike shot. They sampled nearby islands. They just didn't pay enough attention and indeed, it [the radioactive fallout] fell out in the ocean all around."

In its final report, the Joint Task Force made a recommendation that would come back to haunt its authors: "On-site Task Force operations would be greatly simplified if all . . . devices in the megaton range were detonated at locations and under circumstances that precluded evacuation during the detonation phases."

The report also suggested this type of thermonuclear device may be adaptable to a major redesign for weapon purposes. "It is believed that the overall size and weight can be reduced and that the cryogenics system can be simplified to make a usable weapon," the report said.

After the test, there were sixteen separate newspaper accounts describing the Mike shot. Investigation by the government found these "eyewitness" accounts came from letters written by personnel who had been in the task force. All were eventually identified and admitted they had been lectured about security of the operation but had failed to self-censor themselves.

Two weeks after the Mike shot, the second test in the Ivy series took place. Called King, it was an airdrop test of a prototype new fission bomb using the advanced triggering device based on what had been learned from one of the Greenhouse tests. Exploded on November 16, 1952, at a height of about 1,500 feet, its yield was 550 kilotons—the greatest ever produced from a fission device in a deliverable bomb. It was 35 times the yield of the Hiroshima bomb.

The Mike shot had demonstrated that a uranium-based, fission explosion created temperatures in the million-degree range. That intense heat, in turn, caused a thermonuclear-fusion reaction in isotopes of heavy hydrogen, which is also called deuterium, because it has both a proton and a neutron in its nucleus, whereas the more normal hydrogen isotope has only a proton.

Los Alamos scientists knew that the resultant fusion not only caused an explosive release of energy but also set off the release of high-energy neutrons.

The most plentiful variety of uranium on earth is uranium-238. Unlike uranium-235, which is radioactive, uranium-238 is stable, and under normal conditions remains stable. The one thing that will cause an atom of uranium-238 to split is bombardment by high-energy neutrons.

So, the Lost Alamos scientists came up with the idea of surrounding the fusion device (i.e. Mike) that gives off high-energy neutrons with a blanket of uranium-238. The neutrons released by fusion would be captured by the uranium-238 nuclei, which would in turn undergo fission and contribute to the overall power of the explosion.

In short, the scientists developed a super weapon that was really three nuclear weapons in one—two fission bombs and a fusion bomb. The total energy yield, however, would turn out to be half from fusion and half from fission.

An important byproduct of such a super weapon would be uranium-237, created by the fission of uranium-238. Uranium-237 is highly radioactive. It guaranteed that any bomb, such as the one being proposed, would be much more dangerous in terms of fallout than Mike or any other atomic weapon previously exploded.

In addition to the danger of explosion creating highly-radioactive elements such as uranium-237, the hazard from fallout relates to where the detonation takes place. There is a marked difference in contemplating fallout when discussing a nuclear airburst or a nearer-surface, on-the-surface, or below-ground nuclear explosion.

In a relatively high nuclear airburst, where the fireball does not touch the surface, the radioactive fission products are vaporized. These include products resulting from neutron interaction with materials from the device itself, i.e. the container, and some unfissioned uranium and/or plutonium that are already radioactive.

These vaporized products condense as the fireball rises and cools, becoming small particles and sometimes smoke-like as they are carried up in the mushroom cloud. At some point the cloud stops rising and eventually disperses depending on the winds and weather.

The explosive yield determines how high the cloud rises and also how long it takes for the fallout material to be deposited. Larger detonations, such as those contemplated for thermonuclear explosions, could rise into the stratosphere and not settle down to earth for months, with dispersion over a wide area.

Overall, airbursts—such as those over Hiroshima and Nagasaki—contributed little threat for radiation exposure to persons on the ground other than those directly under the burst. They would, however, have short-lived exposure should released neutrons actually reach the ground.

Surface or near-surface nuclear explosions would present a far greater problem. More radioactive debris would be created because more earth materials from the surface would be affected by neutrons generated by the explosion. The extreme heat would vaporize materials, including those from whatever exists below the explosion, and they all would be carried up within the fireball. As the fireball rose and cooled, the particles formed would tend to be larger, with radioactivity embedded inside them or coated on their exteriors.

The largest radioactive particles would tend to fall within the crater or nearby. One estimate was that nearly 80 percent of radioactive debris would fall within the first day, but that would depend on winds, weather, and the height of the cloud.

As the Crossroads Baker shot had shown, surface or below-surface bursts on or beneath seawater would generate radioactive particles, mainly of salt and water. But large-yield surface bursts over shallow lagoon waters, coral islands, or—worse yet—city harbors would form the most complex combination of land-surface-and-water-surface radioactive, particle-size, potential fallout that, if blown toward inhabited land, could have long-standing and devastating effects.

20 · PREPARING FOR THE BRAVO TEST

■ **AT THE TIME** of the November 1, 1952, Mike shot, planning was already underway for Operation Castle, the next set of tests which would lead to an American H-bomb, since US scientists believed they were on the right track. Preparation for testing a bomb design with such enormous potential destructive power would take almost two years. In June 1952, the Joint Chiefs of Staff had already set out a formal requirement to the AEC that they produce a thermonuclear bomb in the megaton range that would be deliverable by an available aircraft within the next two years.

Between the summer of 1952 and spring 1954, scientists at Los Alamos and Lawrence Livermore worked on theoretical calculations, while weapons designers and mechanical engineers, acting in parallel, worked to design a metallic casing that would later be manufactured into actual hardware—the bomb.

In January 1953, Bikini was secretly selected as the site for Operation Castle's three thermonuclear bomb tests. Inside the federal government, the Navy, joined by the AEC, proposed the US expand the danger area to commercial shipping around that atoll from that which had existed for earlier atomic tests.

The high commissioner for the Trust Territories, Elbert D. Thomas, raised a question. In a February 5, 1953, letter to his boss, James P. Davis, director of the Office of Territories in the Interior Department, Thomas wrote that "use of Bikini Atoll and extension of the contingent Danger Area present problems that are difficult for me to resolve because of my responsibilities to the people of the islands," meaning the Marshallese.

Mentioning specifically the people of Bikini and Rongelap, Thomas said the proposed extension of the danger area would include two-thirds of Ailinginae Atoll, which although not "inhabited . . . is regularly used and harvested by the people of Rongelap and contributes a substantial part of their living."

He said there would probably have to be remuneration for the land, but that money would not replace the land on which the Marshallese were dependent. He foresaw the Rongelap people, along with the already displaced Bikinians, looking more to the US government for continued support for what would have to be taken from them.

As a result of this objection, the danger area was changed, "based on information from the Department of Interior," according to a March 1953 letter in the Energy Department archives. "A consideration in establishing the Danger Area was to fix the boundaries *so as not to include unnecessarily any land areas* that were inhabited by natives," according to the letter [emphasis added]. "If the danger area had included such inhabited atolls as Rongelap and Utirik it would have required that the natives of those atolls be evacuated and that a permanent home be found for them elsewhere."

Concern over costs and finding a temporary or perhaps permanent home for the Rongelap and Utirik Marshallese set the danger area boundary; not concern for possible fallout from the upcoming tests.

Gordon Dunning, of the AEC Division of Biology and Medicine, would state that "the main objection to evacuation is the high cost and the logistic problems presented in supporting such an operation." Joint Task Force Seven, headed by Army Maj. Gen. Percy W. Clarkson, had been created to run the Castle tests. Clarkson and his colleagues agreed that financial austerity was afoot at the time and even affected the number of ships and aircraft that could be available should evacuation be necessary.

Simple steps on behalf of the Marshallese were not taken, including placement of dosimeters on those atolls they occupied. On the other hand, the

American military service personnel, then manning the weather station on Rongerik, possessed dosimeters and film badges that could be used to measure their total radiation exposure.

In April 1953, the AEC announced publicly that Bikini would be used again "to accommodate the rapidly expanding program of developing and testing new and improved nuclear weapons." At the same time, the US United Nations Mission in New York announced that Bikini Atoll "and its territorial waters have been declared closed for security reasons."

One month later, a notice to mariners was sent to foreign nations through the Navy Hydrographic Office calling attention to the danger area around the enlarged Pacific testing region and setting the eastern edge of the new area roughly eighty miles from Bikini and just a mile short of Ailinginae Atoll.

An October 31, 1953, letter sent from Adm. Herbert G. Hopwood, then-chief of staff to the commander of the Pacific Fleet, to Joint Task Force Seven Commander Clarkson on "Radiological Hazards in the Marshall Islands Area During Operation Castle," spelled out shortcuts taken that would eventually prove shortsighted.

It directed that "no special efforts will be implemented by JTF 7 [Clarkson's Joint Task Force Seven] in support of the following safety measures." First was to limit "atomic cloud tracking outside the immediate danger area." Since prior to the detonation the danger area was thought to have been downwind toward Enewetak and further away, Ujelang, a limit was put on use of tracking aircraft upwind, toward Rongelap and Rongerik. This was in contrast to the Ivy tests when almost every Trust Territory atoll was surveyed by airborne monitoring, even though little if any residual radiation had been anticipated.

The second step was to limit "sampling of drinking water on distant atolls." A provision was added that "only in the event of radiological conditions requiring such action" would drinking water sampling—which had been done for the Ivy shots—be done post-Bravo on the outlying atolls.

The third item spoke for itself: "Evacuation of native populations is not planned for JTF 7 effort due to unavailability of task force equipment. However, consideration of populated islands will be one of the major factors influencing the decision to shoot."

At that time, Bikini, the shot site, and Enewetak were considered the only populated atolls inside the danger area.

At Enewetak, 165 miles west of the Bikini test site and swept by the lower trade winds, Joint Task Force Seven had based some 10,000 personnel. Almost 300 miles to the southwest, and also in the path of the lower winds, was Ujelang Atoll, where the Marshallese from Enewetak then lived.

Ninety miles east of Bikini was Ailinginae Atoll, owned by the Rongelap Marshallese whose own atoll was 25 miles farther east. Another 35 miles to the east of Rongelap Atoll, was Rongerik Atoll, once home to the Bikinians but now a US weather outpost. Utirik Atoll lay 300 miles east of Bikini and was home to 167 Marshallese. Kwajalein Atoll, the US permanent Navy base, was 250 miles to the southeast of Bikini.

The only direction in which there were no inhabited atolls closer than five hundred miles to Bikini was north, and Bravo's planners were hoping that the winds between twenty thousand and sixty thousand feet would blow the expected radioactive cloud that way, as they had in the case of the Mike shot sixteen months earlier.

Joint Task Force Seven had stipulated that before the Bravo bomb could be detonated, the winds up to sixty thousand feet must be blowing toward the northwest or due north. According to the Joint Task Force's official history, all atolls except Ujelang and Enewetak, the two lying in the trade winds path, were "considered to be a favorable location with respect to fallout."

To assure that adequate weather information was available, Joint Task Force Seven had established an observation network consisting of seven stations in the Marshall Islands, with additional data to be provided from distant Pacific islands such as Midway, Wake, Guam, Iwo Jima, and Johnston.

The Marshall Islands weather station on Rongerik, was manned by twenty-eight Army and Air Force men who had a small number of tents and aluminum-sided buildings which served as barracks and offices.

Before the Bravo detonation, no detailed preparations were made for the emergency evacuation of any atolls except Ujelang and Enewetak. For the people on those atolls, the Navy element of Joint Task Force Seven was to provide ships if an emergency arose, and the Pacific Fleet commander had agreed to furnish extra ships from Kwajalein if they were needed. In addition, a representative of the Joint Task Force met with Trust Territory and Navy personnel at Kwajalein Island before the Castle series began to make certain that interpreters and Trust Territory officials, as well as an amphibious plane, would be available if an evacuation of any Marshall Islands atoll became necessary.

■   ■   ■

Rongelap had been evacuated before the 1946 Crossroads tests, when much smaller yield atomic bombs than those built for Bravo had been detonated. The Marshallese living there during Bravo would not be removed, although planning estimates had forecast a thermonuclear explosion of six megatons—the equivalent of six million tons of TNT. That size yield would be at least three hundred times greater than the atomic bombs used at Crossroads, which were a mere twenty-three kilotons.

Since Bravo was to be a surface explosion with unknown radioactivity implications, Navy Radiological Defense Laboratory personnel, who had been studying the ability of the Navy to operate in a nuclear environment, wanted to try improved techniques to measure the expected fallout. They proposed floating one hundred buoys in the Pacific Ocean equipped with measuring devices in a vast circle whose edge would be fifty miles from the blast.

At first, Adm. Felix B. Stump, the commander in chief of the Pacific Fleet, objected. He claimed the buoys would create a potential target for his patrol planes, whose radar apparatus could mistake them for submarine snorkels. To counteract this problem, low-frequency radio transmitters were installed in each buoy to send out recognizable signals to the destroyers, patrol craft, and airplanes that would be searching the exclusion area for Soviet submarines.

The Air Force explored the possibility of getting an aircraft that could obtain better high-altitude particulate samples from the nuclear cloud, but in the end Joint Task Force Seven had to settle for two B-36 bombers and again F-84G fighters specially outfitted with collection capability by the nuclear weapons laboratories. They also had a B-36D with pressure gauges for measuring blast and heat effects, an RB-36 reconnaissance plane for high-altitude photography, and three C-54 transport planes to take photos at lower levels. Finally, there were twenty-one H-19 helicopters, which were to ferry ground-sampling teams after the shot and handle other logistical problems.

New radiation-protection measures were prepared for pilots and the air-sampling, cloud-tracking crew members. The F-84G pilots during Operation Ivy had worn lead-cloth suits and lead-covered helmets that were bulky and restricted movement. For Castle, a nylon sleeveless pilot's vest with a chest-size section of fiberglass and lead was developed, with more lead added to the sides for torso protection.

Weighing about six pounds, the vest was even tested by swimmers playing escaped pilots in the pool at Kirtland Air Force Base in New Mexico, where it had been developed. Additionally, the backs and bottoms of the pilot seats in F-84G aircraft were given a coating of lead. A heavier lead vest, weighing about fourteen pounds was also developed for the WB-29 crew members.

Before Bravo, scientists had assumed that any significant fallout from a nuclear explosion would descend within six hours and would occur in a circular area out from ground zero. That had been the experience in the handful of surface, kiloton-level shots at Nevada testing grounds where fallout had occurred. But since there had been no successful fallout measurements made from the Mike test outside the islands of Enewetak Atoll, there was no way to judge how far fallout would travel in a major Pacific surface explosion, or whether it would come to earth in six hours.

So, as a safety measure, planners decided to track the Bravo fallout for twelve hours.

The basic tracking of the fallout cloud was to be done through ground observation and aircraft survey. Monitoring stations on surrounding atolls, whose primary responsibility was meteorology, were also to be called on to supply radioactivity reports based on readings from low-level dosimeters.

To aid in shot planning, and in the Air Force's post-explosion sampling of air particulates, the radioactivity-safety unit of Joint Task Force Seven was to use information on wind direction and velocity, to make projections—air particle-trajectory forecasts—at H-minus twenty-four (twenty-four hours before detonation) and again at H-plus six hours (six hours after detonation).

Before the Mike shot, fallout estimates had been based on the assumption that radioactive debris would emanate from a single point, the mushroom cloud, like salt from a shaker. The Mike experience showed that crosswinds and updrafts created by the explosion caused a radioactive swirl thirty miles in diameter from which fallout could rain down. Using those trajectory forecasts, radiation-safety planners outlined areas where people would be barred for at least six hours after the blast, unless they had special authorization.

Since the trajectory forecasts would depend primarily on what effects the winds had on the radioactive cloud, the major ingredient in safety planning for Bravo was the weather. One reason for the selection of the Marshall Islands as a testing area back in 1946 had been the presumed stability of the winds and weather. A trade wind of ten to twenty knots blew steadily over Bikini and Enewetak toward the west-southwest. Above twenty-thousand feet, the winds blew in the opposite direction more than half the time—a fact that meant little in the 1946–1947 test period because the mushroom clouds from those early weapons did not go that high.

Higher, the tendency for the winds to blow toward the east grew, as did their speed. At forty-five thousand feet, wind speed averaged about thirty-five-knots. Higher still, at the point where the lower atmosphere (or troposphere) meets the upper atmosphere (or stratosphere), the winds switched again to a northerly direction.

In 1954, there were few meteorological records to disclose to planners the vagaries of Marshall Islands weather—and particularly the changing nature of the very high-altitude winds. The information available to the Bravo planners indicated that the most significant variations were to be found between twenty thousand feet and sixty thousand feet.

Based on Bravo design expectations, planners expected the top of the mushroom cloud to reach almost one hundred thousand feet, a height above the troposphere and into the stratosphere. Trapped there it was expected to have "cooled" radioactivity, and to disperse so widely that it would not be a health hazard when the fine particles eventually floated down to earth. There was recognition that particles in the stratosphere could be dispersed over large portions of the earth, and that "global fallout" would create some concern since it could add relatively small but increased levels to natural radioactivity. How much, however, was unknown.

On January 8, 1954, the AEC announced publicly in Washington that it planned to hold the Castle test series but gave no dates or further details. The announcement generated a one-column, page-one *New York Times* story headlined, New Bomb Tests Slated in Pacific, Greatest Hydrogen Explosion May Be Produced, and linked it to upcoming US-Soviet talks on peaceful uses of atomic materials.

The previous night, President Eisenhower, in his annual State of the Union speech, had referred to "our great and growing number of nuclear weapons," and said he was looking at sharing knowledge of the tactical use of nuclear weapons with allies, but he made no mention of the upcoming tests.

Two days after the commission's announcement, the *USS Curtiss*, a Navy seaplane tender, sailed from Port Chicago, California, with the bulk of the Castle thermonuclear materials and test devices, including the Bravo bomb elements. The *Curtiss*'s movements were conducted under wartime conditions. There was radio silence, and all of the ship's outside lights were darkened at night. Escorted by four destroyers, the *Curtiss* followed a route that kept it miles from normal ocean traffic, and while the convoy was within five hundred miles of the US mainland and then Hawaii, and finally Enewetak, Navy planes had provided air cover.

Two weeks after the *Curtiss* left California it arrived at Enewetak. Its thermonuclear materials were unloaded at an assembly building on Perry Island at the southern end of the atoll.

Meanwhile, at a point on the northern part of the Bikini reef near the southwest tip of Namu Island, a one-acre, man-made island had been constructed. It was connected to Namu by a 2,900-foot causeway built across a reef. On this man-made island was a shed to house the Bravo device, along with sensor devices linked to recording instruments.

In mid-February, the Bravo device itself was put on a barge, taken from Enewetak to Bikini Atoll, and installed in its specially prepared shed. Although Bravo was a deliverable bomb, it didn't look like one. It was an aluminum-encased cylinder with flat ends. It was about 15 feet long, 4.5 feet in diameter, and weighed 23,500 pounds. It reportedly contained 200 pounds of uranium-235,

200 pounds of lithium deuteride, and more than 2,000 pounds of uranium-238. It also contained high explosives to initiate the U-235 fission explosion.

Inside Bravo's two-story, cement shed was a range of electronic inner parts associated with the testing instrumentation. Running out from the shed along the causeway to Namu Island were a series of vacuum line-of-sight pipes designed to carry instantaneous information on the operation of Bravo, and particularly the fission and fusion processes in the microseconds before the explosion vaporized both pipes and instruments. The pipes were linked to scientific detection and recording laboratories on Namu.

On towers in the shallows of Namu were reflector shields designed to measure both the neutron output of the explosion and the early stages of the fireball.

Eight times a day toward the end of February, the winds aloft were measured from more than a dozen points within five hundred miles around Bikini. Planning forecasts began forty-eight hours before the proposed shot time. They included winds at the shot site and for surrounding areas at ten-thousand-foot intervals up to ninety-thousand feet.

As of February 7, the winds appeared favorable, but forecasters could see that changes were coming. The trade winds up to 20,000 feet were blowing west; they were blowing northwest at 30,000 feet and due north at 40,000 feet. At 50,000 feet, they were blowing slightly east of north.

Five days before the shot, an advisory message to the AEC and Joint Chiefs in Washington from Joint Task Force Seven predicted that fallout from Bravo would spread in a fan-shaped pattern from northwest to northeast under its then windfall prediction.

Conditions remained favorable the afternoon of February 27, but the winds blowing to the northeast continued.

On the morning of February 28, less than twenty-four hours before scheduled detonation, the winds at twenty thousand feet had swung around and were blowing to the northeast at sixteen knots. The winds from thirty thousand feet up to forty-five thousand feet were also heading northeast. Nevertheless, at 11 a.m. that morning, Joint Task Force Seven's advisory predicted "no significant fallout for . . . the populated Marshalls."

By that afternoon, 1,400 technicians who had helped put together the test equipment had been moved out away from Bikini Atoll. Most officials and the few observers were aboard the *USS Estes*, one of ten ships afloat thirty miles east of Bikini, an area considered to be upwind from the expected local fallout.

At the 6 p.m. weather briefing, the predicted winds appeared less favorable, but the decision to continue as planned was "reaffirmed," according to a later Department of Energy report. Another review of the winds was scheduled for midnight.

The midnight briefing should have raised questions, since the winds at twenty thousand feet "were headed for Rongelap to the east," according to the report. But the predicted wind speed was low enough to ease concern for other local islands such as Bikini. The decision to go ahead again was reaffirmed, at least until the final weather briefing just two hours before planned detonation.

Building 70 had been constructed into the coral reef on the downwind side of Enyu Island and twenty-two miles across Bikini Lagoon from the shed housing Bravo. It contained a radio room, laboratory space, and a large room filled with classified equipment used to monitor, prepare, and fire the Bravo bomb, and then continue to monitor the aftermath.

The building served as the operations center for the nine-man, Bravo firing team. It was surrounded by a moat to absorb any post-shot tidal wave if one were to take place. The building's cement walls were three-feet-thick, steel-reinforced, and covered by ten feet of dirt and sand to keep out radiation from any fallout. Its door was a submarine hatch.

■　　　■　　　■

At dusk on February 28, John C. (Joe) Clark, head of the firing party, boarded a helicopter that took him to a platform near the shed holding Bravo. Clark was the chief so-called "triggerman" for the AEC, having done the job on some forty other nuclear tests in Nevada and in the Pacific. He was also considered somewhat of a daredevil for the two times he climbed towers in Nevada and disarmed atomic devices that had failed to detonate, once in 1951 and again 1952.

Clark, on that February 28 evening, had radioed Joint Task Force Seven's Command Post on the *USS Estes* and requested permission to arm the bomb.

All ships had cleared out of Bikini Lagoon, so permission was granted to complete the preliminary arming process.

To do that, earlier that day Clark and his colleagues mechanically created an electronic circuit capable of carrying the firing signal from Building 70 on Enyu to the bomb. When he was finished attaching the circuits to Bravo, he got back aboard the helicopter and flew back to Enyu along with the military police who had been guarding the bomb since its arrival.

The military police remained on the helicopter, which then flew off to join the fleet at sea, leaving behind the nine men in Building 70. Their next job would be to close the final links in the firing circuit and prepare for detonation planned for 6:45 a.m., March 1.

The bomb itself was ready; but so much depended on the weather.

21 · BRAVO DETONATION

■ **AT MIDNIGHT ON** March 1, 1954, the weather was holding although light winds at low altitudes became variable, raising the possibility that fallout in the immediate area of the Castle Bravo detonation might not go as planned. Most ships that were within twenty miles and thirty miles southeast from Bikini were moved out to fifty miles, just in case the lower winds carried fallout in their direction. The *USS Estes*, which carried the navigation system station that controlled the test's drone and piloted aircraft, remained at its original station because it had to be near the detonation point.

The Enewetak airport control tower announced after midnight that aircraft would soon take off, and forty minutes later a B-36, first of the planes participating in the test, went down the runway at Kwajalein, four hundred miles away, and headed toward Bikini. At 3 a.m., one of firing chief Joe Clark's colleagues inside Building 70 announced the beginning of the countdown, timed to the beep from the world-wide, standard-time station. Scores of military personnel on ships and at weather stations synchronized their watches and time settings on instruments.

At 4:30 a.m., JTF-7 reported there were "no significant changes" in the winds except at Bikini where, at the very low level, they were showing northerly and westerly elements, while at the seven thousand to fourteen thousand foot levels there was a shift toward the east.

At 5:45 a.m., in Building 70, a firing team member climbed a ladder to the roof and put metal plates over the air-conditioning vents. Returning, he sealed the submarine-type hatch door, making the structure watertight and blast proof.

At 6:25 a.m., a B-36 flying at 38,000 feet made a final pass over the detonation site and reported scattered clouds, but no rain. Rain could interfere with viewing the shot and bring fallout down to earth more quickly.

At 6:30 a.m., Clark ordered the button pushed on the automatic sequence timer, and although Clark could use a stop switch to halt the process, for the final fifteen minutes the firing mechanisms were set in operation electronically, and the men in Building 70 just watched lights go from red to green to show experiments and circuits were ready to operate. Some remote-controlled cameras would automatically switch on a half-second before detonation.

From Building 70, the announcement was made, "At the next tone it will be H minus one minute," according to an interview Clark gave in 1957 to the *Saturday Evening Post*. "Thirty seconds . . . Fifteen . . . All except two of the lights were green. Ten, nine, eight, seven, six, five. All was absolutely quiet except for the soft whining of the sequence timer . . . four, three, two, one, Zero."

"I looked at the panel," Clark continued. "All the lights were green, [and] we knew the bomb should have detonated. 'How did it go Al?' I called on the radio to Joint Task Force Director Dr. Alvin Graves, forty miles away on the command ship. 'It's a good one,' he answered."

Bravo had exploded.

The force of the explosion was expected to be six megatons, but in fact it was more than double that. At fifteen megatons, it was the equivalent of fifteen million tons of TNT and a thousand times as large as the bomb that destroyed Hiroshima.

In a few seconds the fireball, recorded at one hundred million degrees, had spread nearly three miles in diameter, then quickly spread to ten miles. The sandspit and nearby reef where Bravo's shed had stood, along with island areas, were vaporized down almost two hundred feet into the sea, creating a crater about one mile in diameter.

It was estimated that three hundred million tons of vaporized sand, coral, and water shot into the air as the fireball rose, and one-hundred-mile-an-hour winds created by the blast pulled additional debris up into the fireball. Within one minute, the fireball had gone up forty-five thousand feet with a stem four miles wide filled with radioactive debris. It continued to zoom upward, shooting through the troposphere and into the stratosphere within five minutes.

Later data showed the cloud bottom was at 55,000 feet, the secondary mushroom cloud bottom was at 114,000 feet, and the upper cloud hit 130,000 feet.

Ten minutes after detonation the mushroom cloud had widened and measured seventy-five miles across just below the stratosphere.

Original projections had Bravo fallout emanating from a fifteen-mile-wide cylinder that could stretch into the stratosphere. Instead, it turned out to be a one-hundred-mile-wide cloud where "debris was carried up and dispersed over a much larger area than was thought possible," wrote Dr. William Ogle, the task force commander of the scientific group who dealt with radioactivity.

■   ■   ■

John Anjain, Rongelap's magistrate, was on the beach by the atoll's lagoon when he saw that light from the west.

"After some time, a big noise came," he once recalled. "People kept doing their normal work."

North of Rongelap and ninety miles east-northeast of Bikini, the Japanese fishing boat *Lucky Dragon* with its twenty-three crewmen was drifting west with the trade winds that morning. A crewman at the stern rail saw a whitish flare in the west that lit up both the clouds and the water. It grew in size and turned to yellow-red and then orange. After a few minutes, the color faded, and shortly thereafter the ship was rocked by the blast of the explosion.

The *Lucky Dragon*'s captain and the fishing master realized that they might have strayed into a nuclear test area. They quickly decided to haul in their nets and head back to Japan, almost 2,500 miles away.

At Rongerik, 135 miles east of Bikini, the US servicemen already outside that morning saw a flash and 11 minutes later felt a blast wave that shook buildings.

■   ■   ■

Clark and the men in the firing crew were sealed in Building 70, unable to see the explosion. They had heard Graves saying that it was a good shot.

They braced themselves, and a minute later, as the air shock passed over, the building groaned but held. After fifteen minutes, a few members of the firing crew stepped outside to look up at the cloud, which by then was twenty miles high and directly above them.

At 7:07 a.m., twenty-two minutes after detonation, Clark began getting radiation readings on his instruments, and as the levels continued to rise rapidly he ordered the men back inside the blockhouse. The portholes were shut, and the men gathered in a corner of the concrete building where they were partly shielded by sandbags. Within twenty minutes, coral that had been vaporized by the fireball and had then condensed with radioactive material began to fall like deadly hailstones on and around the blockhouse. Within an hour after detonation, the readings in Building 70 had reached 250 R/hr.*

"The radiation is building up pretty fast, Al," Clark reported to Graves by radio. The two discussed sending helicopters to get them off Enyu but decided the radiation levels outside made that too risky. It was safer to remain inside the covered blockhouse, where they looked for the most secure place to wait. "We were trapped," Clark wrote.

The control room and radio room had high radiation levels. Readings were lower in the data room, where more sand had been piled on that part of Building 70. "I advised the command ship of our situation. I told them that we had found a room in the blockhouse which seemed perfectly safe unless the fallout level outside got much higher. At the rate of ten milliroentgen-per-hour** we could remain for days without harmful effects. I did advise them, however, that we would man the radio in the control room only every fifteen minutes," Clark wrote.

------

* Two hundred fifty R/hr means gamma radiation exposure of 250 roentgens per hour. That measure of gamma radiation exposure could cause nausea and/or vomiting. Above that could result in acute radiation sickness and require immediate medical care.

** A *milliroentgen* is a thousandth of a roentgen. You get several hundred *milliroentgens* when you get a chest x-ray.

They moved some Army cots into the data room. At one hour after detonation, Clark opened the outside hatch door and at arms length took a quick reading. "It read forty roentgen. I quickly closed the door," he wrote.

Aboard the task force ships thirty miles away, men had listened to the firing team's radio reports of radioactivity as they watched the mushroom-shaped cloud being pushed and pulled in several directions by winds at various altitudes. Some of it appeared to be coming their way. At about 8 a.m., Geiger counters aboard the ships began to record substantial levels of radioactivity.

Fallout, described as "pinhead-sized, white and gritty snow" in a later report, began falling on the USS *Bairoko,* which had been preparing to reenter Bikini Lagoon. A helicopter that had launched from the *Bairoko* for radiological reconnaissance was recalled when it was told that its destination area on Bikini was too radioactive for reoccupation. Launching of four other helicopters was delayed.

All men aboard the *Bairoko* were ordered below decks, and hatches, doors, and portholes were sealed. As part of a planned decontamination process, specially constructed wash-down systems were activated. Each support ship had been outfitted with hoses, pipes, and sprays to wash away radioactive fallout.

At the same time, Task Force Commander Clarkson ordered all ships to head south to get out of the fallout area. A radio message was sent to the blockhouse telling the firing crew that the ships were leaving the area but would be back before sundown so that helicopters could pick the men up before dark. Initially, the firing crew was to have been moved within six hours of the shot, but now they would have to remain in the blockhouse until the radiation levels went down.

For more than two hours, the fleet of support ships sailed slowly south, guiding themselves by radar, since the crews were below decks and the decontamination spray and descending mist from Bravo's cloud made visibility poor.

Some fifty miles from Bikini they halted. The radioactivity level was low enough for the observers to remain there and see what was going to happen next. Most of the three hundred million tons of vaporized matter drawn up into the Bravo fireball was coral, which consists chemically of calcium carbonate. A chemical decomposition accompanied the vaporizing within the fireball, and a calcium oxide, known as unslaked lime or quicklime, was formed.

In the subsequent cooling process, the quicklime combined with droplets of water, which had picked up beta and gamma radiation, to form particles that were adhesive, caustic, and highly radioactive. They ranged in size from a fiftieth to a thousandth of an inch in diameter and made up most of the fallout particles.

There were also substantial amounts of other radioactive elements in the fallout cloud, among them strontium-90, a beta emitter; cesium-137, which emits both beta and gamma rays; and iodine-131, also a beta emitter.

With the Bravo explosion, the US had thus unleashed a set of furies unseen in history, furies for which its own scientists were unprepared, let alone some of the least modernized people on earth.

22 · IMMEDIATE AFTERMATH

■ **AT 11 A.M.,** the Joint Task Force got word from its air-sampling F-84Gs and B-36s that fallout around Bikini had subsided. Accumulation on the ground, however, had created high levels of contamination. The ships themselves had been washed clean, but radiation safety officers found that drains on the flight deck of the small aircraft carrier USS *Bairoko* showed an extremely high reading. Intensities on the deck of the small carrier reached five rem/hour.* Absorbing one hundred rem over a short period of time would cause radiation sickness, whose symptoms are diarrhea, nausea, and vomiting.

At about 9 a.m., the crewmen of the *Lucky Dragon* were still pulling in their fishing lines when white, sand-like ashes began falling on the deck. A bit of it got in the eyes of one crewman and hurt, so he and others put on sunglasses. The particles began accumulating on the deck and blew into the pilot house, where the captain was at the wheel. Curious crewmen put the granules on their tongues and found they had no taste. One crewman put some in an envelope to be saved for good luck.

At 9:49 a.m., according to a Defense Nuclear Agency report drafted later, a Navy patrol aircraft from Kwajalein that was sweeping the sector eastward

---

* A rem (roentgen equivalent in man) is a measure of absorbed dose of radiation in human tissue.

from the detonation point for transient shipping found it had become "heavily contaminated and aborted its mission." The aircraft "sent a message stating that heavy contamination (0.5 to 1.0 R/hr) should be expected." That message did not reach the Joint Task Force radioactivity center for five days, and there was never an explanation for the delay.

At that time, according to the report, the aircraft's logs showed it was "approximately 65 miles due east of ground zero," and it had aborted its mission and headed back to Kwajalein just before the *Lucky Dragon* could have come into view.

For two hours, the fishing boat had continued to head north toward Japan, and then the fallout, or "white storm" at sea, ended. By that time, the captain was speculating that the strange material was coral ash from an American nuclear test. He remembered reading the warning—avoid Enewetak, three hundred miles to the southwest—before he sailed from the Japanese Maritime Safety Board. He increased his speed, convinced that if his ship were found close to an American test, it would be sunk.

In the past, some ships had been escorted out of test areas by US destroyers. Some Japanese fishermen also believed the Americans had sunk a fishing boat that disappeared in the Marshall Islands in 1952, at the time of the last nuclear tests.

■   ■   ■

Later on the morning of the explosion, John Anjain and five other Marshallese men set off in boats to collect coconuts on Jaboan Island, which was on the reef beyond the west end of Rongelap Island. At about 1 p.m., radioactive ash began to fall in that area. Some of it stuck to the skin and eyes of the men, causing itching and, occasionally, sensations like mosquito bites.

Back on Rongelap, Anjain's wife, Mijjua, was gathering pandanus leaves to weave a mat, and their son Lekoj, who had just learned to walk, was playing around the house. Mijjua recalled that as the powder began to fall, she heard an airplane overhead and thought that the Americans were dropping medicine to stop damage from the bomb that morning.

Tima, one of Rongelap's best fishermen, had gone out in his canoe that morning to fish the lagoon, and the descending ash turned the water around him yellow. As he headed home, the dust continued to fall on him and on the fish he had caught.

At 11:30 a.m., Billiet Edmond, a schoolteacher on Rongelap, had dismissed classes. He and his students went outside and were confronted with falling "powder-like particles," he said. "The once blended green and yellow leaves of our coconut trees, breadfruits, and pandanus gradually took on the white color of the falling stuff."

There had been little rainfall in the two months before March 1—spring is the dry season in the Marshall Islands, and the water in Rongelap's community cisterns had started to run low. In February, John Anjain and the village council had agreed to ration the communal water, allowing each person one pint a day. The families could continue to use their personal water supplies, accumulated in barrels from the rainfall on their own roofs, as they wished.

Late in the afternoon of March 1, in the midst of the fallout, rain fell briefly, dissolving the radioactive ashes on the roofs and carrying them down drains and into the water barrels that provided water for each household. At the town cisterns, rainwater carried the radioactive matter from the tin roofs into the island's communal water supply.

The children, who had played hard along the beaches during the day, were tired and thirsty by the time the rainfall ended. All of them drank the radioactive water, as did some of the adults.

The women of Rongelap regularly wore dresses; the men wore pants but usually no shirts. Adults wore sandals or went barefoot. Children were always barefoot, and the youngest wore no clothes at all.

In the heat and humidity, they perspired, and as the radioactive fallout came down it stuck to their hair and their bodies, gathering particularly at the folds of their necks. Fallout that was accumulating on the ground stuck to their feet and caught between their toes.

The fallout continued into the evening. It fell on fish drying on wooden racks outside houses. When people ate that fish for dinner, it had a bitter taste. Tima ate some raw fish that he had caught that afternoon, and it, too, was bitter.

Some people didn't eat dinner. A few told Anjain that they felt sick in their stomachs and blamed the falling white powder.

That night, Lekoj had radioactive, coral limestone dust in his hair and on his body. He had played that afternoon in the sand along the lagoon and had literally been rolling in fallout. Mijjua tried to wash the white limestone dust from his hair, but it held fast in part because the islanders used coconut oil to keep their hair shiny.

The fallout over Rongelap, which had started as a mist in late morning, had turned heavier around 1 p.m. and continued for more than ten hours, ending sometime before midnight.

John Anjain also remembered that the green leaves of the coconut palms and the pandanus trees were covered with white fallout. On parts of the island, the fallout was an inch and a half deep on the ground. It covered the roofs of houses and along the lagoon beach. On the reef, pools of water turned yellow. When the moon broke through the clouds that night, its light caused the white powder to glow like snow.

■   ■   ■

The primary radiation hazard to the human body comes from beta rays, which cannot travel far in part because the beta particles which emit them are absorbed by dense material, such as clothing. They can cause problems if the isotopes that emit beta rays land on exposed skin.

The outer layer of human skin—the epidermis—is constantly being worn away and replaced by new cells from the epithelium, which is the base layer of skin cells. As these new skin cells rise to the surface and become the outer layer, they get flatter and tougher.

Beta particles can penetrate the skin as far as the basal layer, destroying skin cells and, at the level of basal cells, killing and sterilizing the cells that promote new growth. If no new skin cells or basal cells are produced, the second skin level, containing capillaries and blood cells, is exposed as the epidermis wears down; it becomes sore and tends to bleed and get infected.

Since cell replacement in the skin takes from ten days to two weeks, that amount of time normally passes before the effects of exposure to beta particles become apparent. These effects—beta burns—show up as blistering, sores, and skin infections.

When radioactive material, such as the Bravo fallout, lands on the head and emits beta particles into the scalp, the particles again penetrate to the basal cells, which produce cells at the base of hair follicles. When these cells are killed and the reproduction is halted, the result can be epilation—temporary or permanent loss of hair. Because regeneration takes about two weeks, that amount of time must pass before hair loss takes place.

Radioactive matter taken into the body and bloodstream also creates special health hazards.

Normally, each ingested chemical element, whether radioactive or not, follows a certain course through the human body. Thus iodine is picked up by the thyroid and concentrated there. Calcium is drawn to the bone structure. Strontium and barium, though not elements usually ingested, are chemically much like calcium, and so their radioactive particles taken into the body are drawn to bone tissue.

Two radioactive elements not ordinarily found in the human body, but produced in fission explosions, are cesium and plutonium. They are not chemically analogous to calcium but do nevertheless find their way to the bone structure. Other radioactive elements that do not resemble the chemical constituents of the body will, if ingested, be eliminated from the body by natural biological processes. The danger from those radioactive isotopes which do remain in the body is heightened because they tend to concentrate in specific cells, tissues, or organs that frequently are the most sensitive to radiation.

Radioactive iodine (iodine-131), drawn to the thyroid, remains there— emitting gamma rays which can destroy thyroid tissue or impede its growth. Iodine-131 emits half its radioactive energy in eight days, and so can spend almost its entire radioactive life in the thyroid before being naturally eliminated from the body.

The smaller the gland or organ, the greater the effect of radiation on it. The thyroid of a child, which is only one-half to one-third as large as the thyroid of an adult, would be subjected to a higher amount of radiation proportionately from the intake of radioactive iodine.

Radioactive strontium, barium, cesium, and plutonium, once drawn to the bone marrow, impede the formation of white blood cells, among the neutrophiles, which are cells important in combating certain types of bacterial infections. The blood platelets also decrease with the ingestion of these radio-active elements.

Since blood platelets have a key role in blood clotting, excessive bleeding and hemorrhaging may follow after someone absorbs radioactive strontium or other bone-seekers. Because most of these radioactive particles emit gamma and beta rays, their effects upon surrounding cells and tissues take place slowly over long periods and may not be apparent for days, weeks, or even years.

◼   ◼   ◼

On Rongerik, hours after the detonation, Warrant Officer J. A. Kapral, who was in charge of the military-manned weather station, reported seeing a "haze [that] closed in like a cloud and a dust was deposited on buildings and flat surfaces." Although the Rongerik unit had a gamma-ray monitor, little attention was paid to it until around 3 p.m. when the device went off scale at some 100+ millirems (100+ mr/hr) per hour.*

Kapral ordered a radio message to be sent to the Joint Task Force radiation safety office. It was routed through a communications center in Enewetak for delivery to the command ship *USS Estes,* where it was received some three minutes later. However, since the message was not clear, it was shrugged off. There was also a sense that the Rongerik unit was not reliable, since its wind readings the night before had come in garbled and were, at the time, considered useless when accuracy was imperative.

Further casting doubt on the Rongerik message was a cloud-tracking aircraft that believed it was in the Rongerik area and reported no airborne contamination. It was learned days later that the cloud-tracker was in error about its location, and in fact had never passed near Rongerik.

Warrant Officer Kapral took matters in his own hands. He told his men not to panic, that the high radiation level was probably temporary. He ordered them to get out of shorts and T-shirts and put on long pants and long-sleeve shirts. If they went outside, they were to wear hats and shoes.

---

* One thousand millirem equals one rem.

By then the fallout had accumulated, and to one soldier looked like "light ashes."

A message to the Rongerik weather unit from its detachment commander came in that afternoon telling the men to cease all operations and remain inside metal buildings.

■   ■   ■

At Bikini Atoll just before sunset, the radiation level had dropped to 20 R/hr. Inside Building 70, the firing crew had prepared to be picked up by wrapping themselves completely in bed sheets, cutting holes for their eyes, to keep the radioactive fallout, which was like dust, from their bodies.

Three helicopters from the *USS Bairoko*, which then stood ten miles outside the lagoon, were sent to get the firing crew on Enyu Island. When the men heard the helicopters over the building, they ran to their jeeps and drove the half-mile to where landing mats had been anchored. The helicopters hovered above them the whole time, and then followed them to the mats where they landed for the pickup.

The operation took less than five minutes. Aboard the helicopters, the firing crew members took off the sheets and all were checked by radiological safety officers. Twenty minutes later they were on the *USS Bairoko*, where they showered and were given a thorough radiation check. "None of us had received any harmful amount of radiation," Clark wrote.

At 6 p.m., the Rongerik detachment radioed its regular scheduled weather message to Enewetak. By then the group knew that they were being subjected to something potentially dangerous. After dinner that night, some Army men examined the fallout material under a microscope. They later sent in a post-shot report to their superiors: "Under microscope, sand particles appeared solid— fallout particles looked like crystals. Material not soluble in water on microscope slide. Crystals had rough edges."

In the Army's van on Rongerik, fallout material of about thumbnail size was placed on a cathode-ray tube, causing part of the tube to glow a pinkish-white color. Fallout particles on an ordinary piece of paper glowed in the dark.

At 9 p.m., the fallout was still coming down, so Warrant Officer Kapral sent a second message to Enewetak, this one carrying the marking "EMERGENCY."

At roughly the same time, the Joint Task Force Command aboard the *USS Estes* received measurements from earlier that afternoon from another airborne element that found secondary fallout north-northeast of the Rongerik area at 100 mr/hr. Although that was not seen as cause for alarm, it was believed by some that Rongerik and Rongelap would need to be evacuated. The question remained just how quickly it needed to be done.

An error in communications took place that evening, according to later Defense Department reports. The Joint Task Force Command prepared a message to Rongerik's detachment saying the fallout posed no significant hazard—but that message was not sent until 5 a.m. the following morning. Meanwhile, the Rongerik "EMERGENCY" message had been passed on to the Test Services Unit on Enewetak, where officers took it seriously. They responded with a message that all operations on Rongerik should be suspended, and the men should move into metal buildings, which they already had done.

This message was sent just after midnight. At the same time, officers of the Test Services Unit decided to send two radiation safety advisors aboard a plane that was scheduled to fly the following morning to Rongerik from Kwajalein. The two safety advisors left Enewetak at 3 a.m. on a military aircraft to Kwajalein so they could catch the morning resupply plane to Rongerik.

At Joint Task Force Command, a new set of night-time airborne surveys had been requested. However, due to communication bungling, those orders were not transmitted from the *USS Estes* radio room for some twelve hours.

The delay would prevent task force commanders from understanding the full extent of their fallout problem.

The Baker test shot was detonated underwater, creating a geyser that picked up the battleship *Arkansas* and at the same time produced a six-mile-wide, highly radioactive cloud. As the column of water collapsed, driving the battleship to the bottom of Bikini Lagoon, a doughnut-shaped tidal wave of deadly spray and steam spread out over the surrounding target ships creating a fallout problem never dealt with before.

In September 1945, J. Robert Oppenheimer and then-Maj. Gen. Leslie Groves, who was in charge of the Manhattan Project, returned to the site of the first Trinity atom bomb test in July 1945 at the Alamagordo Bombing Range in New Mexico.

The bunker, Building 70, was operations center for the nine-man Bravo firing team and twenty-two miles across Bikini Lagoon from the detonation point of the device. It had cement walls three-feet-thick, steel-reinforced and covered by ten feet of dirt and sand to keep out radioactive fallout. Nevertheless, only one room was deemed safe after the fallout began, as the team had to wait all day until nightfall before they could escape to helicopters that took them to safety.

Stored in a two-story shed on its own artificial island, Bravo was a deliverable thermonuclear bomb encased in an aluminum cylinder. It reportedly contained two hundred pounds of uranium-235, two hundred pounds of lithium deuteride, and more than two thousand pounds of uranium-238. Despite its fifteen-foot length, its nickname was "Shrimp."

The *USS Estes* was the command ship for the Bravo test and at the time of detonation was thirty miles east of Bikini.. Aboard were the Task Force Commander Army Maj. Gen. Percy W. Clarkson and other officials and observers. The destroyer carried navigation equipment that controlled drone aircraft and the communication systems that bungled early reports of fallout on servicemen on Rongerik Atoll.

The restored *Lucky Dragon No. 5* has been in its own exhibition hall in Tokyo's Yumenoshima Park since June 1976 to remind visitors of the 1954 "radioactive rain" that came down on the crew. Located outside the hall is a memorial to Aikichi Kuboyama, the crew's radioman. Engraved on the memorial are his words, "Please make me the last man who died from an A- or H- bomb."
CREDIT: Daigo Fukuryu Maru Exhibition Hall

At fifteen megatons (equivalent to fifteen million tons of TNT), Bravo's explosive power was more than twice the six megatons expected and one thousand times more powerful than the Hiroshima atomic bomb. It vaporized an estimated three hundred million tons of sand, mud, coral, and water in a fireball and mushroom cloud that within five minutes went through both the troposphere and into the stratosphere. It created three cloud levels—one at fifty-five thousand feet; another at one-hundred-fourteen thousand feet, and the upper cloud at one-hundred-thirty thousand feet.

The **Marshall Islands** consists of 29 atolls spread across a sea area of over 700,000 nautical square miles. The total land area is about the size of Washington, D.C.

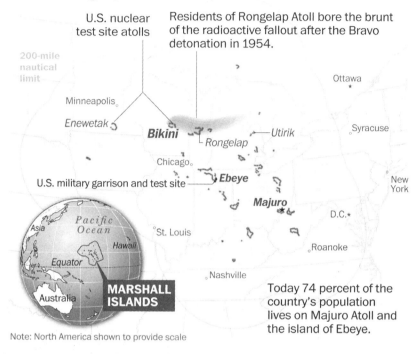

U.S. nuclear test site atolls

Residents of Rongelap Atoll bore the brunt of the radioactive fallout after the Bravo detonation in 1954.

200-mile nautical limit

Ottawa

Minneapolis

Enewetak

Bikini

Rongelap

Utirik

Syracuse

Chicago

U.S. military garrison and test site — **Ebeye**

New York

Majuro

D.C.★

Pacific Ocean

Asia

Hawaii

St. Louis

Roanoke

Equator

Nashville

Australia

MARSHALL ISLANDS

Today 74 percent of the country's population lives on Majuro Atoll and the island of Ebeye.

Note: North America shown to provide scale

Before the United States conducted its first test in the Marshall Islands in 1946, there were 167 people people living on Bikini Atoll. Twenty-nine who were displaced are still alive.

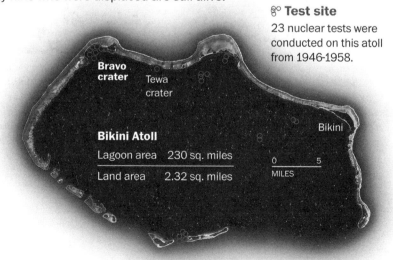

Test site

23 nuclear tests were conducted on this atoll from 1946-1958.

Bravo crater

Tewa crater

Bikini

Bikini Atoll

Lagoon area	230 sq. miles
Land area	2.32 sq. miles

0   5
MILES

CREDIT: Laris Karklis, *The Washington Post*

Utirik Island village on March 3, 1954, as seen from the raft bringing the first radiation monitors to shore.

With their PBM seaplane in the background, two American radiation monitors took readings on Utirik Island on March 3, 1954, days after radiation fallout first fell on Utirik Atoll nearly three hundred miles from the Bikini detonation.

Rongelap Island village was deserted when radiation monitors arrived a week after the Bravo detonations.

More than a month after the fallout, on April 20, 1954, a radiation survey party came ashore on Rongelap Island to determine radiation levels on the island.

23 · EVACUATION

■ **ON RONGELAP, THE** morning of March 2, John Anjain called a meeting of the island council. The villagers walked to the meetinghouse down paths still covered with radioactive white powder. Many islanders said that they or their children were feeling nauseated, and a few reported itching and burning skin. Jabwe, the island's health aide, whose training had been provided by the US Trust Territories government, was asked what could be done.

The night before, Jabwe had eaten fish that had been covered with fallout, and he was himself feeling sick. Others had lost their appetites; a few vomited and had diarrhea. Jabwe told the council that he thought the white powder was the source of their troubles and suggested that water collected the previous day not be drunk. Few of the people listened, however, and most continued to drink water during the day.

That evening, the Rongelap people spent a second night exposed to the fallout that lay on the ground, that was dissolved in their drinking water, that was encrusted on their food and cooking utensils, and that was caught in their hair, skin, and clothing.

■   ■   ■

At the Joint Task Force weather headquarters on Enewetak the morning of March 2, the major question was how much fallout had taken place on the inhabited atolls east of Bikini and what human exposure and absorption of the radiation had occurred.

Was evacuation necessary?

Aircraft that had followed the radioactive cloud after the explosion had reported back that snow-like fallout had been blowing eastward the previous day and had accumulated on the ground on Rongelap and Rongerik.

At ten minutes after midnight on March 2, an order had been sent from the weather unit headquarters to Rongerik to halt all activities except for radio communications and to keep all the men either in the mess hall or in other buildings. The Rongerik detachment was also told that a PBM-5A seaplane from the Naval Air Station at Kwajalein was being sent that morning to begin an evacuation of the entire detachment.

On Rongerik, radioactive ash covered all the buildings and trees. The twenty-eight Air Force and Army men there had breakfast prepared from canned food and juices that were untouched by the fallout. After breakfast, Kapral ordered the aluminum-roofed mess hall washed down inside and out with sea water.

At 9:45 a.m. on March 2, a bulky Navy PBM-5A amphibian aircraft carrying Air Force Capt. Louis B. Chrestensen and other radsafe officers circled over Rongerik, swooping down over the atoll at five hundred feet and then lowering to twenty-five feet, in order to take measurements of the radiation. It showed a level of 200 mr/hr at 500 feet, and at 25 feet, 350 mr/hr.*

Chrestensen and his colleague decided to use their plane to evacuate the weather unit from the atoll, and to request a second aircraft to assist in that operation. The aircraft climbed to five thousand feet in order to message that decision to their weather unit headquarters at Enewetak and to the Task Force

---

* One millirem per hour (mr/hr) equals one thousandth of a rem. The rem is a unit of effective dose. With an exposure of twenty-five rem, there can be a clinically observable blood change.

Command on the *USS Estes.* The message requested permission to carry out those tasks.

According to a later Defense Energy Agency report, "Some garbled and ineffective transmissions ensued, followed by a message denying permission for evacuation and then permitting it." In the face of that confused set of responses, Captain Chrestensen decided to go ahead on his own, and the plane landed on the Rongerik Lagoon at 11:30 a.m.

On shore, Chrestensen briefed the detachment personnel about what he knew, then directed Warrant Officer Kapral to select eight men for evacuation— saying the rest would be picked up later. Kapral decided to choose the first eight by last names in alphabetical order. Meanwhile, Chrestensen and his colleagues did a quick radiological survey of the immediate area.

Inside the metal roofed building where the unit had moved, the reading was 600 mr/hr. Outside the same building, at waist height while standing on a pierced steel planking platform, it was more than twice as high, 1.8 R/hr (1,800 mr/hr); and while standing nearby on a sand surface it was 2.4 R/hr (2,400 mr/hr).

With the evacuees aboard, the amphibian took off from Rongerik Lagoon at about 12:30 p.m. and arrived at Kwajalein at about 2:00 p.m. Since no other plane was available, the PBM-5A returned to Rongerik around 4:00 p.m., and some 45 minutes later took off with the remaining 20 men.

The final message Chrestensen sent to the Task Force Command that evening was a recommendation that Rongelap be surveyed as soon as possible, since it was expected that the Marshallese there had received as much fallout as the detachment on Rongerik.

■   ■   ■

After the lunch on March 2, Task Force Commander Clarkson had begun a special conference aboard the *USS Estes* to discuss what was known about the fallout situation in the inhabited atolls with the senior radiation-safety officers, the group commanders, and Dr. William Ogle, the scientific director.

One hour into the conference—at 2:30 p.m.—Chrestensen's earlier message from Rongerik had finally come through to the *USS Estes*, providing Clarkson and others with the first direct knowledge of the unexpected fallout on that atoll. Almost at the same time, the first report from one of the morning fallout monitoring flights arrived, indicating that at ground level on Rongelap Island the radiation level was 1,350 mr/hr, and on nearby Ailinginae it was 400 mr/hr.

Clarkson ordered the evacuation of Rongelap and additional monitoring flights undertaken to see about possible contamination of other Marshallese atolls. He also sent an Air Force Grumman amphibian with two radsafe monitors late that afternoon to check before dark the surface conditions at Rongelap. The two radsafe officers arrived by seaplane, stayed on the island for only ten to twenty minutes, according to one report, and never spoke to the Rongelap islanders because they did not speak the Marshallese language. They reported back that the average reading at 5 p.m. in Rongelap's living area was 1,400 mr/hr.

Spending just ten hours in that environment would begin to affect an individual's blood cells, and longer exposure would lead to more acute reactions, depending on age.

Since no special emergency evacuation assets had been provided, it was decided that the task force's destroyer escorts would be assigned as needed for the evacuations. The *USS Philip*, a thirteen-year-old Fletcher class destroyer that had seen battles in both World War II and Korea, was chosen to go to Rongelap to pick up the Marshallese.

The task force commander also arranged for a Navy PBM Mariner seaplane to be ready early the next morning, March 3, to fly to Rongelap with Trust Territory Representative Marion Wilds and his Marshallese interpreter, Oscar DeBrum, where they would meet up with the *USS Philip* and assist with the evacuation of the people.

At 7 p.m., information regarding ground contamination at Utirik Atoll arrived. Taken earlier that afternoon, it was put at 240 mr/hr. Utirik was more than 300 miles from Bikini, and so a tentative decision was made to send a different destroyer, the *USS Renshaw*, to evacuate the Marshallese there.

At 9:45 p.m. that night, the *USS Philip* left Bikini for Rongelap under the command of US Navy Cdr. G.W. Albin Jr. The destroyer's crew had itself already experienced Bravo's fallout. Some twenty-one sailors aboard the *Philip* had suffered lesions classified as beta burns from the March 1 shot.

John Bianco, a young *USS Philip* sailor at that time, recalled in a 2012 article in the *Newsletter for America's Atomic Veterans*, that an hour or so after the

Bravo detonation, "A fine mist began to cover the ship, and it looked like it was snowing." He said the crew was ordered below decks, and the ship's ventilation system was shut down. For hours it felt like the crew was in a toxic sauna.

The eighty-six-year-old Bianco finished up by stating, "Over the years, I often found myself—in my dreams—seeing that flash, hearing the roaring thunder, seeing those bones in my shipmates, watching that boiling mushroom cloud rising above our ship, and feeling the radioactive fallout."

■   ■   ■

The decision to bring the Marshallese to Kwajalein created more than just problems in logistics and possible radiation sickness. The Bravo test was highly classified, and the sudden appearance on Kwajalein of Americans and Marshallese who had been over one hundred miles from the site of the explosion could call public attention to the tests and, in effect, notify the Russians of this new phase in the American nuclear weapons program.

A decision was made in Washington to keep the fallout incident secret, and that was passed on to Admiral Clarke, commander of the Kwajalein Naval Base.

At Kwajalein, a barracks area isolated from others was hastily readied to receive the Rongerik military personnel. Of the first eight men to arrive, one showed a radiation reading of 250 mr/hr, another 90 mr/hr and the six others only 5 mr/hr. At that time, 3.9 rads or 3,900 mr/hr was the permissible limit for nuclear test participants. Their initial blood counts were in the normal range.

The basic treatment for external fallout exposure was to wash thoroughly. The initial eight Rongerik men immediately began to take showers, and after only one shower the two with the highest original readings dropped to 10 mr/hr while the rest remained at 5 mr/hr.

Meanwhile, tests were made of the radiation film badges they had. When developed, they indicated accumulated levels of radiation exposure. Several of the badges had been worn, one had been hung in a tent, another was in the refrigerator of a barracks, and the last one was outside the mess hall.

One worn badge showed accumulation of forty rads, and that, plus the initial pre-shower measurements, convinced the task force commanders that a full evacuation of all the exposed atolls should be undertaken.

The first Rongerik group continued to take showers every few hours during the rest of the day. Meanwhile the seaplane was sent back to Rongerik, and the remaining twenty detachment members were picked up four hours later.

When the second group arrived at Kwajalein at about 7 p.m. on March 2, they were sent to the isolated barracks area, measured for radiation, and sent to the shower room. Since they had spent more time exposed to the radiation on Rongerik than the earlier group, their initial radiation readings were higher.

One airman showed a level of 200 mr/hr, and two more were at 150 mr/hr. After five showers that evening, their readings were still substantially above those of the eight men who had arrived that morning.

■   ■   ■

It was 7:30 a.m., March 3, when the *USS Philip* dropped anchor in the lagoon off Rongelap Island. The Navy PBM carrying Trust Territory Officer Wilds and translator DeBrum was already there. The plane, which had been in radio contact with the *Philip*, had already made a reconnaissance flight around the atoll, and after landing settled some one hundred yards off the beach near where the destroyer would anchor.

Commander Albin, along with his executive officer and a four-man, radiation monitoring team, set out from the destroyer in a motorboat, first to the PBM to pick up Wilds and DeBrum, and then on to the beach.

John Anjain met them there in his role as Rongelap's magistrate. The monitoring team immediately began taking measurements and collected samples of the water in the cisterns. They found readings of 1,000 mr/hr to 300 mr/hr throughout the village. Two days later, Albin wrote in a report to his superiors, "On the basis of initial readings it appeared obvious that evacuation was definitely in order."

Based on procedure set by the Joint Task Force Command, the request to the Marshallese for evacuation of Rongelap had to come from a Trust Territory official, but Wilds would not take that step until it was approved by the Marshallese themselves. At a meeting with the Rongelapese, with DeBrum interpreting, Wilds explained that it was in their best interests to leave the atoll and that the United States had sent the Navy destroyer there for that purpose. Anjain would later recall that Wilds said the Rongelapese would die if they remained on their island. Anjain explained to the assembled group that those who felt sick would be provided with medical care by the US when they reached Kwajalein.

When the group approved departing, each person was told to bring a small handbag as the only baggage, since the monitor readings had shown high radiation dosages in sleeping mats, palm baskets, and other personal belongings. Wilds made these requests through Anjain, who passed them on to the people.

For the Marshallese, it was not just that they were leaving behind their homes and personal belongings but also the community's common assets, their one hundred chickens, ten pigs, two dogs, and a considerable amount of drying copra, which was their main product for trading for other foods and commodities. Anjain was particularly upset about leaving the copra, the community's source of livelihood. He also worried about his thirty-foot sloop which was used to connect Rongelap to other larger atolls. The boat was used to carry copra to markets in exchange for food and medicine in between those times that a Trust Territory ship arrived. To meet Anjain's concerns, the sloop was moved and, using two anchors, was left in a lee placed just off the island in what was considered a safe spot.

There was no discussion about when the Rongelap people would return, but the belief was that it would be much like what had happened in 1946, when they left during Operation Crossroads and returned in just a matter of weeks.

Albin noted in his memo, "Since the people were not given an estimate of the duration of their evacuation, the concern over the above items will no doubt increase as the absence from their homes grows longer."

He later recommended that aircraft periodically check conditions on Rongelap, but added that the animals "could be of much value for scientific research," and suggested they be transferred at a later date for study.

It was decided that a number of the elderly and any who were sick or with the highest radiation levels would be taken on the PBM and flown to Kwajalein, while the remainder would go on the *USS Philip*. Anjain picked sixteen for the seaplane, ten of them women including Luior, who was eighty-three. Jabwe, the young health aide who had announced he felt ill a day earlier, was also among those chosen.

The aircraft took off from Rongelap's lagoon at about 10:00 a.m.

The forty-eight remaining Marshallese were brought to the *Philip* using the ship's small boats. All together the group had spent some forty hours being exposed to Bravo's radioactive fallout.

Commander Albin decided to separate the children and women from the men because of the limited space on the destroyer. The men were given temporary shelter under a canvas tarpaulin rigged up on the destroyer's deck between the smokestacks. The women and children were put in the torpedo room, which was a short distance from the officers' bathroom, which was then reserved for their use. The Marshallese men used the crew's bathroom. Cots were set up in both areas to be used for sitting and sleeping.

To ensure privacy and assist the Marshallese in any request, a continuous sentry watch by petty officers was set up at both locations.

A routine was established whereby a decontamination team guided the Rongelap people to the crew's washroom at the rear of the destroyer where they were given showers. Their old clothes, which had high radioactive indications, were put into covered metal cans while lightly contaminated clothing was sent to be washed in the ship's laundry in a strong soap solution. Dried and pressed, most clothing was returned to their owners four hours after the ship had sailed.

In the interim, the Marshallese men, women, and children were given clean clothes, most being dungaree trousers and T-shirts donated by the crew.

After decontamination, sailors gave all the children milk. For meals, the Marshallese went through the normal mess line and were served food prepared for the crew. Meat was not popular, and the adult Marshallese asked for more soup and vegetables, while the children wanted more ice cream.

With all aboard, the *USS Philip* sailed from Rongelap to nearby Ailinginae Atoll, where another eighteen Rongelap people—fishing and drying copra—had been temporarily living. Anjain explained to them the need to leave the atoll, and they agreed to join the rest on the destroyer.

It was 6:00 p.m. when the *Philip* left Ailinginae for Naval Station Kwajalein. That night, a dozen additional cots and two stretchers were set up as beds in the torpedo room for the women and children. On deck, the men were given kapok life jackets to serve as both blankets and mattresses.

At 8:30 a.m. the next morning, March 4, the *Philip* arrived at Kwajalein.

■   ■   ■

At Kwajalein, the Naval Station's commander, Rear Admiral Clarke, had arranged for an emergency decontamination center to be hastily established in an older barracks area of the base—two adjacent buildings for sleeping, a storage area for clothes, a large shower room, and an outside walking area enclosed by a wire fence.

Burlap bags had been placed on the fencing so that no one outside the area could look in; nor could the Marshallese look out.

The plane with the first sixteen Rongelap inhabitants had arrived at Kwajalein about noon on March 3. They were immediately moved to the new decontamination center, showered, their clothes taken to be washed, and given new clothes contributed by the Navy.

When the remaining sixty-six arrived the morning of March 4, they too were taken to the decontamination center. Some complained of itching and burning of their skin and eyes, an apparent result of exposure to radiation. Others spoke of queasiness, nausea, and loss of appetite, which may have come from a combination of seasickness and perhaps the unusual food they had been given while aboard the *Philip*.

At the time it was believed that the whole-body exposure of the people from Rongelap was about 130 R (Roentgens), while those picked up later was about 80 R.

Based on their initial examinations, no one was described as being acutely ill.

■   ■   ■

Utirik Atoll, home of 154 Marshallese when the Bravo fallout came, was 320 miles east of ground zero on Bikini and 200 miles east of Rongelap. In the late afternoon of March 3, a Navy PBM seaplane flew over Utirik and conducted an air-to-ground radiation survey which recorded 60 mr/hr readings.

A decision was quickly made by the Joint Task Force that evacuation had to be undertaken and the destroyer *USS Renshaw*, which had set sail toward Utirik a day earlier, was ordered to begin the operation early the next day. In addition, a seaplane from Kwajalein with Trust Territory representatives, an interpreter, and two public relations civilians was ordered to arrive at Utirik early on the morning of March 4 to aid in the evacuation.

■   ■   ■

Many of Utirik's inhabitants had seen the strange morning light on March 1, and later had heard a rumbling sound. No one on the atoll had known what it was, and no one there had paid much attention when, just before dawn on March 2, a faint mist began to fall on the atoll and continued for several hours.

The *USS Renshaw* had used its travel time to prepare to receive the Marshallese. Plans were made to set up an awning over the fantail, and areas were roped off to ensure privacy for the expected guests. The crew set aside its washroom and bathroom at the rear of the ship to be used for decontamination. An outside saltwater shower was rigged up, and the crew was asked to supply clean clothes for the expected Marshallese.

Arrangements were even made should the "natives" bring their pigs, chickens, and dogs.

The *Renshaw*, captained by Commander L. H. Alford, arrived off the atoll at 6:30 a.m., March 4, but because the channel through the coral reef into the Utirik Lagoon was too narrow, the vessel had to anchor on the seaward side of the main inhabited island. The destroyer's executive officer, Lt. Cmdr. V. H. Easton, along with a radiation-safety officer and a signalman set off in a rubber boat for shore with a hospital man and others aboard. They had trouble crossing the reef, and finally were aided by the Marshallese, some of whom swam out and helped the group finally reach the beach.

Meanwhile, the Navy PBM from Kwajalein arrived and landed in the lagoon. When the Navy executive officer, riding in a Utirik outrigger canoe with cooperating Marshallese, approached the plane, it taxied away, took off, circled the atoll, and landed again outside the lagoon near where the *Renshaw* had anchored. The pilot later claimed he had been advised by radio that there would be a problem disembarking in the lagoon and had not realized the Navy exec was in one of the outriggers.

In the interim, the first ground survey had been taken and indicated readings of 100 to 130 mr/hr throughout the village, including on the bodies of some Marshallese. Drinking water had less contamination, thanks to roofs that were over most reservoirs.

The formal decision to evacuate Utirik was made quickly. The Utirik chief, whose name was Compass, was told that he had to prepare his people for evacuation. A meeting was held in the village, and the Navy exec, along with the Trust Territory official, spoke through the interpreter about the need for the islanders to leave their homes. Although the interpreter was asked not to frighten the Marshallese, "the natives really moved though it is not believed he [the interpreter] shook them up too badly," according to a later report by Commander Alford.

After some consultation among the Americans, it was decided that the livestock would be left behind, since the contamination levels were low enough that the inhabitants could be "reassured that their possessions and animals would be safe until their return," according to Alford's report.

The Utirik people were told they could take only two bundles each, and soon many, carrying woven bedding mats, began moving toward the beach area set aside for the evacuation, which began at 11 a.m.

Using a life raft shuttle that could carry fifteen people, and beginning with women, children, and older people, the operation was about half completed by noon when the wind rose and the surf began to kick up. A second raft was employed. Although two loaded rafts almost overturned, the entire group reached the *Renshaw* without any injuries.

In all there were forty-seven men, fifty-five women (two or three of whom were pregnant), and fifty-two boys and girls under the age of sixteen. Aboard ship, after each was monitored, the average readings were 7 mr/hr. That was below the level found ashore, and the discrepancy was later attributed to the need for all to wade out through water to get to the rafts.

"Ten loads of about 15 people each were required to complete the evacuation of 154 natives," Alston would later write in his report. "The last raft load left the beach at about 12:45 p.m., leaving as forlorn a set of dogs as you have ever seen."

The destroyer set sail for Kwajalein at 1 p.m., with an estimated arrival time of dawn on March 5.

While some children went through the decontamination showers after coming aboard, it was decided to delay doing the adults until after lunch. Alford reported, "They didn't eat very well, perhaps from the excitement or maybe they just don't like meat loaf. They did better on the bread, mashed potatoes and oranges."

A problem arose when showering came up. The Trust Territory officials had suggested segregation between the men and women's sides, but the crew could only get 10 percent of the adults, mainly the aged, sick, and infirmed to go through. The majority also resisted putting their clothes through the laundry, and the women balked at putting on dungarees when offered. In the end, it was decided that since the radiation levels were so low, the Marshallese, themselves, would decide who would shower and change clothes.

By supper, all was quiet on the *Renshaw*. Dinner was boiled fish with rice, tomatoes, and lima beans, and more to their liking. Ice cream and cookies were a great success, as was a sweetened grape drink, plus colored hard candy left over from Christmas.

Evening aboard ship closed with a movie about which, Alford wrote, "There was not the slightest reaction of any kind from any of them the whole time." Looking back, he added, "It should be remembered that most of these natives had never been off the atoll, and as far as it is known had never seen a movie."

At dawn, the *Renshaw* entered Kwajalein harbor. After a breakfast of hot cakes, bacon, bread, and jam, the Utirik people disembarked to waiting buses at 5:10 a.m. Through an interpreter, Chief Compass was asked what he had seen, and with a gesture indicated a large explosion. Asked what he thought it was he said, "The world, we think she start over again."

▪   ▪   ▪

Significant exposure to radiation depresses some elements of the blood, particularly white cells which protect against infection and platelets that prevent bleeding. With large amounts of radiation, such depression may lead to death from infection and bleeding.

The Japanese at Hiroshima and Nagasaki were exposed to direct gamma radiation from detonations more than 1,500 feet above ground, with some additional neutron radiation being contributed from fission products resulting from the bomb's explosion. However, there was no significant radioactive fall-out from those two atomic bomb detonations, although there were of course thermal burns to individuals from the heat and extensive trauma from the blast.

On the other hand, the Marshallese—more than one hundred miles away from the Bravo explosion—felt no direct heat or blast effects. Their radiation effects were solely due to fallout, which in their case involved penetrating gamma radiation exposure to the whole body; depositing of radioactive fallout on the skin; and internal absorption of radioactive materials through both eating of contaminated food and liquids and the inhalation of radioactive dust in the air.

Dr. Conard would later write that in the two days after exposure, two-thirds of the Rongelap people "experienced anorexia and nausea, a few vomited and had diarrhea. The majority had itching and burning sensations of the skin in exposed areas." Only one of the eighteen who had been on Ailinginae had these symptoms, and none of the symptoms were noted in the Americans on Rongerik or in the Utirik people.

As the evacuees arrived in Kwajalein, they were initially in the hands of the Navy base's medical team, which had little experience in handling such a situation and no way to estimate the whole-body radiation dose received by any individual.

Based on previous studies of individuals and animals exposed to radiation, the team decided to focus on testing the blood of the exposed people. "They initially decided that clinical findings, particularly the degree of depression of blood elements, would be the index used to estimate the seriousness of the effects of the exposure," according to Conard's reconstruction of that moment. That decision proved to be the right choice.

The twenty-eight servicemen from Rongerik, examined on Kwajalein, showed normal blood counts. One, who claimed to be feeling ill, later decided it was psychological after being reassured his exposure to radiation was minimal.

The Rongelap and Utirik evacuees each were given physical examinations, including complete blood counts. In addition, their medical histories were taken. The mixed results hardly told the story of what was to come.

24 · KEEP IT SECRET

▪ **BACK IN THE** United States, the public was unaware of what was really going on in the South Pacific, while those Washington officials "in the know" took the traditional governmental approach and tried to keep all the facts from coming out.

They had an unexpected break that diverted public attention.

Beginning with news bulletins on the afternoon of March 1, 1954, and with wide coverage each day thereafter, the dominant news story in Washington and around the country was about a shooting at the Capitol. Four Puerto Ricans, three men and a woman, had entered the US Capitol. At around 2:30 p.m., the men arose in the south gallery of the House of Representatives and fired at least twenty-five shots at some two hundred of the legislators below them on the House floor.

With the uproar going on in the city, the AEC that afternoon in Washington released a brief press announcement: "Lewis L. Strauss, Chairman of the Atomic Energy Commission, announced today that Joint Task Force Seven has detonated an atomic device at the A.E.C.'s Pacific proving ground in the Marshall Islands. This detonation was the first in a series of tests."

The next day's front page of the *New York Times* was dominated by a five-column, triple-decker headline which read: FIVE CONGRESSMEN SHOT IN HOUSE BY THREE PUERTO RICAN NATIONALISTS; BULLETS SPRAY FROM GALLERY.

Well below the front page fold of the *New York Times*, under the House shooting story, there was a much shorter two-column headline that read: ATOMIC BLAST OPENS TEST IN THE PACIFIC. With a byline that read only "Special to The *New York Times*," the article reported that Strauss "did not make clear whether the 'atomic device' was a fission or thermonuclear (hydrogen) type," adding, "There had been unofficial indications, however, that a variety of hydrogen weapons or devices would be tested in the next several weeks."

The story pointed out "the extraordinary security precautions applied to the current trials," and also noted that "the number of observers invited to these tests was believed to be smaller than for any comparable weapons test in the past." The *Times* reported that Strauss's announcement was designed to head off "leaks from seamen and servicemen from the Joint Task Force," adding that it was through such "leaks" that word initially came out about the first thermonuclear Mike test in 1952.

On March 4, 1954, at a time when the servicemen from Rongerik and the Marshallese families from Rongelap were arriving on Kwajalein and put under the care of doctors, the AEC tried to make sure there were no leaks about those events. The commission sent a cable to General Clarkson, the Joint Task Force commander, which stated again that nobody on the scene should "make anything public on these matters."

That night on Kwajalein, Admiral Clarke and his staff announced to personnel at the base movie theater, and at the enlisted men's and officers' clubs, that no news was to leak out that "natives" from outlying Marshallese atolls had been brought to the base. They all knew what had been going on. Many had seen the bright light from Bravo in the sky days earlier, or had seen or heard about the two destroyers that had brought more than two hundred evacuees to the base.

Not everyone agreed with keeping it all secret.

On March 4, Dr. Alvin C. Graves, the Los Alamos scientist who was serving as scientific director of Joint Task Force Seven, sent an "Eyes Only" message to the AEC's Gen. Kenneth Fields, then the agency's director of military operations, saying he was uncomfortable with the idea of hiding the exposure of the Marshallese. He wrote he was "very much concerned [about] the recent decision not to make a release on evacuation of natives unless forced to do so." He wrote, "I hope that the fact that these natives are not United States citizens, but wards of the Government was given appropriate weight. . . . I should regret very much the impression that we are being furtive in our actions with regard to these people."

On March 6, with the Utirik evacuees added to the Marshallese already at Kwajalein, Fields replied to Graves and the task force leaders that he was "being pressed by higher authority [in Washington] for an explanation regarding the circumstances that led to the exposure of the natives." In short, more details were needed and that "your request for such release [about the exposure to radiation of individuals] must be weighed by highest authorities before a decision can be reached." Meanwhile, from the AEC's point of view, the need was to wait "until we know reasonably confidently whether or not serious illness or worse is going to result."

▨   ▪   ▨

Original planning for all scientific investigations to be conducted in conjunction with Operation Castle, including medical, began in early 1953, more than a year before the Bravo shot actually took place. Biomedical studies had been done at earlier nuclear tests, and one for Castle was suggested in 1953. It was to test the medical effects of neutrons on mice, and was to be titled Project 4.1, since Project 4 was the designation used for biomedical experimental projects in earlier tests.

At a planning session that had been held at the Los Alamos National Laboratory on March 3–4, 1953, the mice project was dropped because it was decided that in order to measure neutron exposure, each mouse would have to be placed so close to the high-yield thermonuclear explosion that the heat effects of the detonation would incinerate them. As a result, at the time of Bravo's detonation, there was no Project 4 biomedical investigation underway associated with Operation Castle.

Three days after the Bravo shot, and apparently in response to Clarkson's earlier request for medical assistance, a meeting was held in the office of the

Navy Surgeon General Lament Pugh. Attending were Dr. Stafford Warren, past director of the AEC's Department of Biology and Medicine (DBM); the then-current DBM Director Dr. John Bugher; and Dr. Charles Dunham, then-AEC medical program director and representatives of the Armed Forces Special Weapons Program (AFSWP).

Beginning in 1953, the Joint Chiefs of Staff had given AFSWP responsibility—as a participant in Joint Task Forces set up for nuclear tests—for "weapons effects phases of development or other tests of atomic weapons." As a result, AFSWP had in the past assisted Los Alamos in designing effects experiments. Therefore, it was not unusual for AFSWP Cmdr. Maj. Gen. Alvin P. Leudecke to join in establishing the separate study of the Marshallese who had been exposed to the radioactive fallout from the Bravo test.

Also directed to be at the meeting was Navy Commander Dr. Eugene P. Cronkite, who at the time was assigned to the National Naval Medical Research Institute in Bethesda, Maryland, just outside Washington. Cronkite had, in 1953, handled biomedical experiments associated with nuclear tests carried out that year in the Nevada nuclear testing site.

The purpose of the meeting was to organize a medical team to take care of the exposed Marshallese and military personnel, but at the same time there was apparently a belief among some present that this could also provide an opportunity to study the effects of radiation on them. Cronkite wrote later that he was told to organize the team. As a result, he quickly needed laboratory equipment plus AFSWP people to make up the team, and they all had to be ready to leave within forty-eight hours.

Among the group of twenty-five that Cronkite pulled together were Dr. Conard, then a commander in the Navy Medical Corps who had been at Crossroads, and Dr. Victor P. Bond, then at the Naval Radiological Defense Lab. Their participation would be just the beginning of what became for them a three-decades-long association with the Marshallese and the fallout from the Bravo test.

The next day, March 6, Clarkson promised "all possible assistance including necessary assignment of class one priorities" be given to Cronkite, whose medical group would manage the proposed study.

Clarkson had spent March 5 and 6 on Kwajalein. He had been briefed on what was going on and had spoken through an interpreter to some of the

Marshallese. In a cable back to the AEC, he said the "health situation of the natives is satisfactory and physical examinations are within normal expected variations." Most important, he wrote, "There were no symptoms of radiation sickness as of 5 March."

As is often the case with overseas commanders and government bureaucrats, Clarkson downplayed the possibility of trouble when reporting back to Washington. Perhaps any problems would simply go away.

The Marshallese had asked Clarkson when they would be able to go back home, because earlier they had been told it would be in four to six weeks. However, Clarkson cabled Washington that a final answer had to await surveying Rongelap and Utirik after testing ended.

He also wrote of the concern the Marshallese had about the chickens and pigs they had left behind and suggested it would be "many times more expensive" to resupply them with food "than the cost of replacement at a later date."

The interpreter reported to Clarkson, "Some of the heads of families had left money underneath their huts. This is their only possession of any consequence. Care will be taken to insure that any re-entry parties do not disturb the natives' belongings." Based on his Kwajalein experience up to that time, Clarkson also warned Washington that "It is out of the question to keep news of this event from leaking out," citing as his sources, Kwajalein's commander, Rear Admiral Clarke, his staff, and the local Trust Territories representative.

"There is no mail censorship," Clarkson wrote, adding that with some two hundred American families living there along with people who may be involved in caring for the evacuees, it could not be kept secret, especially with "constant traffic on MATS [Military Air Transport Command] aircraft bound for Hawaii."

On March 8, Cronkite and his medical team arrived in Kwajalein, and that same day he was given a letter that was classified "Secret and Restricted Data," among the highest security classifications at that time.

The letter described what had been designated "Project 4.1—Study of Response of Human Beings exposed to Significant Beta and Gamma Radiation due to Fall-out from High Yield Detonations." It said the project's objective was "to study the response of human beings in the Marshall Islands who have received significant doses due to the fallout from the first detonation of Operation Castle," i.e. Bravo.

The letter also advised Cronkite, "Due to possible adverse public reaction, you will specifically instruct all personnel in this project to be particularly careful not to discuss the purpose of this project and its background or its findings with any except those who have a specific 'need to know.'"

Cronkite later said his initial responsibility was "to examine and treat the Marshallese and American servicemen who had been exposed." But by creating a situation where medical treatment was to be done in conjunction with research on radiation effects, US officials had established a package that would eventually give critics the ability to charge that the Marshallese were being used as guinea pigs because, in fact, they provided a unique opportunity for understanding the effects of low-level radiation on humans.

Critics would later point out that the team conducting the study did not ask the Marshallese for their consent or explain to them that a study was being conducted.

Cronkite's team established examination and laboratory facilities in a Kwajalein building adjacent to the Marshallese living quarters and quickly got down to work. They took medical histories of each Marshallese with the help of translators. Daily they took temperatures, blood samples, urine samples, checked white cell counts and platelet levels, examined skin for radiation burns, and monitored pregnant women.

Clarkson assured Admiral Clarke and the Trust Territories representative "that the Joint Task Force would stand any expense from Task Force funds over and above normal Naval or T.T. [Trust Territory] expenses," which were expected initially to run in the neighborhood of $50,000.

■　　　■　　　■

On March 8, Herbert J. (Pete) Scoville Jr., the technical director of AFSWP (the US Armed Forces Special Weapons Project), led a survey team on what was to be a highly secret mission: To take measurements of the residual radiation of the atolls that had been evacuated, which also meant the outer islands that made up the entire Rongelap Atoll.

The purpose was to get better data on the amount of radiation exposure the Americans and Marshallese had received. Walmer Strope, a radiation protection researcher from the US Naval Radiological Defense Laboratory, was made part of the team because of his background in measuring fallout and his experience using equipment necessary to gather the data.

The survey team traveled on the destroyer *USS Philip* and used a whaleboat to reach Rongelap Island at the southernmost end of the atoll. There, they took samples from the Rongelap Village cistern water, as well as soil and leaf samples. To collect soil, they used giant ice cream scoopers they had borrowed from the Navy mess hall. "The mess hall did not want them back after we had used them, so I brought one home and still use it to scoop ice cream," Strope wrote later in a privately published memoir.

The average reading in the center of Rongelap Village was 280 mr/hr, while near the central cistern it was 300 mr/hr. Near the northern cistern it was 350 mr/hr.

The team moved up to the islands north of Rongelap Island. On the first one, measurements were "about double the levels on the home island we had just left," Strope wrote, meaning "the radiation rate on the land had indeed doubled in little more than a half mile."

They stopped at another island where a survey member looked at his meter before jumping into the water. He then ran into the jungle and then raced back and across the beach to where the boat had backed off into the lagoon. The surf was up to his knees when he flopped over the gunwale into the boat and said, "I had to change scales . . . It's 800 mr/hr in there!" That was the highest reading yet.

Since the land was so "hot," it was decided to spend the night back on the destroyer and go to the atoll islands further north one day later to see if the radiation had "cooled down" a bit.

Next day, they headed for the northernmost island of Rongelap Atoll. If experience so far had been any guide, this would be the "hottest" island in the atoll—and it was Strope's turn to go ashore.

He would write about it later: "I shifted the scale setting on my meter to 0 to 5 R/hr (equal to 5,000 mr/hr). I was dressed in a jumpsuit, hood, and booties. We came up on the beach and I vaulted over the gunwale into the surf. In a couple of steps I was free of the water. I turned on the radiac and read over 1 Roentgen (1,000 mr/hr). The whaleboat had already backed off the beach.

I jogged up the beach and into a stand of coconut palms. I noticed that the undergrowth looked dusty. There was also a smell that reminded me of the biblical reference to 'fire and brimstone.' It was not a sulfuric smell like hot lava. It was more like an empty saucepan smells when one inadvertently leaves it on the stove until it is red hot. It seemed to me I had smelled that brimstone smell once before—in the fallout at the Nevada Test Site.

"I held the radiac at the prescribed three feet above the ground and read 2.2 R/hr (2,200 mr/hr). I turned and fled to the beach. I was knee-deep before the boat got back to pick me up. We backed off a respectful distance and discussed this latest reading. I took it about 100 hours after the Castle Bravo shot. The fallout had arrived at Rongelap about 7 to 10 hours after the shot. Our curve of radioactive decay indicated that when the fallout arrived, the level must have peaked at about 45 R/hr (45,000 mr/hr). Although it began to fall thereafter, the exposure to unprotected persons during the first day would have been lethal. Thank goodness the natives lived on the south side of this atoll, rather than on the north side. Otherwise, we would have had a much greater tragedy, the deaths of innocent people."

Back at Kwajalein, Scoville and Strope talked with the Navy doctors treating the evacuees, including Lieutenant Commander Conard and Victor Bond. The blood tests had convinced them that none of the Marshallese had received life-threatening exposure. They were interested in the radiation levels of the soil, water, and plant samples because the people had eaten and drank contaminated food and water, and it was being reflected in their blood and urine.

Of particular interest was the radioactive iodine, which concentrated in the thyroid gland, especially in children, where the gland is much smaller than in adults and can have greater effect.

· PART II ·

LONG-TERM PROBLEMS

25 · SECRET'S OUT

■ **NOTHING HAPPENS IN** a vacuum. We often forget that fact when looking back at historical events.

As important as the nuclear weapons tests were in 1954, there were other events on the Bravo test day, March 12, 1954, that were considered more important by the White House, the media, and therefore the public than that nuclear explosion in the Pacific.

In the United States, that day's attention was almost totally focused on Sen. Joe McCarthy (R-WI) and his fight with the Army, which had grown out of McCarthy's allegations that the military service branch had coddled communists within its ranks.

The *New York Times* that morning had twelve stories on its front page, flooding its readership with many more issues and events than morning newspaper readers would encounter more than half a century later. The *Times* led its page one with a four-column headline, MᴄCᴀʀᴛʜʏ Cʜᴀʀɢᴇꜱ Aʀᴍʏ 'Bʟᴀᴄᴋᴍᴀɪʟ'; Sᴀʏꜱ [Aʀᴍʏ Sᴇᴄʀᴇᴛᴀʀʏ Rᴏʙᴇʀᴛ] Sᴛᴇᴠᴇɴꜱ Sᴏᴜɢʜᴛ Dᴇᴀʟ Wɪᴛʜ Hɪᴍ; 'Uᴛᴛᴇʀʟʏ Uɴᴛʀᴜᴇ,' Sᴇᴄʀᴇᴛᴀʀʏ Rᴇᴘʟɪᴇꜱ.

McCarthy, whose red-baiting campaign against alleged communists in government had been going on for years, was beginning to face his first major opposition from the Eisenhower administration, boosted by the media. In the days that followed the Bravo test, *Time* magazine's March 8 issue featured McCarthy on its cover. In its lead story, the magazine described the senator, during his two weeks of attacking the Army, as having "built the smallest of molehills into one of the most devastating political volcanoes that ever poured the lava of conflict and the ash of dismay over Washington."

In his notes from a March 8, conversation with Eisenhower, White House Press Secretary James Hagerty recorded that the president had called McCarthy a "pimple on the path of progress." Hagerty also noted that "Ike really made up his mind to fight Joe from now on in—all to the good."

On March 9, television viewers were riveted by that evening's edition of Edward R. Murrow's *See It Now* on CBS. The entire thirty-minute program was devoted to using McCarthy's own words to show the harm the Wisconsin senator had done to the nation by alleging communists had infiltrated the government and that the Democratic Party and its leaders were to blame.

In closing the program, Murrow said, "The actions of the junior senator from Wisconsin have caused alarm and dismay amongst our allies abroad, and given considerable comfort to our enemies. And whose fault is that? Not really his. He didn't create this situation of fear; he merely exploited it—and rather successfully. Cassius was right. 'The fault, dear Brutus, is not in our stars, but in ourselves.'"

As soon as Murrow signed off, calls swamped the CBS switchboard, and in less than three hours the network had received 1,046 congratulatory telegrams and 13 protests; 2,211 supportive phone calls—including one from former First Daughter Margaret Truman—and 149 against.

Though other politicians and journalists had courageously challenged McCarthy before, it was the Murrow broadcast that people remembered as the start of McCarthy's downhill track, which was just picking up steam and would dominate Americans' attention in mid-March 1954.

At Eisenhower's March 10 press conference, the first seven questions were related to McCarthy and to a planned speech by Vice President Richard Nixon to answer questions raised by Adlai Stevenson about how the Republican administration and party leaders were dealing with the Wisconsin senator's allegations.

While all these events were occurring, the leak of information about the Bravo test fallout—which Washington had feared and Joint Task Force Cmdr. Maj. Gen. Percy Clarkson had warned about—was about to take place.

It began March 3, two days after the detonation, with Marine Corporal Don Whitaker, who was stationed on Kwajalein and wrote the first of two letters to his mother, Mrs. G. R. Whitaker in Sharonville, Ohio, just outside Cincinnati.

"I was walking back to the barracks from chow, and it was just getting daylight, when all of a sudden the sky lighted up a bright orange, and remained that way for what seemed like a couple of minutes," Whitaker's first letter began. "About ten or fifteen minutes later . . . we heard very loud rumbling that sounded like thunder. Then the whole barracks began shaking as if there had been an earthquake. This was followed by a very high wind."

The second letter, dated March 5, contained the following: "There were two destroyers here to-day bearing natives of one of the Marshall Islands that was within seventy-five miles of the blast. They were suffering from various burns and radioactivity."

Mrs. Whitaker sent her son's letters to her hometown Cincinnati newspaper which published a story about them on March 10. When picked up by wire services, the story created an obvious public relations problem for the US government.

AEC Chairman Strauss would later describe Whitaker's letters as a "colorful account," but a partial and inaccurate appraisal of "new dangers" posed "from radioactive fallout from high-yield nuclear explosions."

However, the published story based on Whitaker's letters forced the AEC to put out a press release late in the afternoon of March 11 that echoed Strauss's words that it was inaccurate in terms of facts then known.

The AEC press statement read, "During the course of routine atomic tests in the Marshall Islands, 28 US personnel and 236 residents were transported from neighboring atolls to Kwajalein Island according to a plan as a precautionary measure. These individuals were unexpectedly exposed to some radioactivity. There were no lesions. All were reported well. After completion of the tests, they will be restored to their homes."

The next day's March 12, *New York Times* page one was dominated by a four-column, triple-decked headline reporting that day's big news, that the Army had charged Senator McCarthy and his top counsel Roy Cohn with seeking favorable treatment for their colleague G. David Schine, who had been drafted into the military service.

Located in the lower left of page one was an Associated Press story about Strauss's release, headlined, 264 Exposed to Nuclear Radiation After Nuclear Blast in the Pacific. It carried the AEC announcement that 28 Americans and 236 natives were "unexpectedly" subjected to "some radiation" from a recent test in the Marshall Islands, but that "there were no burns," i.e. lesions, and all those exposed were "reported well."

The article further explained those exposed had been moved to Kwajalein Island as a precautionary measure, and "after completion of the tests they will be returned to their homes." The AEC reference to the test as "routine" led the AP writer to interpret that to mean "the detonation in question was not a hydrogen bomb."

That press release was just the beginning of a series of misleading government statements about the Bravo test and its aftermath, all initially justified—in the name of national security—to hide that the US had a powerful, deliverable hydrogen bomb. There was another, less apparent, reason: to limit public concern that such nuclear tests, even when held out in the Pacific, carried a worldwide risk of affecting human lives and the environment.

From this first AEC press release dealing with the Bravo fallout, the withholding of information troubled those who knew better. For example, Dr. Conard, who was at Kwajalein when that AEC release came out, later wrote, "Our group felt this announcement was misleading and inaccurate since, at that time, some effects were being observed."

Just one day earlier, on March 10, Joint Task Force Commander Clarkson had sent a classified message back to the AEC reporting that "Erythema [a rash or lesion that could be caused by exposure to radiation] can be very difficult to detect on native type skin." He added that there was a great variance in white blood counts post-detonation among the Rongelap evacuees, and they would be "retained in Kwajalein for intensive study by AEC-DoD research team. No decision as to time to return home."

Clarkson also wrote that the Utirik people would be transferred to Ebeye, the island six miles from Kwajalein, "when the [Trust Territories] high commissioner approves, and studied intermittently." In fact, they were moved to a tent city on Ebeye, where they continued to be fed and clothed by the Americans. Ebeye was where Marshallese who worked at the Kwajalein Naval Base lived in extremely crowded quarters.

Five days after their initial examinations, only eight of the twenty-eight American servicemen from Rongerik showed any contamination at all, and that was at a very low level. Shortly thereafter, all were flown to Tripler Army Hospital in Honolulu for further examination by Army physicians.

However, the initial medical examinations of the Rongelap evacuees, according to a later report, had already shown "a significant amount of penetrating radiation to the entire body . . . and that extensive contamination of the skin and possible internal disposition of radioactive materials had occurred."

Decades later, in his 1992 report entitled "Fallout," Conard—who had spent twenty-five years monitoring the Rongelap people—wrote, "Beginning about 10 days after exposure [which would put it at March 11], radiation burns of the skin began appearing in the Rongelap people. These so-called 'beta burns' appeared as dark pigmented spots on the scalp and on parts of the body that had not been covered by clothing." The top skin layer then peeled away, leaving a depigmented area. At the same time some of those burns became infected.

To combat the beta particles, the exposed people were told to shower more often and to use detergent soaps and a heavy brush. When scrubbings, particularly of the hair and scalp, caused tenderness, cloth towels were used.

Accompanying the burns were itching and a burning sensation. Burns on the scalp led to loss of hair. Lesions on the feet, caused by barefoot contact with the fallout-littered ground during those two days on Rongelap, made walking painful.

Worst of all, "the majority of the Rongelap children had these burns," Conard said. New blood samples were taken every second or third day, and sick call was held twice a day.

"Significant exposure to radiation depresses some elements of the blood, particularly the white cells which protect against infection and prevent bleeding,"

Conard wrote, adding that such depression began within the Rongelap group in the first weeks. In some it fell to one-half to one-quarter normal levels.

Again, he noted, "The greatest drop occurred in children."

Conard later wrote that in those early days, "Serious consideration was given to the possibility that further medical assistance, such as the use of a hospital ship might be necessary." The Pacific Fleet surgeon, Adm. Bartholomew Hogan, assured him that if needed, naval assistance would be available.

By that time, it was clear to the US medical team that the eighty-two Rongelap people—not those from Utirik or the Americans from Rongerik—represented the real problem, and the next ten days would show how serious their condition would be.

The concern of the American doctors was nothing compared to the mixture of fear and confusion felt by the Marshallese. They had been hastily taken from their homes because of a strange white powder, which they had begun to refer to as poison.

At the Kwajalein Naval Base, though isolated, most of them were seeing trucks, cars, and airplanes for the first time. In addition, they were being fed three Navy meals a day—a gargantuan amount of food by Rongelap standards. Many had felt sick. Doctors who had come all the way from the United States were constantly examining them, taking their medical histories, and drawing blood.

Those Marshallese given medicine were told they were being treated for their various illnesses, but rarely was a translator present to explain what tests were being conducted or for what purpose. Marshallese were given pills to take with no accompanying explanation as to why they were supposed to take them

The effects of radiation terribly frightened the Marshallese, as they would anyone. John Anjain's wife, Mijjua, had superficial burns around her neck, and the couple's one-year-old son, Lekoj, had burns and blisters all over his body, and some of his hair fell out.

Tima, the fisherman, lost most of his hair and had deep lesions on his neck and behind his ears. The beta burns that appeared were more severe than expected, and they did not respond to treatment with ointments and calamine lotion.

Dry scabs formed on the superficial burns, and healing for some was rapid. But areas of irregular pigmentation remained—on dark-skinned Marshallese, light-colored patches appeared where the burns had been. In cases of the more serious burns, there were severe skin breaks accompanied by itching and pain.

Doctors were watching carefully for signs of major infection or blood hemorrhaging among the victims. Their concern was based on those blood tests conducted when the Rongelap people first arrived at Kwajalein.

Two weeks after exposure to the fallout, the levels of white blood cells and the platelets were continuing to fall, particularly in the case of children under five. With the blood counts of the victims showing increasing effects of gamma-ray exposure, the medical team decided to check on the amounts of other radioactive material that had been taken into the body.

On March 16, the first urine samples were taken, mixed, and tested. The level of strontium-90, a radioactive form of strontium, was high. Radioactive strontium is taken up into bone, and over time the radiation released can damage nearby soft tissue including bone marrow, the most important source of red blood cells. Depleted red blood cells can lead to excessive tiredness, blood that does not clot properly, and lowered resistance to disease.

These results led to the Americans sending a team back to Rongelap to pick up animals still there. Scientists would conduct experiments on them to determine radioactive intakes.

It was becoming apparent that the exposure of the Rongelap people to radiation and their subsequent unknowing ingestion of radioactive isotopes were both a medical problem and an opportunity for the American doctors to discover more of how human bodies react to such an event.

26 · TALES FROM THE
UNLUCKY DRAGON

■ **WHAT AEC CHAIRMAN** Lewis Strauss and his commission officials did not know, when they released their March 11 press statement, was that the *Lucky Dragon*—the Japanese fishing boat that had been ninety miles off Bikini the morning of the Bravo test—at that moment was making its way back to Japan and its home port of Yaizu.

Although the boat was in constant radio communication with Japan all the way home, the radioman never said anything in his conversations about what they had witnessed or the strange white mist that followed thereafter. Several of the twenty-three seamen aboard had shown early symptoms of radiation sickness—vomiting, diarrhea, and nausea—and a number of them had felt the skin sensations that indicated beta burns.

On March 12, as they neared home, the crewmen began playing around, and the radioman tugged at the boatswain's hair. A clump of it came out in his hand. Others began grabbing, and soon the boatswain was bald from his left ear to the top of his head. Other crewmen found that their hair was falling out, and the radioman recalled that loss of hair had occurred among survivors of the atomic explosion at Hiroshima.

Several crewmen believed that they had been subjected to the after-effects of an American atomic test, and they decided that upon their return to port they would go to the Yaizu Hospital for some explanation of their ills.

On March 13, the day before the *Lucky Dragon* reached home port, Japanese newspapers were running front page articles, based on the March 11 AEC press release, reporting that 236 people in the Marshall Islands had been affected by atomic radiation. No one in Japan or the rest of the world yet knew about the twenty-three Japanese seamen.

At 5:30 a.m. on the morning of Sunday, March 14, 1954, the *Lucky Dragon* slipped through the early morning mist and docked quietly at the pier in Yaizu, ending what would become its history-making, fifty-two-day journey.

That morning's sale of fish was already over when the boat was tied up at the dock. So, the *Lucky Dragon*'s catch would remain aboard until the next day's auction. However, each crew member was permitted to take six pounds of tuna home, his bonus for the hazardous trip.

The boat's owner, businessman Kakuichi Nishikawa, was there to greet them, as were its regular captain, Shimizu, and a fishing union representative. The owner, who had been worried after receiving earlier messages describing lost fishing lines and a limited catch, began questioning fishing master Misaki and the crew as to what had happened.

Crew members mentioned the atom bomb test they thought they had seen, and Shimizu confirmed they were right, having read about it in Japanese newspapers. Misaki then described the illnesses of some crew members along with some of their burns. He and the owner decided all crew members should go to Yaizu Hospital to be examined. It being Sunday, they had to wait until afternoon before the arrival of the doctor on duty, a surgeon named Dr. Ooi.

When Dr. Ooi heard their story—the flash of the explosion, the ashes falling on the boat, their burns and loss of hair—he took blood samples in order to do white cell counts.

The results showed some decrease from normal, but nothing dramatic. Those with burns were treated with a paste, and Dr. Ooi told them not to worry and come back the next day for a more complete examination.

But Nishikawa and Misaki were worried about the crew members who had shown the worst symptoms. They returned to the hospital late Sunday afternoon and got Dr. Ooi to refer two of the crew—engineer Yamamoto and deckhand

Masuda—to Dr. Masao Tsuzuki, a radiological expert at the University Hospital in Tokyo.

Dr. Tsuzuki had just been put in charge of a new Japanese Institute of Radiological Sciences and recently had published a book in which he had proposed that radioactive fallout from testing could cause radiation sickness.

In a note the two seamen took with them, Dr. Ooi wrote they were among twenty-three crewmen on a boat who "seemed to have been taken with radiation sickness(?) on March 1. They are supposed to be suffering from the atomic cloud of an H-bomb." He also included some ashes from the deck in the note's envelope.

Meanwhile, it was 2 a.m. the morning of March 15 when the *Lucky Dragon*'s tuna were unloaded by crewmen who had slept aboard the ship, while others had gone to their homes. The ten-ton catch, sitting on iced trays, had to be put out on the pier for auction shortly after dawn.

There were some twenty thousand pounds of contaminated fish, primarily tuna but some shark. A small portion would be sold locally, but most would be shipped by train and truck to Osaka, Kyoto, Nagoya, Kobe, Fukuoka, and Tokyo.

■　　■　　■

On the morning of March 16, 1954, Japan's largest newspaper, *Yomiuri Shimbun*, broke the story about the *Lucky Dragon*. To this day it remains the greatest world scoop in the history of the 190-year-old newspaper.

The background of its publication is worth recounting since it reflects the best kind of what we old-timers call shoe-leather journalism.

In 1954, *Yomiuri Shimbun* not only had the greatest national circulation in Japan but it was also part of the Yomiuri Group, the country's largest media conglomerate. The Group owned Nippon TV, the nation's most popular private television network, and the Yomiuri Giants, known as the New York Yankees of Japan, which the company established in 1934.

The tycoon behind the company's growth was Matsuaro Shoriki, an amazing character of the 1920s. He started out as a university-educated, judo champion, turned secret policeman, who then bought a struggling left-wing newspaper with funds from a right-wing politician and turned it into a dynamic, but conservative, media giant.

Shoriki was pro-American in the 1930s, showing it when he brought baseball to Japan. However, as Japanese imperialism rose in the 1940s, Shoriki was named head of the Imperial Rule Assistance Association, an organization created as a vehicle to merge Japan's disparate political parties into one. Shoriki's association forced together all private associations and youth groups under one, fascist, quasimilitary structure designed to support Tokyo's expansionist goals.

At the end of World War II, many of the association's leaders, including Shoriki, were arrested and charged with war crimes. Shoriki spent twenty-one months in jail until released by US occupation forces, which dropped the charges against him—which the occupiers claimed had come from unhappy leftist journalists on his newspaper.

By 1951, Shoriki, then sixty-two, began a political career supported by conservative politicians. He had regained his media empire, and with help from well-placed politicians, officials, and US military commanders, he established Nippon TV.

After Eisenhower's Atoms for Peace speech at the United Nations on December 8, 1953, *Yomiuri Shimbun*'s January 1, 1954, edition carried the first of a pro-nuclear energy series of articles headlined, WE FINALLY CAPTURED THE SUN: DOES NUCLEAR POWER BRING HAPPINESS TO MANKIND? Its publication was followed by other positive stories about nuclear power.

The newspaper's continued promotion of nuclear power, which Shoriki supported, led to unproven rumors, going back to the owner's release from prison, that he had made some deal to work with US intelligence agencies.

Against that background, it's a wonder his newspaper printed what would lead to a flood of stories that would inflame worldwide anti-American and anti-nuclear emotions.

Yomiuri's initial reporting about the *Lucky Dragon* began with a seventeen-year-old, high school student, Keiji Kobayashi. A tech major, Kobayashi had become fascinated with nuclear power thanks to the *Yomiuri* articles on the heels of Eisenhower's speech at the UN. It also happened that Mitsuyasu Abe, a twenty-three-year-old local correspondent for the *Yomiuri Shimbun* the past two years, was a boarder at the Kobayashi home.

Kobayashi's mother had, by chance on the afternoon of March 15, had tea with two fishing company employees who told her stories they had heard that day about the *Lucky Dragon* crewmen. They had talked particularly about the strange dust that fell and the burns that developed and caused one sailor to go to Tokyo University Hospital. At dinner that night, the high schooler listened to his mother relate those stories. He tied them to articles he'd read about the H-bomb test and decided to contact Abe, who had been covering a murder case in a nearby town.

Abe, excited about what he had been told, quickly returned to Yaizu that evening. Soon he gathered enough material about the fishing boat and its crew being near Bikini, ashes falling, and the sailors showing burns that were bad enough that a Yaizu doctor had sent two crew members to Tokyo University Hospital. Abe wrote a short article, which he sent that evening to *Yomiuri*'s regional office where it was immediately sent on to the main office in Tokyo.

Luckily, the night editor that evening was the same person who had edited the newspaper's original atomic power series. He was well aware of the recent AEC release about Marshallese being accidentally exposed to fallout from the Bikini test.

Also on duty in Tokyo that night was a *Yomiuri* reporter named Murao, who had worked on the atomic articles. He raced to Tokyo University Hospital with a photographer to try to find Masuda, the *Lucky Dragon* crewman. The hospital staff refused to help, but by going ward-to-ward into the night, Murao was finally able to locate the man from Yaizu with a blackened face, who may have been exposed to atomic radiation.

Ralph Lapp, in his 1957 book *The Voyage of the Lucky Dragon*, described Murao's "heart pounding like a hunter who has sighted big game" when he finally came upon the sleeping Masuda, who looked "like something from another world." The crewman's face and ears were covered with a white ointment, leading Murao to eventually use the phrase "ashes of death" in his part of the story.

Awakened, Masuda recounted everything he and his crewmates had been through, but when Murao tried to return to the hospital room sometime later with his photographer, it was locked with a nurse preventing anyone to enter. At least, Murao later told Lapp, no other reporter could get the story.

Back in Yaizu, Abe got a local photographer and went to the pier where they took flashbulb shots of the *Lucky Dragon* for the next morning's newspaper. Abe's attempt to talk with other crewmen at local bars proved less successful.

Nevertheless, *Yomiuri*'s Tokyo editors were so excited with the story they had put together that they held it until their final morning edition, so it could not be matched by any of their competitors.

When it finally ran with a picture of the *Lucky Dragon*, the headline across the front page—here translated into English—read: JAPANESE FISHINGMEN ENCOUNTERED ATOMIC BOMB TEST AT BIKINI, 23 MEN SUFFERING FROM ATOMIC DISEASE, ONE DIAGNOSED SERIOUS BY TOKYO UNIVERSITY HOSPITAL, H-BOMB?

■　　■　　■

Few stories in 1954 had the immediate and lasting impact on the Japanese and US governments, plus people around the world, than the story about the *Lucky Dragon* and its crew in the final morning edition of the March 16, *Yomiuri Shimbun*.

Among the first alerted by it were a set of Japanese scientists, who had been trained in radiology and sensitive to the effects of radioactive materials—as were many of their colleagues because of the continuing impact of the Hiroshima and Nagasaki bombings on their countrymen. Professor Takanobu Shiokawa, who taught chemistry at Shizuoka University, lived about a half hour's drive from Yaizu. He had already read the story about the *Lucky Dragon* that morning when he was called by an official of the local Prefectural Sanitary Division and asked to go examine the crew and ship.

He went to his office, picked up some Geiger counters, pocket dosimeters, and other instruments, and with his assistant and a prefecture official drove to the port. His Geiger counter would measure intensity of radiation at a point in time; the dosimeter would provide the accumulated amount of radiation to which it has been exposed over a time period.

Arriving at the docks in Yaizu, Shiokawa was surprised at the Geiger counter's rapid buzz as he approached and then went aboard the *Lucky Dragon*. He showed real concern when he later examined the first crew member. As he brought the device to the head of fishing master Misaki, it zoomed to its highest count.

Considered a gentle and modest individual, Shiokawa was shocked by the levels after his first examinations finished. He called for the ship to be moved to a dock on an eastern jetty, far away from the public fish market, and suggested a police guard be established to prevent people from boarding the fishing boat.

He was given a small sample of the radioactive dust by one of the crewmen who had kept it in a folded-up, waterproof paper. Leaving the *Lucky Dragon*, Shiokawa then went on to examine the rest of the crewmembers, some at their homes and others at Yaizu Hospital.

He left Yaizu with the belief that his measurements showed there was definitely internal radiation exposure in the crew, but he did not publicize those first impressions.

That same morning, two hundred miles west of Yaizu at Osaka, Japan's second largest city, a thirty-seven-year-old physicist named Dr. Yasushi Nishiwaki had also read with interest the *Yomiuri Shimbun*'s story about the *Lucky Dragon*.

Nishiwaki had an accomplished past, having studied experimental physics at Osaka Imperial University and having been a research fellow at the aviation laboratory at Tokyo Imperial University, after which he returned to Osaka and his alma mater as an assistant professor. During World War II, he conducted experiments on thermal diffusion of uranium enrichment as part of the Japanese Imperial Army's nuclear development research for a weapon.

Under American postwar occupation of Japan, such nuclear weapon research had been banned. So Nishiwaki turned to medical research of radioactive materials. He joined the faculty of Osaka City Medical College and taught radiation biophysics. In the late 1940s and early 1950s, he was part of the first Japanese American exchange program and traveled to the US where he studied radiation biophysics at the University of Pennsylvania and Columbia University.

In 1948, when he was thirty, Nishiwaki met Jane Fischer, who at the time was teaching English at a Presbyterian mission high school in Osaka. She was the idealistic, twenty-three-year-old daughter of a Missouri Presbyterian minister and a recent graduate of Millikin University, located in Decatur, Illinois. Four years later, in February 1952, they were married and settled in Osaka.

Intrigued by the Yomiuri story, Nishiwaki was surprised to find that other newspapers had nothing about the radiation-exposed Yaizu fishermen. Nonetheless, he immediately responded when later that morning Osaka public health officials called him to go to the city's central market to examine tuna, which had been purchased from the catch of the *Lucky Dragon.*

Using a Geiger counter, he found that wet skin of the contaminated, unsold *Lucky Dragon* tuna registered one hundred times the radiation level of other tuna. Market officials estimated contaminated fish had been sold to roughly one hundred people. The unsold *Lucky Dragon* tuna that remained was confiscated and buried.

Nishiwaki took samples of the fish back to his laboratory and found some of the fish interior meat was even more radioactive than the skin. City officials, informed of the findings, set out to find those who had earlier bought from the *Lucky Dragon* catch.

When the Osaka and other evening papers came out with the news, the Japanese public at large stopped buying tuna and almost all other fish. There were radio broadcasts making "emergency announcements" warning that more than twenty thousand pounds of "atomic tuna" had been shipped from Yaizu to other cities, and a nationwide panic ensued.

That evening, Nishiwaki and his wife talked over what he had found and decided they had to pursue it further by going to Yaizu to examine the boat itself and members of the crew. They took an overnight train and arrived at the port city early on the morning of March 17.

At the harbor they found the *Lucky Dragon* unguarded at its distant dock space. Nishiwaki's Geiger counter went off as they approached, so he and his wife put on protective clothing before going aboard.

Although the lower deck appeared to have been washed, there were still strong signals of radiation. Nishiwaki collected radioactive dust from the upper deck of the steering room, and he noted that the cabin of Kuboyama, the chief wireless operator, was located at the highest part of the boat.

Examining the roof of the upper deck above Kuboyama's cabin, Nishiwaki found radioactive dust that hadn't been cleaned, opening up the possibility that the wireless operator could have received the highest amount of external gamma-ray dosage during the voyage home.

When Nishiwaki examined some crew members later at Yaizu Hospital, he found Kuboyama's skin burns appeared smaller than some of the other fishermen's burns. But having found radioactivity in their blood, urines, and feces, he felt there was "good reason to believe that the crew had inhaled or ingested at least some of the radioactive dust during the voyage in addition to receiving external radiation," according to a paper he later wrote.

Yaizu city officials on the afternoon of March 17 asked Nishiwaki for advice on what should be done. What was missing was not just determination of the dose the crewmen had received but also what radioactive elements had been released with the H-bomb explosion. Nishiwaki knew both those factors were needed to determine the treatment to be given to those who had been exposed.

Nishiwaki decided to write a letter to AEC Chairman Strauss, asking questions on behalf of himself and other Japanese doctors and scientists. In it he said that "in order to minimize possible radiation injury and damage to human subjects in Japan, we need to know immediately in detail the possible types of radioactive elements contained in the radioactive contaminated material."

He turned the letter over to an American wire service reporter, having been told that doing so was the fastest way to have it reach the US. However, the wire service reporter's editor in Tokyo blocked its transmission on the grounds that Nishiwaki was "an alarmist," so it never reached the AEC.

By the afternoon of March 17, the price of fish in Japan had plunged by nearly 50 percent, and sales of sushi were nearly zero. In Tokyo, a load of so-called "atomic fish" arrived and almost immediately had been buried in a nine-foot-deep hole.

Other Japanese newspapers had immediately assigned reporters to the story, which had naturally shocked and aroused the one nation against which nuclear weapons had previously been used.

In Tokyo, the English-language *Japan News* and other newspapers began referring to the twenty-three crew members as "the third group of Japanese victimized by nuclear weapons following Hiroshima and Nagasaki." Others called the event "the third US atomic attack on Japan."

27 · WASHINGTON RESPONDS

▪ **IN WASHINGTON, WHICH** is thirteen hours behind Tokyo, first reports about the *Lucky Dragon* were received at the AEC through commercial channels late in the evening of March 15. Chairman Strauss was on his way to the Pacific to observe the next test shot, so on the morning of the sixteenth the other three commissioners asked AEC General Manager Nichols to arrange technical assistance to the US Ambassador John Allison in Japan, who was looking for help.

The Kansas-born Allison was a long-time Foreign Service officer with twenty-four years of prior service. Still, he had quickly recognized his unpreparedness for this complex matter, which had far reaching international implications.

In Tokyo, the morning of the sixteenth, the Japanese Foreign Office officially informed the US embassy in Tokyo that the *Fukuryu Maru No. 5* [the *Lucky Dragon*] had arrived on March 14 with "23 crew members showing signs of radiation exposure suffered during the United States March 1 atomic test." The initial notice stated that no warning of atomic tests had been received by Japanese government, although Japan's Foreign Minister Okazaki later that day told the Japanese National Diet's Lower House Foreign Affairs Committee that the Foreign Office had discovered it had been warned back in October 1953 about the coming tests.

One of Washington's first concerns was finding out the details of what had occurred vis-à-vis the *Lucky Dragon*, but attention was equally focused on maintaining security about the nature of the H-bomb that had been tested.

Few in Washington looked at it from the Japanese point of view. Instead, as one senior AEC official would later put it, "Initial reaction was disbelief, that this was just a propaganda stunt, that there would be nothing to it."

AEC and State Department officials—seeming to ignore that US occupation of Japan had ended almost two years earlier—decided the Tokyo embassy had to get the Japanese Maritime Safety Board to take control of the *Lucky Dragon* and turn it over to representatives of the US Navy's Far East Command. Thereafter, the US Navy would limit access to the boat, prevent representatives of other countries from having access to it, and thus protect American security interests. The Japanese were also to be to be reassured the US would decontaminate the vessel.

This proposed approach was put in a cable sent to Allison in Tokyo from Washington at 6:35 p.m. March 16, which was 7:35 a.m. March 17, Tokyo time. The instructions read, "Vitally important reasons US security that access vessel be restricted and controlled every extent possible through Japanese Government cooperation. May be helpful this connection offer undertake full responsibility decontaminate vessel. Also desire do all possible investigate circumstances injuries received by crew members."

Meanwhile, the AEC directed its Division of Biology and Medicine to draw up what a commission memo that day described as "a firm statement" about whether it would be safe to eat fish caught near the test area. The AEC memo noted "scientific tests are still inconclusive," and though it was felt "there is little danger of radiological poisoning hazard . . . tests will take 10 years to establish this fact as a scientific certainty."

Nonetheless, the memo concluded, "AEC will issue a statement and we will be informed as soon as it is formulated."

Another decision was made to arrange for Dr. John Morton, the sixty-seven-year-old, white haired, head of the Atomic Bomb Casualty Commission

(ABCC), then stationed in Hiroshima, to contact the US embassy in Tokyo and offer the commission's aid and any medicine needed by the Japanese.

The ABCC had been established in November 1946 by President Truman to run an ongoing investigation into the continuing, long-term effects of radiation among the survivors of the atomic bombs dropped on Hiroshima and Naga-saki. The ABCC was not popular among the Japanese because it only collected data about the bomb victims while providing them with no real medical assis-tance. Beginning in 1951, its activities had started closing, as Congress limited its funding to some $20,000 a year.

Morton, in Hiroshima, got the call about his new mission near midnight on March 16, but would not arrive in Tokyo until March 18, accompanied by Air Force Col. Arthur Meeks, another expert in radiology.

Under the headline NUCLEAR DOWNPOUR HIT SHIP DURING TEST AT BIKINI—US INQUIRY ASKED and datelined "Tokyo," the March 17 morning *New York Times* story focused initially on Japanese police attempting to "find and remove from public sale today some 12,000 pounds of fish landed from a vessel showed with radioactive ash from recent atomic tests at Bikini Atoll." It reported that Japanese scientists had tested the fish and deemed it "to be dangerous to human life." The story cited the incident as headline making in the Japanese press and subject of an inquiry in the country's parliament. "It seemed likely that it would become a new focus of spreading anti-American feeling here," according to the *Times* story. The *Times* also raised questions about the exact location of the ship when the radioactive ashes came down, whether inside or outside the "closed area."

The *Times* report noted that the Japanese claimed their fishing boat was eighty miles off Bikini when the incident occurred, which would be outside the danger zone.

The newspaper also quoted Rep. Sterling Cole (R-NY), then-chairman of the House-Senate Joint Committee on Atomic Energy, who said that based on the Bravo test, the US now had an H-bomb that it could deliver anywhere in the world.

That item became the basis of the first question for President Eisenhower at his press conference the morning of March 17. United Press's Merriman Smith, "dean" of the White House press corps at the time, asked if the president would discuss Cole's statement.

Ike responded, "No, I wouldn't want to discuss that." He added, honestly, that he hadn't seen Cole's statement and did not recall "what we have released . . . [and] I would say that was a question not to be discussed until I was more sure where I am standing."

Other questions followed. They concerned the French fighting in Southeast Asia, America's getting involved in Indochina, and the US and NATO confronting the Soviet Union in Europe.

At the very time that Eisenhower was speaking to the press, quite a different meeting was going on at the AEC's General Counsel's office. A cable had come in from Ambassador Allison which stated that the American embassy in Tokyo had put out a press release expressing concern over the incident. Allison also requested permission to state that in the future if investigation showed the US was at fault, proper compensation would be made.

The AEC-run meeting members discussed what liability for fallout on the Japanese seamen existed for the US government under American law. Someone noted during the discussion that similar claims against the US government were limited to $1,000, although American law did not apply to claims arising in a foreign country. However, Bikini was part of the US-administered Trust Territory.

A participant suggested that "it might be politically desirable to settle Japanese claims independently of legal liability," so the question arose as to what funds were available for such a purpose. A State Department lawyer present said he would inquire. He found that funds for "emergencies in the diplomatic and consular services" could be used, but only between $25,000 and $30,000 would be available for that purpose.

At the White House, the president had an "Emergency Fund National Defense," but any request to use that money would have to be made through the director of the Bureau of the Budget. Late on March 17, the State Department issued a formal statement that referred to the "regrettable" incident and said the US government was conducting an investigation with the cooperation of Japan on how the event occurred "despite the careful precautions taken, including warnings over a wide area."

That same day, Joint Task Force Seven on Enewetak was preparing for five more thermonuclear tests to follow Bravo, four of which were to be in the megaton range. Meanwhile, at a meeting of AEC and State Department officials, the decision was made to enlarge the danger area to shipping by eight times around Bikini. As an additional precaution, the US Navy was directed to keep all commercial shipping five hundred miles north of the testing sites.

On the morning of March 18, the AEC and State Department cabled Ambassador Allison that "the US would see that appropriate compensation was made to injured [Japanese] parties."

Looking back more than a dozen years later, Dr. Charles L. Dunham, AEC's director of Biology and Medicine, made an interesting comparison between how the Marshallese and the Japanese had responded to the US under similar circumstances.

Marshall Islanders, Dunham said at a 1967 meeting with colleagues, were "a small group of native people who are quite literate but who weren't well educated, and I think this is the distinction to make. They were a possession of the Germans, later the Japanese, and then the United States.

"I think they do not really love the United States . . . [they think] we're going to be wards of somebody. The US has been pretty good. . . . So when something had to be done and they were moved [from their homes for the tests], they took it all very quietly and were totally cooperative."

He continued: "In contrast, of course, are the Japanese. A highly sophisticated people, just as sophisticated as we, who had this extra sensitivity to the whole phenomenon of radiation, and who had been a beaten people who were very worried about their relations with the United States and with the world as a whole, but who were just beginning to sort of feel their oats a little bit."

■   ■   ■

Japan's Dr. Tsuzuki, who had prime responsibility for care of the two *Lucky Dragon* crewmen at Tokyo University Hospital, had a troubled history dealing with Americans.

Within days after the bombing of Hiroshima and Nagasaki, the Japanese government had sent medical and scientific teams to the two cities to collect data during the critical first month. Dr. Tsuzuki held the rank of rear admiral in the Japanese Navy during World War II, and after the war was director of the Japanese National Research Council. He had led the Japanese teams sent to the two cities. Studies produced from data the Tsuzuki teams collected were later largely kept from publication by the American military occupation force. US desire for secrecy when it came to the atomic bombs' nuclear effects was probably the main reason, but Tsuzuki believed it was related to his former military service.

One story claimed that when US investigators visited Hiroshima on September 8, 1945, less than a week after the surrender papers were signed, they met with Tsuzuki. At the time they were given an earlier paper Tsuzuki had written about his well-known, earlier radiation experiments and their effects on rabbits. Tsuzuki was said to have commented, "Ah, but the Americans, they are wonderful. It has remained for them to conduct the human experiment."

Nine years later, Tsuzuki was still suspicious of the real reason for American offers of medical assistance, this time for the exposed *Lucky Dragon* crew members.

Meanwhile, on March 18, the envelope with the small sample of ashes from the *Lucky Dragon* that the crewmen had turned over to Tokyo University Hospital doctors was given to Professor Kenjiro Kimura—a distinguished, internationally respected, Japanese chemist. Kimura was a specialist in radioactivity and prior to World War II had built the first cyclotron outside the United States. Not only that, in 1938 he had discovered and described the creation of uranium-237, having bombarded uranium-238 in his cyclotron with fast neutrons.

Later, Kimura had studied the soil samples from Nagasaki after the atomic bomb had dropped. Based in part on what he had found and what he had read in newspapers, Kimura had concluded there was plutonium in that bomb.

Kimura analyzed the dust from the *Lucky Dragon* to determine its radioactive components so that Tokyo University Hospital doctors could decide on medical responses to the *Lucky Dragon* seamen's radiation sickness.

At the same time, throughout Japan, the tuna market continued to be in turmoil. The Japanese, at that time, got 90 percent of their protein food out of the sea, so it was no surprise their great concern about radiation made the emerging situation that much worse. Additionally, in spring, Japanese consume in excess of a million pounds of sashimi or raw tuna fish a day. Just over a month away was the emperor's April 29th birthday, a time when fish delicacies such a sashimi were part of the ceremonial tradition.

Then, a new jolt struck when the American-based Van Camp Company told Japanese exporters to hold up shipments to the US of 1,500 tons of Japanese tuna it had already purchased.

The Tokyo government had to act. Foreign Minister Katsuo Okazaki appeared before the Japanese Diet's House of Councilors budget committee on March 19, and announced that he would seek compensation from the US government for the burns the *Lucky Dragon* crewmen had received. He also wanted confirmation from Washington that the fishing boat had been operating outside the specified danger area at the time of the Bikini Atoll H-bomb test.

■   ■   ■

Following directions from Washington, Dr. Morton on March 19 visited the two crewmen at the Tokyo University Hospital. He assured Dr. Tsuzuki and his colleagues that the US was ready to assist with antibiotics.

Without realizing it, Morton had taken sides in an internal dispute that already existed between Japanese medical experts at the National Institutes of Health, headed by Dr. Rokuzo Kobayashi, and the staff at Tokyo University Hospital, headed by Dr. Tsuzuki. They were at odds over the proper treatment of the exposed crewmen, and the conflict led to strained relations between Washington and Tokyo.

Morton and others, months later, confessed in reports to Washington about being "naïve" regarding "evidence of rivalry among various Japanese medical groups," but they focused their major ire on "the rantings of the hysterical, sensation-seeking, irresponsible, sometimes mendacious Japanese press."

After examining the two crew members, Morton briefed Sen. John Pastore (D-RI), a member of the Joint Committee on Atomic Energy, who was in Tokyo on his way home from Kwajalein, where the Senator had also seen some of the Marshallese. As Morton later wrote, he warned Pastore that a medical opinion at such an early moment was dangerous, and "trying to minimize radiation injuries suffered by citizens of the only country ever to be under attack of a nuclear weapon" would be "seized upon by the paranoid, hysterical Japanese press and thereby create an atmosphere of suspicion, distrust, and contempt."

But that was exactly what happened. Morton forgot his own advice. Before leaving for Yaizu, where he planned to examine other crew members, Morton met with Japanese and other reporters. He said he found the two *Lucky Dragon* seamen "in better shape than I had expected," and added that they "are improving and will be better. . . . In two or three weeks to a month the men will recover."

At the same time, Senator Pastore held a press conference in Tokyo and repeated Morton's optimistic view about the fishermen's prospective recovery. The remarks of these two Americans contrasted with Japanese press reports up to that time, which had repeatedly written about the long-term danger of exposure to radiation.

In addition, some Japanese newspapers were already demanding "all information" on the makeup of the fallout. The US embassy reported to Washington on the Yomiuri newspaper speculating that Japanese scientists feared American scientists will use "them," the seamen, "merely to gather information, meanwhile refusing to give essential scientific facts of the bomb type necessary for treatment of the injured."

■   ■   ■

Inside the government in Washington, where many officials had been caught up in McCarthyism and the serious nuclear arms race with the Soviet Union, heightened fear of communists became another factor affecting the US reaction to the *Lucky Dragon* affair.

Unknown to the public, and a deep secret within the American government, was knowledge of a major Soviet defection in Japan just two months earlier on January 24, 1954. The defector was Yuri Rastvorov, a lieutenant colonel in the NKVD Soviet security force, who had been undercover in Tokyo as a third secretary in the Russian embassy. Initially, the story was that Rastvorov had mysteriously disappeared, but he emerged six months later in the United States—after the *Lucky Dragon* had made headlines.

At the time Rastvorov had defected, he was the highest-ranking Soviet intelligence officer to do so. The CIA debriefing of him had been tightly held while the *Lucky Dragon* uproar with Japan raged on.

Rastvorov provided his CIA interrogators—one of whom he later married—with details of Moscow's in-country Japanese agents, some in the government, in labor unions, and others among North Korean refugees. He also described widespread pro-Soviet propaganda activities in Japan. Then-CIA Director Allen Dulles, in a later thank-you letter to the defector, said Rastvorov provided "considerable counterintelligence information on Soviet espionage and subversion activities in Japan and some of the earliest retrospective data on how the Soviet Union had actually planned the Korean War and how Stalin successfully pressured China to enter that conflict."

Rastvorov's defection added to American military and intelligence agencies' preexisting concerns about security surrounding any sensitive information exchanged with the Tokyo government. A State Department paper prepared around that time noted, "There is no law in Japan today which defines treason, espionage or state secrets, or which provides for political screening of public employees." As a result, "The problem for the communists is not so much winning over a Japanese majority as it is the penetration of the Japanese government and society and the infiltration of its key positions."

Based on these concerns, some US government officials—with the McCarthy charges fresh in mind—were reluctant to share what they considered sensitive nuclear information, such as the makeup of the Bravo fallout, with Japanese scientists, doctors, and government officials.

Activities in Japan related to the radioactive fallout further fed security concerns among American officials worried about the Communist infiltration in Japan. On March 19, local Japanese Communist Party officials, including some former members of the Diet, visited the families of the *Lucky Dragon* seamen to offer assistance. That same day, a group of doctors associated with a Communist organization showed up at the Yaizu Hospital wearing Mao Tse-tung badges and promised to raise money for the affected seamen.

On March 22, 1954, days after the *Lucky Dragon* had surfaced, Defense Secretary Wilson's assistant for Special Operations, retired-Marine Corps Gen. Graves B. Erskine, warned the White House national security staff that

"Communist propagandists will make the most of the excellent opportunity they now have to exploit." They had already taken advantage of Nagasaki and Hiroshima to fuel anti-American sentiment in Japan, and Erskine noted in his memo that, "In effect, the present situation involving radiation burns on Japanese fishermen and the panic caused by confiscation of fish cargoes due to radioactivity gives the Communists an opportunity to sow the same seeds . . . with much greater effect."

Erskine recommended, "A vigorous offensive on the non-war uses of atomic energy would appear to be a timely and effective way of countering the expected Russian effort and minimizing the harm already done in Japan."

■   ■   ■

On Kwajalein, the Rongelap Marshallese continued to show hematological depression, but there was no evidence of bleeding or increased susceptibility to infections. As a result, blood transfusions were not considered, nor were antibiotics.

Urinalyses showed radioactive strontium and iodine levels were higher but not above permissible levels. Only the radioiodine, which concentrates in the thyroid gland, would eventually prove troublesome.

AEC Chairman Strauss had flown out to the Pacific test site on March 13 to attend the Romeo test shot, which had been moved up from April so he could be present. The project engineer on the test of the Romeo device was Harold Agnew, a Manhattan Project physicist who in 1945 had flown in the B-29 chase plane and filmed the Hiroshima bomb going off. In 1970, he would become director of Los Alamos National Laboratory. He had nick-named the Romeo device "Runt," as compared to "Shrimp," the nickname for the Bravo device.

Runt, according to Agnew, used unenriched lithium in the lithium deuteride fusion fuel and became the design that was finally weaponized as the Mark 17 thermonuclear bomb. A copy of it now resides in the National Nuclear Museum of Science and Industry in Albuquerque, New Mexico.

Despite Strauss's presence, three shot dates planned for Romeo—March 13, 15, and 21—were postponed due to weather, mostly because of unfavorable upper winds. Each postponement required the resetting of equipment, made more hazardous because the changes had to be done in the radioactive environment that still remained from Bravo.

While on Enewetak, Strauss sent Eisenhower Press Secretary Hagerty a report on the Bravo briefings he had received. He described the results of the test as very important to the weapons program, and that the radiation injuries suffered by the Marshallese had been exaggerated. Based on reports about the Japanese fishermen, which had not been verified, Strauss said that expansion of the danger area for shipping would allow the next tests to continue as planned.

The Romeo test postponements were underway at the time of President Eisenhower's March 24, morning press conference in Washington. He was questioned about the McCarthy hearings and whether an upcoming conference in Geneva could halt fighting in Indochina. Then, a reporter asked whether there would be more information coming from the AEC about atomic weapons.

"I do believe this," Eisenhower told reporters, "that entering the atomic age, you people have legitimate questions—and all America and possibly the world—affecting this whole development. You have a right to ask them at places, specifically to me or the White House or other places, when information can be given without definitely jeopardizing the security of the United States." He didn't say the test Strauss had gone out to see had been postponed three times, but he did promise, "After Admiral Strauss comes back from the Pacific . . . he [the President, would] review this whole question with him [Strauss] again and determine, if we can, what is the scope or the limits of the things of which I can talk."

Near the end of the press conference, Eisenhower was asked by CBS correspondent George Herman about US policy on continuing atomic testing since "some anti-American newspapers in Japan and other countries" had described the radioactive fallout from the Bravo test on the Marshallese and Japanese fishermen. Eisenhower made the next day's headlines when he admitted that the Bravo test was "something . . . we have never experienced before and must have surprised and astonished the scientists." He added that the US "has to take precautions that never occurred to them [the scientists] before." He claimed not to know the details, but following the Strauss approach to play down the known effect, Eisenhower said news reports about Bravo's fallout "were far more serious than the actual results justified."

Later, on the afternoon of March 24, probably in response to the president's press conference, the AEC released a public announcement about the enlargement of the Pacific test area, which—as expanded—covered 570,000 square miles, twice the size of Texas. The AEC did not announce that US Navy vessels and American aircraft were regularly patrolling the area in advance of the delayed Romeo test shot. There was to be no repeat of the *Lucky Dragon* episode, if the US could prevent it.

■   ■   ■

In the Soviet Union, which knew it was behind in development of an H-bomb, initial worldwide publicity about the Bravo bomb test spurred Moscow's scientists to try new ideas.

Following the March 5, 1953, death of Stalin and the arrest months later of minister of Internal Affairs and head of the nuclear program, Lavrentiy Beria, the Politburo and Central Committee had become more involved in nuclear planning. Although work on a true H-bomb had continued, it went on in a more decentralized fashion.

In the spring of 1954, the Soviet leadership knew it had fewer than fifty fission nuclear bombs when Moscow and the world learned of the US Bravo test of an impressive H-bomb.

On March 12, 1954, Georgi Malenkov, as chairman of the Council of Ministers of the Soviet Union, responded to the Bravo test by saying that the government in Moscow wanted to ease world tensions and prevent a return to the Cold War because "that policy is the policy of preparing for a new world war which with modern weapons means the end of world civilization."

Russian scientists working on the H-bomb project responded to Malenkov's speech with a published article that tried to follow Eisenhower's earlier proposal of atoms for peace. Veniamin Malyshev, Kurchatov, and two other Soviet scientists had concluded that they were free to follow Malenkov with a statement about the insanity of nuclear war. They wrote that a thermonuclear bomb can "destroy all surface buildings in a city of multimillion population," but that a broader use of such weapons would be "enough to create impossible conditions for life all over the globe." They were calling attention to radioactive fallout.

Molotov and Nikita Khrushchev, both then moving within the Politburo to gain power, subsequently denounced the speech made by Malenkov, who later retracted it. That step marked the beginning of the end of Malenkov. Less than a year later, Khrushchev and others forced him to resign as Council of Ministers chairman.

■   ■   ■

In the wake of Bravo problems, Dr. Merril Eisenbud, the AEC's director of its Health and Safety Laboratory and the commission's specialist in radioactive fallout, flew to Japan at the request of Dr. Morton, arriving at Tokyo's airport at 10:30 p.m., March 20.

He was surprised to be met by a crowd of some thirty reporters and photographers and behind them hundreds of demonstrators, who had been generated by earlier Japanese press stories about his coming. The situation required American military police to station themselves at the foot of the airplane's ramp where they were then able to hustle Eisenbud into a waiting US embassy car.

What Eisenbud did not know was that a small American pharmaceutical company, seeking publicity for an over-the-counter burn lotion, had put out a story about sending a miracle drug to Japan to help the *Lucky Dragon* crew members. That item had gotten into Japanese news media, along with the announcement that an AEC expert was also on the way, implying Eisenbud was carrying the drug.

Although his flights from New York had taken forty hours, and he had not been able to have a night's sleep for two days, Eisenbud was nonetheless taken directly to the embassy for a conference that went on for two hours. He was told about the rapidly developing situation, the anger of the Japanese people, the conflict between Japanese doctors, the tuna market collapse, and the emerging crisis between Tokyo and Washington.

He then had several hours rest and a chance to read cables exchanged between the Tokyo embassy and agencies in Washington about the situation.

They included suggestions by American officials that the exposed fishermen had been spying on the US nuclear test.

The next morning, Eisenbud met with Dr. Kimura, who had been studying the radioactive ashes from the *Lucky Dragon*. As Eisenbud later told colleagues at a 1967 closed, government-sponsored meeting, Kimura "had already analyzed the debris and had detected uranium-237, which led him to the conclusion that there must have been a . . . reaction which involved the fast fission of U-238." Eisenbud told his 1967 audience that Kimura had mentioned "a very sensitive fact in our weaponeering, and here I was sitting with a man who had deduced something in a couple of days that was known to very few people in the United States."

The use of U-238 was one of the secrets of the Bravo bomb that Eisenbud was supposed to protect while carrying out his role as a radiation specialist sent to assist the American embassy and the Japanese.

Eisenbud's dilemma at the time, he later said, "was in trying to be helpful and at the same time trying to protect information that some people [in Washington] thought should be held secure."

Within two days, Eisenbud found himself in a late-night meeting with Dr. Tsuzuki, in which he had to assure the once-stymied Japanese that, unlike his earlier post-Hiroshima/Nagasaki experience with Americans, this time he would be free to publish his studies of the exposed Japanese crewmen.

Eisenbud also learned that the Japanese government wanted to cooperate and settle matters quickly, but its officials first wanted an apology from the US. In addition, they were upset that Senator Pastore and other Americans, who had seen the seamen's injuries, were claiming there was nothing major wrong with them or the Marshallese, and the whole uproar was "a hoax."

As a result, the Japanese did not want the exposed seamen to be examined by American doctors. However, days later, when Eisenbud visited Yaizu and went aboard the *Lucky Dragon*, he was able to take away a strip of wood from the upper deck that still contained radioactive dust. Expecting not to be able to see the crewmen, he was surprised when Yaizu's mayor insisted he visit the hospital. There, Eisenbud was permitted to examine some of their thyroids and take urine and hair samples along with skin scrapings, all of which he sent back to the US for examination.

Eisenbud also met with Gen. John Hull, who had replaced Gen. Douglas MacArthur as Supreme Allied Commander in the Pacific. "By the time of our conference," Eisenbud later wrote, "the true importance of the Bravo fallout seemed obvious to me: Thermonuclear weapons had the ability to contaminate tens of thousands of square miles with lethal amounts of radioactivity."

Although he told Hull he was not in position to discuss the military implications of the Bravo fallout, "it was quite obvious that thermonuclear weapons were far more destructive than had been anticipated by the military planners with whom I had been discussing this possibility for the past two years."

Hull's response to Eisenbud's analysis was not about the military implications of Bravo but rather the security danger for the US of allowing the Japanese to keep possession of the *Lucky Dragon*. Hull believed the tiny amount of radioactive fallout particles still on the boat could reveal classified information about design of the hydrogen bomb. He even suggested he was prepared to seize the boat if higher authorities felt that was necessary. Eisenbud then checked with General Clarkson's Joint Task Force Seven intelligence unit and was told taking control of the fishing boat was unnecessary.

■   ■   ■

Back in Washington, out of public view, the Eisenhower White House was preparing to deal with what it anticipated would be a propaganda battle sparked by the worldwide media attention to Bravo test fallout.

Journalists often have a hard time understanding that at times people inside government, familiar with the news cycle, spend a good deal of effort anticipating media demands when it comes to a controversial subject. This was one of those times.

On March 24, the White House Operations Coordinating Board had a preparatory session entitled: "How can we prevent Communist exploitation [of the nuclear testing program]; what policy guidance needed for top officials; include contingencies of death resulting from injuries and widespread loss to Japanese fishing industry."

Prior to that, Defense Secretary Charles Wilson had sent the board a memo in which he had written: "It is reasonable to suppose Communist propagandists will make the most of the excellent opportunity they now have to exploit and develop their 'peaceful' intentions regarding the atom, as compared with what is apparently going on in the current US tests."

Wilson had continued, "Present situation of radiation burns on Japanese fishermen and the panic caused by confiscation of fish cargoes due to radioactivity give the Communists the opportunity to sow the same seeds throughout Japan with much greater potential effect and we must assume they will, without in fact, limiting their effort to the parts of the world nearest the [Pacific] proving ground."

As a response, Wilson suggested, "A vigorous offensive on the non-war uses of atomic energy like a [US] decision to build a reactor in Japan or Berlin or any of tangible implementation of the president's [peaceful uses of atomic energy] speech."

The working group meeting on the twenty-fourth had recognized they would have two press releases, but decided where possible when dealing with the *Lucky Dragon* and its seamen, they should appear to originate in Japan. The board had already dealt with questions raised by Japanese newsmen, such as one posed to the AEC as to "whether it was safe to eat fish caught near the general test area."

The board had been told current tests on fish were "inconclusive," although little radiological hazard was expected. Nonetheless, the truth was that it would take ten years to establish that fact with scientific certainty, so the AEC was trying to formulate an answer to handle the issue. It was agreed work should begin on a formal note of regret to the Japanese government, but it would be "so worded as to refer only to the unfortunate accident, and without any implication of legal responsibility on the part of the US government," according to a summary of the March 24, working group meeting. The note could also make reference to the already announced US assurances of some kind of financial payment.

American officials in Washington would not make official announcements, other than those totally relevant to the American audience, such as any reports by the Food and Drug Administration (FDA) on its monitoring for radioactivity of Japanese fish shipments coming to West Coast ports.

Where possible, the working group decided that releases should come through the recently-established US-Japan Coordinating Committee, a panel of Japanese and American scientific, governmental, and military officials working together to investigate the entire incident. It was noted that the Japanese government had set up its own Atom Bomb Injury Investigating Committee, which the US hoped would resolve coordinating problems.

A contingency plan was to be developed "in the event of the death of any of the radiation-affected Japanese fishermen," according to the White House memo. It was to include arrangements for autopsies to be done jointly by US and Japanese doctors, so that a joint announcement could be made "on causes of death."

The working group concluded, "Official actions should tend to minimize the incident, and to undercut Communist propaganda."

On March 26, the working group decided that the formal note of regret for the incident should not be delayed until the US-Japan Coordinating Committee reached its findings. However, that note should be made public simultaneously in Tokyo and Washington, perhaps accompanied by a statement by Secretary of State Dulles at a press conference.

■   ■   ■

On March 27, on Bikini, the second thermonuclear test in the Castle series, Romeo, finally took place, almost two weeks behind its revised schedule. The Joint Task Force finally had the favorable upper-wind patterns it required.

Bravo's lesson had been learned.

The Romeo device was detonated from a barge that had been anchored over the Bravo crater—another step designed to cut down on potential fallout since there was no coral beneath it to be picked up by the fireball.

Originally scheduled to go off at eight megatons (equivalent to eight million tons of TNT), Romeo, using the unenriched natural lithium, went off at eleven megatons, not quite as powerful as Bravo. "It worked gangbusters, and so we weaponized that [version]," Agnew told an interviewer in 1994.

Romeo's fallout moved to the northwest, away from populated areas. It was found later that large areas of the ocean surface were significantly radioactive after the barge shot.

No immediate public disclosure of the Romeo test was made.

That day on Kwajalein, Dr. Eugene Cronkite's medical team concluded its sixth blood test of the Rongelap people and found in some that the platelet and white-cell levels were still sinking dangerously low. Dr. Conard and his colleagues again began to consider whether to send them to Hawaii or California where intensive-care facilities were available.

That very same day in Washington, Eisenhower met with his Science Advisory Panel, looking at how the US could defend itself from a surprise Soviet nuclear attack. One of the participants was Dr. James R. Killian, president of Massachusetts Institute of Technology.

Asked at the meeting's end to come up with a plan, Killian suggested creation of a new group to study nuclear weapons and intelligence technology. That group became the Killian-led Technological Capabilities Panel, nicknamed the "surprise attack panel," which eleven months later produced recommendations for what eventually became some of the major nuclear weapons systems employed in the Cold War period.

Rep. Chet Holifield (R-CA), chairman of the Joint Atomic Energy Committee, who had observed the Romeo detonation along with Chairman Strauss, sent Eisenhower a note after returning to Washington. In it Holifield said it was "imperative that the people know the effect of these weapons, that they may be able to more realistically evaluate the gravity of international tensions." He called on the president to use "plain words" rather than scientific explanations.

AEC Chairman Strauss had his own, strongly held ideas about what he had learned from watching the Romeo shot, and what should be told to the public about the thermonuclear tests. Back in Washington, he first met with Secretary of State Dulles and told him, "Nothing was out of control, nothing was devastated." Dulles reportedly responded that from his point of view—with Japan, Great Britain, and other countries upset—some things were out of control.

It was late afternoon March 29, with Strauss back in charge, that the AEC released an eighty-eight-word statement in the chairman's name announcing the Romeo test as part of a "thermonuclear series," and for the first time officially describing the event as a hydrogen bomb test. The release also pointed out that in the wake of the *Lucky Dragon* incident, the enlarged test area had been searched, and "no shipping was discovered."

Two days earlier, *Red Star*, a Soviet military newspaper, had published a story describing roughly how Russia's version of a hydrogen bomb would work.

On March 30, Strauss met for an hour with Eisenhower to provide a report on both the Bravo and Romeo tests. In that meeting, according to Strauss, the president expressed concern about the *Lucky Dragon* crewmen, the Marshall Islanders, and the exposed American servicemen. The AEC chairman told Eisenhower that the *Lucky Dragon* had been within the warning range, something Ike would repeat in his memoirs. But on that same day, March 30, Eisenhower's own Operations Coordinating Board (OCB) reported internally that the Japanese fishing boat was outside the danger area by fourteen miles.

With Hagerty present, it was agreed that Strauss would prepare a statement on the Pacific tests and take part in the president's press conference set for the next day.

Strauss's statement was to "ease fears that [the Bravo] bomb had gotten out of control," according to Hagerty's notes on the meeting. The White House would support Strauss, downplaying the "accident" and his "soothing" of world opinion when it came to the radioactive fallout.

The AEC's public confirmation of a US hydrogen bomb test had ramifications across the Atlantic, where Churchill, appearing before the House of Commons, was questioned aggressively by the opposition Labor Party members who wanted Washington pressured to end thermonuclear testing.

The seventy-nine-year-old prime minister, shaken, had to confess, "I haven't got them [the facts]" about the American testing. A request for details about the US activities, he said, might result in "a blunt refusal" because US government officials "are prevented by their own legislation from divulging secret information about them [the tests]."

Churchill did repeat what he had been told by American officials about the Bravo fallout, saying he understood "that the injuries suffered by persons outside the [Pacific testing] area . . . are neither serious or lasting."

When a Labor member pointed out that consultations were needed on the hydrogen bomb "in view of the fact that it is from British airfields that American hydrogen bombers may take off," Churchill replied, "That aspect of the general situation is one which is never absent from my mind. . . . I am in almost hourly correspondence with the Government of the United States."

By that time, sixteen of the *Lucky Dragon* crew had been transferred from Yaizu Public Hospital to the First National Hospital in a residential section of

Tokyo, joining the others in the Japanese capital who were already at University Hospital. Tensions over radiation in Japan had risen so high that if the seamen had gone by train from Yaizu, there was fear other passengers would object. So the US military had provided a C-54 for the flight to Tokyo.

In Japan, on March 30, 1954, Dr. Masanori Nakaizumi at University Hospital had issued a statement which read, "Judging from white blood cell counts, the condition of the twenty-three patients is gradually deteriorating. Blood transfusions are continuing." In an appearance before the Japanese Diet's Health and Welfare Committee, Dr. Masao Tsuzuki said, "I have been silent until now out of consideration of the patients, but 10 percent of the twenty-three crew members may die."

28 · MAKING THINGS WORSE

▪ **PRESIDENT EISENHOWER'S MARCH 31, 1954,** press conference, in what was then called the Old Executive Office Building next to the White House, was a monumental event of worldwide impact.

Some 233 reporters, cameramen, and staff members had packed the Indian Treaty Room by 10:30 a.m. that morning when Ike, with the dour AEC chairman beside him, opened the session saying, "As you can suspect, ladies and gentlemen, from the picture-taking this morning, we are trying a little bit of an innovation."

He continued with an unusual tongue-in-cheek introduction: "There has been some slight interest shown in the tests recently conducted in the Pacific, and for this reason, I brought along with me this morning the expert in that field. After I take a certain share of the press conference time, I am going to turn the rest of it over to him. Of course, this will also give me a unique privilege of seeing someone else in this particular spot."

After talking about taxes, housing, foreign trade, and then answering a handful of questions covering the Far East, use of American troops, the Israel-Arab dispute, the Soviet explosion of thermonuclear test devices, and keeping good people in government, Eisenhower said: "That is my last question, and now Mr. Strauss is going to take over. I didn't realize that time had gone."

Strauss opened by reading a prepared statement that began with a justification for testing.

Soviet nuclear tests, Strauss said, had challenged American military leadership since "our sole possession of the weapon, which had been a major deterrent to aggression, had been cancelled." The United States, he said, had to regain nuclear superiority, and he indicated that the hydrogen bomb was the way to do so.

After acknowledging that the Mike test in November 1952 was "the largest man-made explosion ever witnessed to that date," he acknowledged that the Russian test of August 1953 showed Moscow, too, had "a weapon or device of a yield well beyond the range of regular fission weapons."

In short, the US was in a race with the Soviets in getting a deliverable hydrogen bomb. Or as Strauss put it, further testing was needed since "we now fully know that we possess no monopoly of capability in this awesome field."

At that point, Strauss turned defensive and began to make a series of misleading and at times inaccurate statements that would in time come back to haunt not only him but Americans in Japan and elsewhere.

He said press news reports about the March 1, Bravo test shot had been "out of control" and "exaggerated." He admitted the test was "in the megaton range," and "the yield was about double that of the calculated estimate." In fact, it had been almost triple the estimate and, at fifteen megatons, was unexpectedly outside the margin of error, not, as Strauss put it, within the margin that could be expected with a new weapon.

Much more damaging was what he said in reference to the fallout and exposure of the fishermen on the *Lucky Dragon*, which Strauss mistakenly called "the Fortunate Dragon."

He repeated to reporters the inaccurate claim that the Japanese fishing boat "must have been well within the danger area," something Eisenhower's own OCB had already determined was incorrect. The Tokyo government four days earlier had presented US Ambassador Allison with an aide-memoire that included the *Lucky Dragon*'s location, its course, and other details, but had emphasized that the fishing boat had not "received warnings, by radio message or other means, while in the area before the accident occurred."

That, however, was only the beginning.

Strauss said the 236 "natives"—the Marshallese—exposed within the radio-active fallout area were "promptly evacuated," when it had actually taken two days before those on Rongelap were picked up, and another two days had passed before those on Utirik were taken to Kwajalein.

Referring to his visit the previous week to Kwajalein, Strauss said the Americans who had been on Rongerik had not suffered any burns and "could be returned to duty, but they are still being kept on Kwajalein for the benefit of extended observation."

As for the "236 natives," Strauss said, they "also appeared to me to be well and happy," totally ignoring the burns they had suffered, the loss of hair, and the lowered white blood cell levels that were beginning to show up. On that very day, blood tests on Kwajalein had shown white blood cell levels falling to about a fourth of the normal level for at least 10 percent of the Rongelapese. It would be ten days before blood regeneration would begin to appear.

Strauss claimed, "Today, a full month after the event, the medical staff on Kwajalein have advised us that they anticipate no illness [for the Marshallese], barring of course disease which might be hereafter contracted." That was a totally misleading statement, since thousands of miles away, Dr. Conard and his colleagues were considering moving some exposed Marshallese to Hawaii for better care.

At that point, Strauss paused, and then ad-libbed, saying, "And as a matter of fact, we have more natives than we started with. One child was born while I was there, and four more are expected. They named the child—a little girl—after my wife." When a reporter broke in and asked the name of the child, Strauss said, "Alice," his wife's name.

The AEC chairman, it was later learned, had presented the new mother with a gift of ten pigs, which represented a fortune for a Marshallese, but he did not tell that to the reporters, ending what was the only light moment in the entire press conference.

Getting back to his prepared statement, Strauss claimed his information about the Japanese fishermen was sketchy since "our people have not yet been permitted by Japanese authorities to make a proper clinical examination." That did not stop him from making another totally misleading statement, that "skin lesions observed [on the Japanese fishermen] are thought to be due to the chemical activity of the converted material in the coral rather than to radioactivity, since these lesions are said to be already healing."

As we shall see, this latter claim would be particularly irritating, not just to the Americans on Kwajalein and in Japan but to the Japanese themselves.

Strauss identified Dr. Morton and Dr. Eisenbud as the US specialists on the scene in Japan. He then concluded his prepared remarks by characterizing as "irresponsible and utterly false" comments he had read that suggested "the incident involving the fallout on inhabited areas was actually a planned part of the operation." Such comments, he said, "greatly distressed me" and were "doing a grave injustice to men who are engaged in this patriotic mission."

It is worth noting that such a view—that the Marshallese were being used as guinea pigs—is one still held by many of them and their heirs more than sixty-seven years after the event.

Strauss's prepared statement did not generate the immediate worldwide impact of the March 31, Eisenhower-Strauss joint press conference. Instead, the headline news arose from the AEC chairman's exchange in the question period, after Richard L. Wilson, bureau chief of Cowles publications, asked Strauss if he would "give a general description of what actually happened when an H-bomb went off?"

Strauss paused, and started saying, "The area of the blast would be about . . . " When Eisenhower broke in and said, "Why not depend on all these pictures they are going to see?"

The president was referring to a scheduled press preview of an AEC/Defense Department film about the 1952 Mike thermonuclear test shot that was to follow the press conference, but be embargoed for release April 7, almost a week later. The film spoke of the damage a bomb similar to Mike would create if Washington were hit, with ground zero being the Capitol building. The film would show complete annihilation west to Arlington Cemetery, east to the Anacostia River, north to the Soldiers Home, and south to Bolling Air Field.

Taken aback by Eisenhower's intervention, Strauss acknowledged that the 1952 film was going to be shown but avoided an immediate answer, saying if he "described it specifically, [that] would be translatable into the number of megatons involved, which is a matter of military secrecy. The effects, you said the effectiveness—I don't know exactly what you meant by that, sir, so I don't know how to answer it."

Wilson persisted, eventually asking, "What happens when the H-bomb goes off, how big is the area of destruction in its various stages; and what I am asking you for now is some enlightenment on that subject?"

Aware of what the soon-to-be-shown film described, Strauss apparently felt free enough to say, "Well, the nature of an H-bomb, Mr. Wilson, is that, in effect, it can be made to be as large as you wish, as large as the military requirement demands, that is to say, an H-bomb can be made as large enough to take out a city."

That set off a chorus of "What?"

Strauss continued, "To take out a city, to destroy a city."

Someone shouted, "How big a city?"

"Any city," the AEC chairman responded.

"Any city, New York?" was then asked.

"The metropolitan area, yes," Strauss said, "given a general description of what actually happened when the H-bomb went off."

Strauss ignored the various impacts of an H-bomb going off, such as radioactivity and fallout, but would later change the transcript from "to destroy" a city to "put out of commission," as what he should have said.

■　　■　　■

H-BOMB CAN WIPE OUT ANY CITY, STRAUSS REPORTS AFTER TESTS; US RESTUDIES PLANT DISPERSAL

That was the *New York Times'* page-one lead headline spread across five columns on its April 1, 1954, newspaper. "Vast Power Bared; March 1 Explosion was equivalent to Millions of Tons of TNT," read the subhead on the story.

Similar headlines appeared across the country. H-BOMB CAN WRECK N.Y., led the *Washington Post* and *Times Herald*; H-BOMB CAN DESTROY ANY CITY: STRAUSS, *Chicago Daily Tribune*; and H-BOMB CAN WIPE OUT ANY CITY, STRAUSS REPORTS AFTER TEST, *Los Angeles Times*.

Although the *New York Times* reported the device was six hundred or seven hundred times as powerful as the Hiroshima and Nagasaki bombs, the Bravo explosive power was in fact one thousand times greater. William Laurence, who wrote the *Times* article, claimed that the Bravo explosion showed that while Moscow had "gained a technical lead in the hydrogen bomb field, as has been recently reported, the Russians are still far behind, despite the four-year lead they had received."

Herbert R. O'Brien, director of New York City Civil Defense was quoted by the *Times* as saying, "If this latest bomb is what they say of it, our system of going to shelters is ancient history." He called for mass evacuation of cities such as New York as the only answer.

Two days after Strauss's well-publicized press conference, Prime Minister Jawaharlal Nehru went before the Indian Parliament on April 2, 1954, and said, "We are told there is no effective protection against the hydrogen bomb and that millions of people may be exterminated by a single explosion. . . . These are horrible prospects and affect us as nations or peoples everywhere, whether we are involved in wars or power blocs or not."

Nehru called on the United Nations to begin negotiations immediately on an interim "standstill" agreement to halt all nuclear testing as a first step toward prohibition and elimination of all nuclear weapons.

After the press conference, the White House received over one hundred pieces of correspondence a day advocating an end to nuclear testing.

Four days after Nehru's speech, Secretary of State Dulles gave Eisenhower a handwritten note that read, "I think we should consider whether we could advantageously agree to Nehru's proposal of no further experimental explosions . . . [that]could be policed—or checked—."

Ike's reply was, "Ask Strauss to study."

Three weeks later, Joint Chiefs of Staff Chairman Adm. Arthur Radford sent Strauss a letter which read, "While it is recognized that certain political advantages might accrue to the United States in making or accepting a proposal for a moratorium on the testing of nuclear weapons, it is believed that any political advantages would be transitory in nature, whereas the military disadvantages probably would be far-reaching and permanent."

Radford added, "In the light of the foregoing, the Joint Chiefs of Staff consider that it would not be to the net advantage of the United States to propose or to enter into an agreement on a moratorium on the testing of nuclear weapons."

Bravo, from a military standpoint, had become important because it led the way to a new family of thousands of thermonuclear weapons—from small tactical ones to multi-megaton strategic sizes. As Los Alamos Director Norris Bradbury would say in July 1954, "Is anyone going to care about using a B-47 [bomber] to deliver kilotons, when three-megaton bombs of the same weight are available?"

At the same time, Bravo's radioactive fallout, following broad media coverage of the *Lucky Dragon* seamen, generated worldwide, public fear and thus growing political interest in halting nuclear testing. As historians Richard Hewlett and Jack Holl described it, Bravo's "sweet taste of success" was accompanied by a "sickening reality: mankind had succeeded in producing a weapon that could destroy large areas and threaten life in thousands of square miles."

■   ■   ■

Strauss's statements at the press conference created another kind of uproar in Kwajalein and Japan.

Japanese TV and radio stations focused on Strauss's claim that the *Lucky Dragon* was in the "danger zone," which it was not. Adding to Japanese anger was Strauss's remark that "chemical activity" rather that radioactive fallout had been the cause of the fishermen's skin lesions. The AEC chairman's "all is well" statement about those exposed was compared to the seamen's Japanese doctors who, at that very time, had disclosed concern about their blood conditions.

Dr. Cronkite, head of the medical team in Kwajalein, could not restrain himself after reading the *New York Times* account that reported Strauss's description of the fallout hazard as minimal. Dr. C. L. Dunham, an AEC doctor who had been with Cronkite at the time, recalled, "I can remember when this hit us. We were at Kwajalein. I could see the expression on Cronkite's face when he read this."

Cronkite later said that after the March 31 press conference he had called Strauss and told him that his statement that lesions on the Japanese fishermen were caused by slaked coral was "a downright lie." Strauss replied, "Young man, you have to remember that nobody reads yesterday's newspaper," Cronkite recalled.

Cronkite possessed firsthand experience studying the Marshallese skin lesions. In the immediate aftermath of Bravo, he had sent samples of the affected Rongelap people's skin from Kwajalein back to the AEC where they were compared to cattle hide taken from livestock in Nevada and Utah that had been exposed to radioactive fallout from US-based test shots.

The AEC scientists determined the lesions on the human skin and the lesions on the cattle skin were the same, according to Cronkite.

Eisenbud said later that Strauss's statements came to him in Japan as a "big shock to Morton and myself." He said the AEC chairman's claim that the Japanese fishermen "must have been well within the danger area" was one of those troublesome to him. Eisenbud said he told media in Japan that "there was no evidence this was so."

Eisenbud had also been angered over Strauss's remark that the fishermen's burns had been the result of caustic action caused by the coral's calcium oxide intensely heated by the fireball and not radioactive fallout from the test shot. For Eisenbud, it implied to the Japanese that he was the source of Strauss's information, since he was reporting back to the AEC from Tokyo. "It certainly didn't come from me, but everybody else thought it did," Eisenbud would later say.

Together, Strauss's two statements "were very offensive to the Japanese and that caused things to deteriorate so far as Morton and myself were concerned." Eisenbud recalled.

Conard said years later, "The fallout material was indeed caustic, though this did not cause the 'beta burns' that later developed." At the time of the press conference, Conard would later recall, "We didn't know whether they [the Marshallese] were going to live or die, or whether we were going to have to request a hospital ship to take them and that sort of thing."

Not publicly known at the time, but just hours before the start of the Eisenhower-Strauss news conference, the Japanese foreign ministry had delivered a note to the US asking that no future atomic tests be held at the Pacific sites between November and March, which is the tuna fishing season. At the same time, a US official had announced that talks had opened between Japanese and American officials on compensation for the Japanese fishermen, although the Americans had yet to admit responsibility for the sailors' exposure to radioactive fallout.

While US officials were slow to admit what they were learning, as additional facts were uncovered more of the truth began to emerge.

29 · OUT IN THE PACIFIC

■ **THE DAY AFTER** Strauss's press conference, Dr. Masao Tsuzuki went on international radio with a fifteen-minute speech on the *Lucky Dragon* seamen's injuries that was translated into English and other languages. Eisenbud recalled that Tsuzuki said on the broadcast that "it was ironic to [have the US] tell him that radiation burns might be lye burns, when he [Tsuzuki] had worked all his professional life with radiation, and had been the first to go into Hiroshima. He made a few unpalatable remarks about the ABCC, and about the Americans using the Japanese as guinea pigs."

In a letter to the US embassy, generated by Prime Minister Shigeru Yoshida, the Foreign Ministry reminded the American officials of Japanese "extreme sensitivity to the subject of atomic casualty," because the people "have twice before been victims of the atomic bomb." Therefore, "their reaction to the incident has been unfavorable to US."

As a result, it was difficult for Japanese officials to meet the request of the US government, which at the time still wanted possession of the *Lucky Dragon*. In addition, the letter stated that Washington's offer of American doctors assisting with the seamen "has not worked smoothly due to the fact patients [are] unwilling [to] submit themselves to care [of] American physicians in fear treatment 'as experimental material, rather than as therapeutic object.'"

Responding four days later, the AEC cabled the US embassy in Japan that their scientists "cannot venture to predict the outcome of the five most severely affected seamen, lacking total cooperation," having only reviewed their blood counts and urine specimens.

The letter warned, however, that it "would be unwise to assume at this point that all will recover. People have been known to live many months and even a year or more with the white blood count as low as 2,000," noting that one seaman was that low. A normal white blood count is between four thousand and eleven thousand, although some laboratories believe less than four thousand could mean the body could not fight infection as it should.

The AEC cable said in conclusion that "a draft statement for use in the event of the death of one of the crewmen is being prepared in the Department. The suggested text of such statement will be communicated to the Embassy in the near future."

On April 20, 1954, a committee of the Marshall Islands Congress filed a petition with the United Nations to protest the American tests and to request compensation for the islanders who had been subjected to fallout. The Americans had indicated they were going to pay money to the Japanese seamen. Why, then, shouldn't the people of Rongelap and Utirik also be reimbursed?

Among those drafting and signing the document was Lejolan Kabua, the high chief or iroij of the atolls in the western chain of the Marshall Islands, including Rongelap and Utirik. He was also a major landowner on both Kwajalein and Bikini. Amata Kabua, the ambitious son of Lejolan Kabua, had also been one of the Marshallese working to get compensation from the US for the Rongelap people.

Amata Kabua already was popular among all Micronesians because at age twenty-five, in 1953, he had hired an American attorney in Hawaii, Earnest E. Wiles, to get compensation for Kwajalein landowners after the US took over the island and converted it into a Navy base and missile test site. Kabua's original demand was for $1,000-per-acre-per-year for the US to lease the land. Negotiations went on for years.

As a teacher on Majuro, Amata Kabua also had signed the UN petition which asked that the nuclear tests be halted, but if they had to be continued, they wanted their people to be moved to safe areas or be instructed in safety precautions should there be any future fallout.

If the exposed Marshallese had "known not to drink the waters on their home island after the radioactive dust had settled on them," the petition said, they could have avoided dangers. Instead, the Rongelap and Utirik people had been suffering from "lowering of blood count" and "burns, nausea, and the falling off of hair from the head. . . . Complete recovery no one can promise with any certainty."

On Kwajalein, the petition read, the exposed Marshallese themselves were worried about what was to happen to their home atolls, and they sought funds in the event they could not return to their homes within a short time. The petition concluded with a statement that it "should not be misconstrued as a repudiation of the United States," for the American control of the Marshalls had been "by far the most agreeable time in our memory."

The petition was sent to the United Nations through Trust Territory officials, and nothing was heard about it—publicly—for several weeks.

Meanwhile on Kwajalein, the blood counts for most of the exposed Marshallese had stabilized, albeit for some Rongelapese at very low levels. One of those whose white blood count and platelet levels had remained low was the Anjains' one-year-old son, Lekoj.

Urine tests, now made on an individual rather than a collective basis, permitted an analysis of radioactive elements in each person's body. Some had higher strontium-90 levels than others. The doctors decided to see if administering a recovery agent might help in ridding a person's body of ingested radionuclides.

Seven of the more highly contaminated Rongelapese were selected, and for five days they were given the agent. The treatment was stopped when it appeared to be ineffective.

On April 21, 1954, a survey party including Marshallese as well as personnel from the Joint Task Force, AEC, and Trust Territory visited Rongelap and Utirik to determine if the two atolls were habitable. The radiation level at Utirik was low enough to permit planning for its inhabitants to return after the Castle test series was completed. Meanwhile, they had to remain where they were because in the aftermath of Bravo, Utirik had been placed in the expanded danger zone.

The situation with Rongelap was different. The survey team found it far too dangerous for anyone to consider living there for at least a year. Instead, the AEC, in conjunction with the Defense Department and Trust Territory, developed plans for the Rongelap people to be transferred to Ejit Island in the Majuro Atoll, 250 miles south of their home atoll. There, a new village of wood and aluminum buildings was to be constructed for them.

Since Ejit had few stands of coconut palms and hardly enough pandanus, papayas, and breadfruit to support the Rongelap people, it was arranged that the AEC would pay an allowance to the people so that they could buy food from nearby Majuro.

It began the decades-long Rongelapese dependence on American financial and food support for their survival.

The issue of the *Lucky Dragon* being a spy ship reemerged in the US at about the same time. Rep. W. Sterling Cole, then-chairman of the Joint Committee on Atomic Energy, announced his panel would investigate whether the Japanese fishermen "may have entered the danger zone for a purpose other than fishing, and may have spied on the nuclear test."

Strauss had already initiated an internal government investigation, and both the CIA and the Japanese Public Security Investigation Agency had been on the case, looking into the lives and backgrounds of the crew, their families, the Japan Communist Party, trade unionists, and members of the organization that had become involved with the *Lucky Dragon* issue.

On April 3, a letter from the Japanese Foreign Ministry to the US embassy in Tokyo reported, "As to the career, travel overseas, education, behavior and political activities of the crew members, investigation is being conducted by Yaizu Municipal Police, but no unusual information has been obtained so far."

It went on, "So far, there is no Communist Party member, nor communist sympathizer among the crew." One local Yaizu newspaper had reported on March 19, that one of the sailors, Shinzo Suzuki, made antiwar, anti-American remarks while in Yaizu Hospital, although it was unconfirmed. He was a POW of the Americans during the last war and was said to understand a few English words.

■ ■ ■

Bernard J. O'Keefe was an electronic engineer who as a Navy ensign had been assigned to the Manhattan Project. He would later prepare the fusing of the Nagasaki bomb and participate in the Crossroads testing. For Bravo, he was part of the arming and firing team with Joe Clarke and the others in the Enyu Island bunker in Bikini when the detonation took place.

In his 1983 memoir, *Nuclear Hostages*, O'Keefe described his view of the mood in America in the aftermath of Bravo. He wrote, "Returning home from the Pacific, I found little comprehension of the enormous amount of destruction possible from an explosion of the magnitude of the Bravo test. Even the furor from the Japanese fishing trawler was treated as an isolated incident, a freak that happened because the vessel was someplace it shouldn't have been. The experience of the Rongelap natives was played down in the tight security atmosphere of the times. . . . The public was unable to grip the great scope of the destruction, unable to comprehend that a single explosion could wipe out an area of 2,800 square miles, unwilling to believe that there was no place to run to and no place to hide."

Much more frightening was the description given by Brig. Gen. Kenneth Fields to a closed, secret, May 24, 1954, meeting of the AEC. Describing the radioactive fallout effects of Bravo if it had been detonated at Washington, DC, Fields showed a diagram where those in the Washington-Baltimore area would have received a dose of five thousand roentgen; in Philadelphia, more than one thousand roentgens; in New York City more than five hundred, or enough to result in death for half the population if fully exposed to all the radiation delivered. Fallout northward through New England toward Canada would have been in the one-hundred-roentgen area, roughly comparable to the *Lucky Dragon* exposures.

Fields's classified document had little distribution beyond those in the room and perhaps another seventy-five people.

■   ■   ■

On May 13, Strauss and Defense Secretary Charles Wilson put out a joint statement announcing the conclusion of the Castle series of tests. The statement
said the tests "were successful in the development of thermonuclear weapons"
and had "contributed materially to the security of the United States and the
free world."

Altogether there had been five detonations after Bravo, four of which were
from barges anchored over deep water to limit fallout.

One day after the Strauss-Wilson statement, the Marshall Islanders' petition
to the United Nations was made public. Henry Cabot Lodge Jr., the American
ambassador to the UN at the time, issued a statement that said the United States
regretted the danger to the Marshallese and admitted that they "apparently have
suffered ill effects from the recent thermonuclear tests in the Pacific proving
grounds." He pointed out that the exposed islanders were getting the same
medical treatment as the exposed American servicemen, and that the United
States "would take precautions to prevent financial damage." Lodge also said he
was "informed there is no medical reason to expect any permanent aftereffects
due to the falling radioactive materials."

As Strauss had done earlier, Lodge said that an unexpected shift of the wind
had caused the fallout to come down on the Marshallese islanders.

A Defense Department report on Castle's military effects, put out five years
later, referred to Project 4.1—the medical survey instituted immediately after
the Bravo detonation—in a brief chapter entitled, "Accidental Exposure of
Human Beings to Fallout." The report described Project 4.1 as being organized
to "evaluate the severity of the radiation injury to the human beings involved;
provide for all necessary medical care; and conduct a scientific study of radiation injuries to human beings." It described the project as representing "the
first observation by Americans on human beings exposed to excessive doses of
radiation from fallout (mixed fission products)." The exposures "far exceeded
the normal permissible dosage," according to the report.

It also read, "The groups of exposed individuals were sufficiently large to
provide good statistics." Although there were no preexposure clinical studies

or blood counts available, "it was possible to study Marshallese and American control groups that matched and exposed population closely with regard to age, sex, and background."

■   ■   ■

On June 4, 1954, a Navy vessel carried the 154 Utirik people, plus three new-born babies, back to their home atoll. For the three months they had been on Kwajalein Atoll, the Utirik people were paid $400-a-month as compensation for the money they might have earned making copra on their home atoll—a figure that they considered too low.

They were also given food and water to tide them over until they could once again provide their own food. Since the Americans had removed all the Utirik livestock and family pets for laboratory study, they were given new chickens and pigs. After three months of living on Ebeye with an American food subsidy, it was hard for the islanders back at home to return to their old eating habits.

The men had become accustomed to such former luxuries as cigarettes and liquor, and the women now enjoyed cooking regularly with packaged flour, Crisco, and prepared doughnut mixes. Meanwhile, the children had grown to love candy, which had been easily acquired on Kwajalein and Ebeye. In the past, some of these things could have been bought or traded for when ships stopped by.

As for the Rongelap people, they remained at Kwajalein until July when their newly constructed village on Ejit was completed.

Life on Ejit turned out to be difficult. The AEC, instead of continuing to supply the islanders with all their food, gave a thousand dollars a month for the entire atoll to buy provisions. In addition, they were paid $240 every three months for the copra they would have made if they had remained on Rongelap. Since there were few coconut or other food trees on Ejit, the people normally ran out of food money before the end of each month and had to borrow from the AEC. The food problem was made more acute by the fact that the original eighty-two exposed people were joined on Ejit by relatives who—for schooling or for work—had left Rongelap before the March 1, detonation.

As word of the American subsidy spread through the Marshalls, the size of the group on Ejit continued to grow since, by Marshallese tradition, the newcomers shared in the food bought by the AEC subsidy. John Anjain, who still served as magistrate, held the copra payments for two, three-month periods so that he had just $2 a person when it was finally distributed. On behalf of his people, he asked the Trust Territory government for more money, but at that time no more came.

In Washington, responsibility for paying for the Marshallese was transferred to the AEC from General Clarkson's Joint Task Force Seven, and thus away from the Pentagon. Although their resettlement still remained in the hands of the commander in chief of the Pacific, the AEC's Division of Biology and Medicine continued medical care of the exposed Marshallese.

While Americans began to forget what happened to the Marshallese, and Washington bureaucrats were left to share responsibility for their welfare, the *Lucky Dragon* seamen who had been Bravo victims still remained uppermost in the minds of the Japanese.

30 · SETTLEMENT

▪ **DURING THE SPRING** and summer of 1954, the twenty-three Japanese seamen from the *Lucky Dragon* remained in Tokyo hospitals. In late May, as examinations continued, blood samples were regularly taken from arms and ears while biopsy needles drew marrow from bones. Doctors found both white and red cells dropping along with platelets, essential for blood to coagulate.

Blood transfusions and dry plasma were regularly administered, and for some of the men their blood cell counts began slowly to rise. They also were given vitamins and antibiotics to prevent bacterial infection and anemia.

Almost two-thirds of the crewmen developed jaundice, a skin decolorization caused by red blood problems, but only one of the seamen was severely affected—radioman Kuboyama. In April, his white blood cell count had dipped to 1,900, while at the same time his red blood-marrow count had dropped off sharply. Jaundice appeared in June, and his condition worsened as the summer wore on.

▪   ▪   ▪

Also in late May 1954, with discussions going on about a future visit of Prime Minister Yoshida to Washington, President Eisenhower took a personal interest in Japanese reaction to the plight of the *Lucky Dragon* fishermen. In a memo to Dulles, the president asked that "some competent staff officer prepare a brief analysis of this situation in terms of what things we can and should do now to improve our prospects in that region."

In response, Robert Murphy, the acting secretary of state, sent a note to the president on May 29, 1954, saying the US embassy in Tokyo believed compensation for the fallout from Bravo was "the most important specific issue to dispose of." At that point, Murphy said, the US had publicly announced it would compensate the injured fishermen "and we have suggested to the Japanese Government the sum of $150,000." He added, "The Japanese will probably claim heavy damages of any indirect type, such as for fish thrown away and lost profits. We feel we should be prepared to raise our initial figure, perhaps to $300,000, to avoid legalistic haggling and gain rapid agreement. We should then pay promptly."

When Ambassador John Allison came to Washington and briefed the National Security Council on June 10, he told them the Japanese expected much more. "He suggested $750,000 as a likely figure, mentioning a possible range of $500,000 to $1 million," according to the meeting's notes.

In the discussion that followed, an AEC official stated, "Commission lawyers had found no legal authority for payments of the type in question." It was agreed not to seek a congressional appropriation because a debate in Congress on the matter was "undesirable," since it would bring radioactive fallout into full public view.

Gov. Harold Stassen, then-White House director of foreign operations, said that if the payment were made in yen, rather than dollars, it could be gotten by shipping wheat to Japan under the P. L. 480 program, which would generate counterpart funds when sold in that country. Attorney General Herbert Brownell Jr. was asked to look into the legality of that approach.

Three days later, however, the Defense Department agreed to produce $1 million as an indemnities payment for damages caused by its testing. The funds would come through reprogramming from other of its accounts. In a July 7 cable, the State Department authorized Allison to begin negotiations with the Japanese for a payment "not to exceed $1 million."

On July 26, Allison reported back to Washington about how the negotiations were going. Foreign Minister Okazaki had told him that within the

Japanese cabinet, the figure of $4.2 million was described as the "minimum acceptable claim," but that he, Okazaki, had insisted that "sum was out of the question."

Instead, Okazaki said he told the Japanese cabinet, "He had persuaded US Ambassador [meaning Allison] to go as high as $1 million, but that was the limit." Allison told Washington he had made no such commitment "but indicated some optimism we could go this high or prompt settlement." Allison also pleaded to Washington, "It is important we be able to conclude agreement immediately, once Okazaki has persuaded his colleagues to sign."

In effect there were two sets of negotiations going on, which is not unusual in such situations. Allison and Okazaki were dealing with each other, and then both had to go back to their bosses in Washington and Tokyo to sell what they felt would be an acceptable agreement.

■   ▩   ■

On August 5, 1954, in connection with the Hiroshima anniversary, a press conference was held. Kuboyama, as the oldest seaman, represented the *Lucky Dragon* crew. He had a wife and two daughters, so he talked of the seamen's marriage problems—their present separation from their families. He told of different situations for those with girlfriends, thanks in part to a published story that the exposed men would no longer be able to have children. Kuboyama also talked about finances, saying he and his colleagues were "borrowing money at the rate of 20,000 yen each month [equivalent then of roughly $60] from the Fishery Union for our families' living expenses," while waiting to see if any reparations would be coming from the US.

In mid-August, Kuboyama's health took an abrupt turn; he required oxygen through a nasal tube while blood transfusions continued. When American doctors came to visit him, Kuboyama objected to giving them time, arguing that the Americans were interested more in research than treatment.

The radioman's condition then became critical. In a September 2, 1954, cable to Washington, Ambassador Allison referred to a Japanese doctor working for ABCC, who had three days earlier examined Kuboyama. That doctor had reported the radioman had "recently recovered from jaundice [but] appeared to have virus hepatitis induced by 73 blood and plasma transfusions" which he characterized as "serious overtreatment."

For a short time Kuboyama seemed to recover, but by September 20, he dropped into a coma. His situation was widely publicized, even in the US where the Associated Press story reported, "Not since the Emperor announced the end of the war, has an incident so stirred the emotions of the Japanese."

On the evening of September 23, 1954, with hundreds of Japanese outside the hospital standing vigil, the forty-year-old seaman died. Ironically, his death came on a national newspaper "holiday," so there was no immediate editorial comment, but the news itself was carried on radio and television.

By the afternoon of September 24, when the first newspapers appeared, Kuboyama's doctor was quoted as saying "radiation sickness" had been the cause of death. All Japanese papers demanded compensation for the seamen, and some added to what had become a worldwide call for a halt in US testing.

According to a US embassy cable, radiologist Dr. Tsuzuki had told a *Yomiuri* reporter, "What I was most surprised to see was that his liver, heart, kidney, spleen, pancreas, and all the other internal organs affecting life had been so greatly changed that I could hardly recognize them. I think it means that radiation infiltrated and destroyed his body completely. I realize keenly how horrible the death ash is."

In the US, Kuboyama's death was not that big of an event. A page ten headline in the September 24, 1954, *New York Times* read, JAPANESE DUSTED BY H-BOMB IS DEAD. The Tokyo-bylined story described the *Lucky Dragon* radio-man as "probably the world's first hydrogen-bomb casualty" whose death "was expected to touch off some bitter anti-American expressions here [in Japan]." A *Times* opinion piece the next day described Kuboyama as Japan's "symbol of hatred and fear of atomic weapons," and quoted a Japanese newspaper editor saying, "Only the death of an emperor could have commanded similar attention in the nation's press."

Officially, the US attributed Kuboyama's death to liver disorder, but the exact cause of that disorder was never specified. In the opinion of one Japanese doctor, the death could have resulted from infectious hepatitis, since hospitals at that time did not use disposable syringes and such cases were common. It could also have stemmed from the degeneration of the liver caused by cell destruction, primarily by radiation.

Most likely it was a combination. Whatever the case, the Japanese and most of the world believed the Bravo test fallout was responsible.

The day after the radioman's death, Ambassador Allison sent the Japanese Foreign Office two notes. One, to Japanese Foreign Minister Okazaki, contained a check for a million yen—about $2,600—made out to Kuboyama's widow for delivery to her. The second note, for delivery to Mrs. Kuboyama, was a personal one from Allison, and it spoke of "the deep sorrow and regret felt by the Government and the people of the United States and personally by me over the death of your husband. There are no words of mine which can assuage your grief. While no sum of money can compensate for your loss, it is the desire of the Government of the United States that something be done to make life easier for you and your children in the future. I am therefore enclosing a check for yen one million payable to you as a token of sympathy of the American people."

The Japanese government provided Kuboyama's family with a $16,000 special payment a day after his death and some $1,500 to each of the remaining *Lucky Dragon* seamen still in Tokyo hospitals. It was to be considered a down payment on the expected money the Japanese government was seeking from the US.

■   ■   ■

Well before the Bravo test, the AEC for years had been paying close attention to strontium-90.

As early as 1949, Dr. Nicholas M. Smith Jr., at the AEC's Oak Ridge National Laboratory, had concluded that ingested strontium-90 was "by far the most hazardous isotope resulting from nuclear detonations." As noted earlier, ingested radioactive strontium is a chemical analogue to calcium and is drawn to human bone tissue.

Smith's finding led to the commission's establishment of Project Gabriel, a semi-secret network of some one hundred monitoring stations—half-domestic, half-foreign—which collected fallout on gummed paper or plastic sheets. Periodically, the sheets were sent to a New York AEC laboratory for study.

By 1951, the right questions were being raised within the AEC about what civil-defense studies about the impact of fallout needed to be undertaken.

Among them were questions on whether atomic bombs or warheads could be used without ultimate harm to our own troops or friendly populations. These studied effects would include immediate as well as long-range ones for present generations, according to an October 15, 1951, internal memo sent to Dr. Stafford Warren, then at the AEC's Division of Biology and Medicine.

Among other questions was: "What are the ultimate long-range effects to future generations from the spreading of large amounts of radioactive materials around various countries (this could include such things as effect on food production, genetics, etc.)."

To measure the human uptake, an additional pilot study was undertaken in 1953 by Dr. Willard F. Libby, a Manhattan Project alumnus then located at the University of Chicago. The idea was to collect human bones worldwide in order to measure any rise in radioactive strontium, as testing of thermonuclear devices was beginning to send fallout around the world.

As a cover story for this classified research, participating scientists were told "it is to be for a survey of the natural radium burden of human bones," according to a later government study. "As for the emphasis on infants, we can say such samples are easy to obtain here [in the US], and that we would like to keep our foreign collections comparable."

To hide the AEC role, samples were to be sent directly to Libby at the University of Chicago.

By June 22, 1954, an internal AEC memo showed Project Gabriel was studying the occurrence of strontium-90 in the US and some twenty foreign countries as a result of nuclear tests in Nevada and the Pacific. Samples of soil, plants, animals, various dairy products, and some human bones were being examined. Resampling, where practical, was underway. Human uptake of strontium-90 and its biological effects were being followed "as conditions permit."

"Our observations to date on the content of strontium-90 in human beings," the AEC memo read, "have not fallen outside the range of our expectations of a year or so ago."

Independently, but at the same time, the Defense Department, through the Armed Forces Special Weapons Project, had created its own worldwide fallout network, collecting human urine, animal milk, and tissue samples.

Work on strontium-90, through Project Gabriel, had been the focus of a classified Biophysics Conference that AEC Biology and Medicine Director Bugher held in the commission's Temporary Building #3 along the Mall in Washington on January 18, 1955.

He told the opening session that Bravo and the other thermonuclear tests had shown "that weapons have taken an enormous leap in energy" that required accelerating data collection on how "such elements as strontium-90 mix into the whole environment, both the inanimate and biological environment. The rate at which such material may reach equilibration in nature, is something that we've not yet determined, but is very fundamental to the whole problem."

Bugher noted a great upsurge was underway in public comments by geneticists about radioactivity's impact on humans, and he commented that "the thing that is wrong is that they are all talking largely from a basis of opinion rather than conclusion."

Libby, still involved with Gabriel but now an AEC commissioner, told the group about existing strontium-90 data and the gaps that existed in furthering the research. He surprised them by saying, "By far the most important is human samples. We have been reduced to essentially zero level on the human samples. It is a matter of prime importance to get them and particularly in the young age group."

He continued: "We were fortunate, as you know, to obtain a large number of stillborn babies as material. This supply, however, as of now has been cut off and shows no signs of being rejuvenated."

Researchers had used a calf that had lived with strontium-90, measuring levels in milk, soil, and his bones, which in some way, Libby said, replicated the experience of a one-year-old to three-year-old child.

However, it had not been satisfactory. "We must be careful," Libby said, "So human samples are of prime importance, and if anybody knows how to do a good job of body snatching, they would really be serving their country."

His statement brought forward a doctor associated with a center that specialized in gathering such samples. He said he could get them from New York City through a medical school and in Houston where, he said, "they don't have all these rules," and they "have a lot of poverty cases."

Libby concluded by saying that in the next few months "we expect 10 to 20 samples per month of humans in the range of one to 40 [years of age]. If you could get three, four- or five-year-old children, that would be good, because they are a little hard to come by."

No mention was made of the Marshallese.

Following Bravo and other powerful Castle tests, such studies of the long-term impact of radioactive fallout had specific new areas and people on which to focus: The surface of the Pacific Ocean as well as land masses, such as Rongelap and Utirik, more than one hundred miles away from the detonations. Months after stories about the *Lucky Dragon* seamen had fired up the Japanese public's concerns, Ambassador Allison told the State Department, "No report of long-range air or sea contamination, no story of food or water pollution, no theory of genetic deterioration seemed too wild for acceptance."

The Japanese government tried to respond to public concerns, and on May 15, 1954, sent a small mariner training ship, the *Shunkotsu Maru*, on a two-month, nine-thousand-nautical-mile, marine environmental radiation survey of the Pacific Ocean outside the immediate test area near Bikini and other Marshall Islands. Aboard were Japanese oceanographers, marine biologists, meteorologists, food scientists, and radiation analysts, who turned the small ship into a virtual floating laboratory.

The US government had mixed feelings about the venture. Japanese scientists invited two American specialists to join them. They accepted and flew to Tokyo. However, after they arrived, the US embassy in Japan, at the suggestion of the AEC and State Department, had them remain there.

In Washington, officials questioned supporting the *Shunkotsu Maru* during its voyage. Internal government documents show that the Defense Department initially refused, on security grounds, to allow the ship to stop at Guam or Marcus Island for provisions, suggesting it put in at Wake Island instead. In the end, Commerce Department officials talked the Defense Department into getting the Navy to provide the necessary supplies on Wake Island, with Commerce handling the distribution.

As with earlier disagreements among Japanese scientists, those aboard the *Shunkotsu Maru* did not produce uniform opinions. When the ship returned to port on July 4, 1954, however, it was greeted by reporters and a crowd of some two thousand.

In a July 7, 1954, summary of newspaper and newsreel coverage to the State Department, Allison said that reports from nine newspapermen on the expedition were "largely factual and moderate in tone." Some pointed out that there was some radioactivity in seawater and marine life northwest of Bikini, and overall the group agreed that plankton was much more radioactive than fish.

Allison also reported that the Japanese press "constructively pointed out (a) radioactive fish [were] eaten [aboard the *Shunkotsu Maru*] during trip; (b) Certain scientists aboard have scoffed at over-nervous dumping of radioactive fish here [in Japan]."

In a later report to Congress, the AEC said the Japanese scientists took water samples at various depths to study radioactivity in the water, in addition to plankton and fish. They found that diffusion of the radioactivity occurred rapidly at the surface, but more slowly at greater depths. That result echoed findings earlier in the US that found tritium—used in thermonuclear explosions—diffused at a half-life of about eighteen years between surface ocean water and the next level below.

A sample of seawater taken by the *Shunkotsu Maru* 1,200 miles west-northwest of Bikini showed radioactive disintegration comparable to disintegration found in radioactive sewage ponds near AEC US installations.

In the summer of 1954, the Japanese data from the Shunkotsu encouraged the AEC's Eisenbud to put together a US radiation survey that in 1955 covered the northern Pacific Ocean area. Using the name Operation Troll, the AEC and Office of Naval Research jointly undertook what became a 17,419-mile voyage that lasted from February 25 through May 3, 1955, using a laboratory installed on the Coast Guard Cutter *Taney*.

Operation Troll scientists found widespread, low-level radioactivity in sea water, plankton, and fish samples consistent with the Japanese findings on the *Shunkotsu Maru*. The radioactivity was "greatly diminished in intensity in accordance with known laws of radioactive decay, and through mixing with large volumes of ocean water," according to a Naval Research report.

"The minute traces of radioactivity being found by the *Taney* expedition," the report continued, "exist in proportions predicted by oceanographers. The activity is much smaller than any which would create a health hazard, according to internationally recognized health standards."

Between March 1954 and March 1955, the AEC also carried out multiple surveys of radioactive fallout's impact on plant and animal life on Rongelap. The most comprehensive were done by the University of Washington's Applied Fisheries Laboratory for the commission.

Dr. Lauren R. Donaldson, who headed the laboratory, had a long history with the AEC, having first served as a radiation monitor for Operation Crossroads. He then planned or led radiobiological surveys following the Bikini and Enewetak tests in 1947, 1948, 1952, and 1954. In June 1954, he had participated in the joint Japanese-American scientific conference that dealt with the fallout on the *Lucky Dragon* and its crew.

In a report filed with the AEC in August 1955, Donaldson's Rongelap survey group reported there had been an "expectable decline in radioactive levels" in the months since the fallout, but that "biological uptake and recycling of radioisotopes was proceeding actively." Coconut meat, which measured 1.16 microcuries in late March 1954, was down to .036 microcuries late in January 1955. There were similar declines in arrowroot and pandanus, other mainstays of the Rongelap diet. Some declines still left high radioactivity readings in fish liver.

Two findings had long-range significance. The new leaf buds on trees, formed after fallout, showed as much radioactivity as the older leaves that had received the fallout directly. That meant the tree roots were taking up radioactive elements from the ground and carrying them to the new growth on the tree.

In addition, researchers were surprised to find that while the heaviest fallout had been at the northern part of the atoll, "the radioactive material is being distributed throughout the atoll, at least in the deeper waters." In surveys made in January 1955, birds were found to be as radioactive in the south as they were in the north.

Further surveys were made in October and November of 1955, at a time when there was talk within the AEC of returning the Rongelap people to their home atoll. By then, interest in strontium-90 had grown, and researchers found that it accounted for only 4 percent of total radioactivity on the atoll's islands. However, they also found that in the coconut crab, which islanders considered a delicacy, strontium-90 represented 12 percent of that radioactivity.

The contamination of the coconut crab would be a continuing problem for the Rongelap people in the years to come.

■   ■   ■

The failing health and eventual death of Kuboyama had stalled compensation negotiations for three months. Within that time, public pressure grew for the Japanese government to take care of crew members of other fishing boats, who claimed to have developed radiation disease. The president of the All Japan Seaman's Union sent a petition to the minister of transportation on October 5, 1954, which read, "In precisely the same way as the crew of the *Lucky Dragon #5*, they suffered damage in the course of sailing and fishing. There is no reason for them to be treated differently from the crew of the *Lucky Dragon #5*."

During an October 23, 1954, meeting at the State Department, Foreign Minister Okazaki told Assistant Secretary of State for Far Eastern Affairs Walter Robertson that although he once thought $1 million would satisfy claims, Kuboyama's death and revived fear of radioactive tuna had changed the situation. He said he believed the cost had grown to $1.5 or even $2 million because of fishing industry losses.

Robertson responded by saying he understood the problem facing the Yoshida government, because the nuclear tests "have produced such unfortunate and unexpected results," according to a memo of the conversation. However, Robertson pointed out that US expenditures of "billions of dollars . . . in our atomic program are not only for ourselves but to serve the dual purpose of protecting the human race from Communist enslavement, and also leading toward the peaceful uses of atomic energy with all their great potentialities."

Okazaki said settlement for $1 million "would really end the Yoshida government," with Robertson responding, "Anything over $1 million may possibly require Congressional action and the resulting discussions may possibly lead to bad reactions in United States public opinion."

In December, the Yoshida government did fall. The new Japanese government formed under Prime Minister Hatoyama Ichiro, who had close connections with the fishing industry. From the time of the Bravo test through August 1954, public expectations in Japan about the amount the US would pay rose to $7 million, far above what was being discussed behind the scenes by the two governments. By the end of 1954, the public figure in Japan had reached near $9 million.

The Ichiro government internally recognized it would never come close to the public expectation, but it wanted to reach agreement quickly. To ease concerns about long-term damage, it raised the radiation level for designating fish nonedible by 500 percent and, at the same time, agreed to drop the idea that the US was legally liable for damages.

In a January 5, 1955, settlement note, Ambassador Allison wrote, the US "hereby tenders, ex gratia . . . without reference to the question of legal liability, the sum of two million dollars for purposes of compensation for the injuries or damages sustained as a result of nuclear tests in the Marshall Islands in 1954." Distribution of the fund was left to "the sole discretion of the Government of Japan."

The *Lucky Dragon* crew, who by then had been temporarily discharged from the hospital and were back in Yaizu, considered that an enormous amount, believing it would be distributed to them. That was because Allison's note specifically referred to "each of the Japanese fishermen involved, and for the claims advanced by the Government of Japan for their medical and hospital- ization expenses."

However, after four months and some fifteen meetings among Tokyo gov- ernment agencies, the Ichiro cabinet finally announced how the money was to be split up. Various elements of the tuna industry—boat owners, processors, brokers, and fish market owners, all of whom had suffered severe financial losses during the radiation scare, were to get the lion's share, roughly $1.7 million. The city of Yaizu received $37,000 to cover extra expenses associated with the *Lucky Dragon*, and $71,000 was paid for the hospital and medical bills of the exposed seamen.

As for the crewmen themselves, a pot of $151,000 was split among them— the older, married ones getting the equivalent in yen of a bit more than $5,000; the younger, unmarried men getting slightly less. For each of them it repre- sented several years' earnings at sea. The Kuboyama family received the most, $15,000.

Less than a month after they received their money, the twenty-two remaining crewmen were released from the two hospitals where they had been for more than a year, an event covered extensively by Japanese media. Their doctors considered their recovery slower than comparable cases from Hiroshima and

Nagasaki. Blood counts were still below normal, and some crewmen continued to be easily fatigued.

Arrangements were made for them to be examined annually by Dr. Kumatori, a situation that changed two years later when the National Institute of Radiological Sciences was established and took over their care. The institute's goals were "to study the effects of radiation on the human body . . . study the medical use of radiation . . . and train experts in the prevention of radiation damage to the human body and in diagnosis, treatment and medical uses of radiation."

One of the crewmen, Oishi Matashichi, wrote in his 2011 book, *The Day the Sun Rose in the West*, that although he took part in these annual examinations, "They never told us the results, so we didn't know what they were."

■   ■   ■

The American payments to the Japanese fishing industry and the *Lucky Dragon* seamen attracted widespread attention.

In the Marshall Islands, rumor had it that the $2 million went only to the exposed crewmen, and as a result, pressures grew for the Americans to make good on their promises to compensate the Marshallese.

Since April 1954, when the Marshallese had unsuccessfully petitioned the United Nations for $8.5 million, the islanders had been waiting for the US government to pay them for the damage done to their land and trees, and for the injuries sustained by the people. Amata Kabua, son of the iroij and, by then, an influential member of the Marshall Islands Congress, was a leader in this effort.

Kabua had later filed a lawsuit against the US government.

Late in January 1955, the Utirik people and the exposed Rongelap people who were living on Ejit were asked to fill out forms describing personal property that had been left behind during the evacuation of the islands.

The AEC paid those who presented claims a set value for all their belongings—$1.80 for dresses; $3.60 for men's pants, and so on. For the entire Rongelap group of eighty-two men, women, and children, such claims totaled $5,030.

31 · TELL THE PUBLIC

■ **INSIDE THE AEC,** the Pentagon, and up on Capitol Hill there was major concern that the public would learn about the fallout patterns created by Bravo.

The Joint Committee on Atomic Energy members reacted immediately to a closed briefing on the Armed Forces Special Weapons Project's report, which estimated a fifteen-megaton weapon would create radioactive fallout dense enough over a five-thousand-square-mile area to be hazardous to human life. The committee appointed its own panel to look at the impact thermonuclear bombs would have on defense of North America.

AEC staff members, fearful that the fallout estimate in the Armed Forces Special Weapons Project's report would leak after the Joint Committee briefing, encouraged Strauss to spell out those concerns to President Eisenhower and other members of the National Security Council.

During the summer and fall of 1954, fear arising from radioactive fallout also got stirred up within a small academic community after having been briefly heightened in the wake of the Bravo test impact on the Marshallese and Japanese.

In June 1954, Dr. Alfred H. Sturtevant, chairman of the genetics department at the California Institute of Technology, told an audience at the annual Pacific region meeting of the American Association for the Advancement of Science, there was "no possible escape from the conclusion that bombs already exploded will ultimately produce numerous defective individuals." He also estimated that "1,800 deleterious mutations" had resulted from fallout up to that time.

AEC officials and particularly Chairman Strauss grew concerned over increased public opposition to future nuclear tests. After all, they'd already fought off the internal administration move toward a temporary test moratorium. A November 22, 1954, newspaper column by Joseph Alsop and Stewart Alsop claimed that the AEC was withholding fallout data from the public and called for the AEC to publish their internal studies "however much this may shock such high priests in the cult of Q-clearance [the classification level for atomic secrets] as Lewis Strauss."

On November 30, 1954, after being saluted in the House of Commons for his eightieth birthday, Winston Churchill spoke publicly during a celebration in Westminster Hall. He emphasized his own concerns about radioactive fallout from nuclear tests, past and future, and the effects they would have on the earth and its atmosphere. The fallout message reached a narrow, but important American readership when, in the November 1954 issue of the Bulletin of the Atomic Scientists, Dr. Ralph Lapp published the first of what would be a series of pieces. A former Los Alamos nuclear physicist who also had worked for the Defense Department, Lapp focused on the fallout from the Bravo test as a wake-up call for civil defense planners.

Shortly thereafter, AEC Commissioner Libby took it upon himself to deliver publicly what up to then was considered a definitive statement on radiation hazards from fallout from the new thermonuclear bomb. His audience, on December 2, 1954, was the US Conference of Mayors then meeting in Washington.

Libby was not an alarmist, although he emphasized the short- and long-term differences between blast/heat from a detonation and the aftereffects of radioactive fallout. He said unprotected people would suffer, but that those in shelters outside the detonation area could survive an H-bomb attack. He implied that testing up to then had not appreciably increased worldwide background radiation.

On December 23, 1954, distinguished Manhattan Project physicist Hans Bethe, then teaching at Cornell University, wrote a letter to Libby saying, "There is real unrest both in this country and abroad concerning the long-range as well as short-range radioactivity and it would, in my opinion, greatly allay the fears of the public if the truth were published. I believe the story . . . could be published without giving away any information about our H-bombs; it is merely necessary to put the permissible accumulated yield in terms of fission."

Bethe continued, "Another question is the publication of the fallout data from our tests. I think it would be highly desirable to publish these too, in particular for civil defense purposes," something Civil Defense officials had been pressing for doing inside the government.

In thinking about public disclosure, Strauss was not worried about the impact of what Churchill, Libby, and Bethe were saying. He feared what he considered wilder claims by scientists such as geneticist Alfred Sturtevant; fallout questions raised in their columns by the Alsops; and continued references in articles about radiation linked to the death of Japanese seaman Kuboyama.

The AEC chairman began to believe that the commission had to release another official statement that would tackle the fallout issue head on, if for no other reason than to prevent interference with an upcoming series of nuclear weapons tests. Strauss had been encouraged by a statement Eisenhower had made at a December 10, 1954, cabinet meeting when the president talked about the "virtue of laying all the facts on the line before there is an inquisition."

Seven days later, at a press conference on December 17, Strauss told reporters that the AEC staff was studying the fallout problem, and he hoped a public statement would be released at some future date. He said that three AEC divisions—biology and medicine; military applications; and research—were studying the fallout hazard. However, that day's press corps almost totally focused on Strauss's statements related to a squabble over what was called the Dixon-Yates contract, where the AEC agreed to buy electricity from two private utilities rather than from the federal government-run Tennessee Valley Authority.

Weeks later, after going through five drafts of what was titled, "The Effects of High Yield Nuclear Explosions," the AEC sent a copy to the White House, where it quickly became part of the agenda at several Top Secret National Security Council meetings under the heading: "Proposed Public Announcement of the Effects, Particularly Fall-Out of Thermonuclear Explosions," according to notes of those meetings.

At a February 3, 1955, National Security Council session, President Eisenhower questioned whether "an announcement at this time on the effects of thermonuclear explosions would have an irritating effect on the international situation. In other words, was there any good reason for keeping altogether still on this subject or, alternatively, on saying a great deal more than was now included in the draft."

Ike answered his own question, pointing out that "it was generally understood that the British, among others, tend to exaggerate the effects of fallout," having recently received messages on the subject from Churchill. The British were going to make their own public statement on February 15, 1955, because then-Foreign Secretary Harold Macmillan believed "there was no adequate defense against thermonuclear and atomic weapons," thereby allowing a needed budget reduction on spending for Britain's air defense systems.

The president said he thought "it would seem undesirable . . . for the people of the United States to learn of these effects from the British Government rather from their own Government, the more so since we would probably be obliged to state that the British exposition was substantially true," according to the notes.

Strauss thereafter proposed the US statement be released "on the 11th or 12th of February," with others suggesting simultaneous release with the British. Eisenhower agreed with the latter approach but left it to the State Department to work out the details.

Strauss opened the following week's February 10, 1955, NSC meeting by bringing the group up-to-date on the latest draft, which he said he had gone over with Secretary of State Dulles. Certain revisions had been made, including a decision "to omit the references to the genetics situation since the material on this subject in the report indicated that we knew very little about it." Also omitted was a map of fallout "on grounds that the legend on the map would be too difficult to read as normally reproduced."

The Civil Defense leadership, Strauss said, had been "screaming for months for some such statement" which they thought would help them "press ahead with renewed energy and zeal on its civil defense program." Others, he added, "worry this would snowball the sense of danger, rather than to reassure the population." Defense Secretary Wilson added, according to the notes, "that as far as he could see, the United States' proposed statement wasn't going to be very reassuring in any event." Strauss responded, "It would be reassuring in comparison with so much of the 'scare stuff' which had recently filled the papers on the subject of fallout."

■   ■   ■

At the very time Strauss's fallout paper was being considered by the NSC, the White House was carrying on discussions about a new national security strategy based on nuclear deterrence teamed with a reduction in the size of the country's armed forces.

At the Pentagon, Defense Secretary Wilson had publicly disclosed reduction of some 500,000 service personnel over the coming four years. At the same time, Reserve forces would be increased by training an additional 100,000 men for six months each year so that by 1959 there would be a Reserve of five million men. "Manpower cuts were keyed primarily to the priority given to air-atomic power and only secondarily to the Reserve system," the *New York Times* explained in its December 18, 1954, edition.

Democrats made "an argument to the American people that we are cutting the Armed Services just for budgetary reasons in an attempt to try to save some money," Presidential Press Secretary Hagerty wrote in a January 4, 1955, diary entry. He added, "Of course, nothing could be farther from the truth and the reasons we are taking the steps we have is so that we can put more emphasis on nuclear weapons, guided missiles and the Air Force."

Hagerty explained that Eisenhower often said in private that, "it would have been impossible for him to invade the Continent in the way he did if the Germans had had the atomic bomb. The deployment of troops in the ports of debarkation would have been blown off the earth with that bomb and the great mass of ships coming across the Channel would have been actually dis-integrated. Consequently, in an atomic attack or an atomic war, it is going to be virtually impossible to move land troops to Europe or to Asia, and we will have to depend upon long-range bombers, guided missiles and the like to carry the attack to the enemy and to stop them from coming over here."

The president had pressed for the Reserve increase, Hagerty wrote, because atomic "bombing of our cities is going to create panic and riots within those cities and it is going to be necessary to have a strong ready Reserve to throw into those areas merely to keep order and to get our production going again."

On February 14, 1955, the day before the planned public release of Strauss's fallout report, the Killian Committee, formed three weeks after the Bravo test, delivered its 190-page, Top Secret study, "Meeting the Threat of Surprise Attack," to a special National Security Council meeting.

Based on projected US technological capabilities, the Killian Committee proposed development of what would become the Polaris submarine-launched ballistic missile; acceleration of the Distant Early Warning line across Northern Canada; more resources for the Intercontinental Ballistic Missile (ICBM) program that led to the Atlas and Titan missiles; and creation of the Jupiter and Thor Short-Range Ballistic Missiles (SRBM).

The Killian panel's recommendations also eventually led to military communications and intelligence satellites and the U-2 high altitude spy plane, a CIA-developed reconnaissance aircraft.

That February 14, NSC meeting concluded with Eisenhower approving a Net Evaluation Subcommittee, to provide "integrated evaluations of the net capabilities of the USSR, in the event of general war, to inflict direct injury upon the continental US and key US installations overseas, and to provide a continual watch for changes which would significantly alter those net capabilities," according to notes of the session.

■   ■   ■

The next day, on Tuesday, February 15, 1955, the AEC report, "The Effects of High Yield Nuclear Explosions," was publicly released.

Strauss included a brief statement which emphasized that testing at the Nevada site had created no hazard, but that the fallout in the Pacific had enabled study of the fallout phenomena, which would be useful if an enemy, meaning Russia, were ever to use such a bomb against the US.

"It should be noted," Strauss wrote, "that if we had not conducted the full-scale thermonuclear tests . . . we would have been in ignorance of the extent of the effects of radioactive fallout and, therefore we would have been much more vulnerable to the dangers from fallout in the event an enemy should resort to radiological warfare against us."

The early part of the report described the blast and heat effects of early atomic bombs detonated in the air before discussing fallout from Bravo and other detonations. "In the air explosion, where the fireball does not touch the earth's surface, the radioactivity produced in the bomb condenses only on solid particles from the bomb casing itself and the dust which happens to be in the air. In the absence of materials drawn up from the surface, these substances will condense with the vapors from the bomb and air dust to form only the smallest particles. These minute substances may settle to the surface over a very wide area—probably spreading around the world—over a period of days or even months. By the time they have reached the earth's surface, the major part of their radioactivity has dissipated harmlessly in the atmosphere and the residual contamination is widely dispersed."

The report then turned to what fallout would occur if the fireball hit the ground. "If however the weapon is detonated on the surface or close enough so that the fireball touches the surface, then large amounts of material will be drawn up into the bomb cloud. Many of the particles thus formed are heavy enough to descend rapidly while still intensely radioactive. The result is a comparatively localized area of extreme radioactive contamination, and a much larger area of some hazard. Instead of wafting down slowly over a vast area, the larger and heavier particles fall rapidly before there has been an opportunity for them to decay harmlessly in the atmosphere and before the winds have had an opportunity to scatter them."

It described the Bravo fallout as looking like snow "because of calcium carbonate from coral," and then noted its "adhesive" quality thanks to moisture picked up in the atmosphere as it descended. In the end it contaminated "a cigar-shaped area extending approximately 220 statute miles downwind, up to 40 miles wide," from Bikini. It "seriously threatened the lives of nearly all persons in the area who did not take protective measures," the report said.

The report then addressed radioactive strontium in fallout as having a long, average lifetime of nearly thirty years, noting it could enter the human body either by inhaling or swallowing. Deposited directly on edible plants, the

strontium could be eaten by a human or animal. While rainfall or human washing of the plants would remove most of the radioactive material, radioactive
strontium deposited directly on the soil or in the ocean, lakes, or rivers could be
taken up by plants, animals, or fish. There it would lodge in their tissue where
it could later be eaten by humans.

The report further noted that radioactive strontium-90 fallout from all
nuclear explosions up to that time—both US and Soviet—would have to
increase many thousand times before it had any effect on humans.

The other radioactive element in fallout described specifically as a threat was
radioactive iodine. Even though the average life of radioactive iodine was only
11.5 days, it was described as a serious hazard because if inhaled it concentrated
on the thyroid gland where it could damage cells, depending on dosage.

Overall, the AEC report emphasized that the amount of radiation received
from all US and Russian tests up to that time to an American resident is equal
to "about the exposure one received from a single chest x-ray."

Reflecting Strauss's concern about the next set of Nevada tests, the report
concluded that the degree of risk they presented to Americans "must be bal-
anced against the great importance of the test programs to the security of the
nation and of the free world. . . . None of the extensive data collected from all
tests shows that residual radioactivity is concentrated in dangerous amounts
anywhere in the world outside the testing areas."

Of course, the report's conclusion was based on what was known at that
time. Future findings of medical examinations of Rongelap people would prove
the report was wrong.

■   ■   ■

The *New York Times* on the morning of February 16, 1955, led its paper with
the headline: US H-Bomb Test Put Lethal Zone At 7,000 Sq. Miles. It
added subheads: "Area Nearly Size of Jersey Covered by Atom Fallout After
Bikini Explosion," and "Strauss Warns That Human Survival Might Depend
on Prompt Protective Steps."

Calling it the AEC's "first official estimate of the perils of a fallout of radioactive materials beyond the point of a nuclear blast," the newspaper reported

that the commission had temporarily called off nuclear tests at the Nevada site, which originally had been scheduled for that day and the next.

The *Times* not only published the entire AEC report, which covered almost an entire inside page, but also presented a map—similar to the one the AEC left out—which showed how the Bravo H-bomb, if dropped on Washington, DC, could cause almost a 100 percent lethality rate from cigar-shaped fallout that stretched from the nation's Capitol to Philadelphia.

Other newspapers that day had major stories. The *Los Angeles Examiner* produced a front-page fallout map with that city as the detonation point. The *Las Vegas Review-Journal* carried a page one headline, H-BOMB FALLOUT TERROR IS TOLD.

Despite the nationwide media coverage of the report, it did not have a major impact on the American public. The *New York Times* followed it the next day with a story inside the paper quoting White House Press Secretary Hagerty saying Eisenhower saw the report as a boost for the president's atoms-for-peace proposal, which was hung up at the United Nations. At the same time, Hagerty said it demonstrated to Americans "how, with adequate protection, they could safeguard themselves and their families."

LIFE, then the nation's largest-circulation, weekly, pictorial, news magazine, ran in its February 28, 1955, edition, a three-page spread under the headline: FACING THE FALLOUT PROBLEM. It focused primarily on the civil defense aspects, providing rules for seeking shelter if there were an attack and how to clean up afterwards. Using illustrations, it showed that cellars provided safety where a family "can wait out the danger period in relative comfort." It went on to show a "much safer" shelter "with a lead-lined door leading from the basement and with several feet of earth packed overhead."

The article also suggested that a foxhole-like shelter could be dug after warning, and before the start of fallout, reporting that even a hastily dug foxhole could provide 95 percent protection if persons had some type of cover—as long as it were shaken occasionally to free it from falling radioactive dust. "Turning over turf is necessary to bury radioactive dust which falls on lawns and gardens," was another bit of advice, with the addition that "Buried dust may make future plants and crops radioactive . . . clothing worn during decontamination process should be changed and buried."

■   ■   ■

The next series of American nuclear tests in 1955 were called Operation Tea-pot and were reset to begin the week after release of the AEC fallout report. They were to include some twelve detonations with a variety of yields to meet different military applications. Some were for development of smaller weapons for tactical applications. Others supported development of warheads for land-based, guided, anti-air nuclear missiles. Some were for nuclear air-to-air missiles, and even one for an atomic demolition mine that was to be exploded sixty-seven feet below ground.

Almost all tests had troops in trenches three and four miles from detonations as observers. After two of the detonations, units were scheduled to march into areas close to where the shot took place.

A few tests were of new designs. Pressure had come from the Joint Chiefs, whose plans called for early field deployments of a variety of nuclear weapons in the face of perceived growing threats from the Soviet Union.

On February 21, 1955, the day before a postponed test shot was scheduled to take place at the Nevada Test Site, Sen. Clinton P. Anderson (D-NM), the new chairman of the Joint Committee on Atomic Energy, wrote to Strauss saying that although his panel had "staunchly supported the Commission in its desire for a continental test site, we have also insisted that every necessary step must be taken to assure . . . that no significant hazard to off-site public health should result from test operations."

Anderson, whose New Mexico constituents were nearby the testing area, raised concerns that weather over the Nevada site varied substantially, making it difficult to hold to schedules because of fallout worries. Therefore, he asked for a report that would "consider whether only very small yield devices should be tested there [at the Nevada Test Site], leaving all substantial shots for the Pacific where they can be precisely scheduled."

"Please do not misunderstand me," Anderson continued, "I do not advocate taking any real risk with public health and safety; rather, assuming that the present criteria are necessary for public safety, I am raising the question of whether we can use the Nevada Test Site efficiently for anything other than the test of very small yield devices."

At 5:45 a.m. the next morning, February 22, 1955, the windows in Las Vegas, Nevada, seventy-five-miles southeast of the Nevada Test Site, were rattled by the two-kiloton, Moth nuclear test shot, a Los Alamos Laboratory prototype for a fission air defense weapon. Though Moth's yield was far below what was to come, the impact of this very small shot reached Washington quickly in the form of Anderson's letter. Strauss had the commission discuss it at a meeting the very next day.

Actually, Anderson had been at the Nevada site for the Moth shot. "He was out there. He didn't see anything because he left before the shot," Strauss told the meeting, adding that the senator's letter represented a big change in Anderson's attitude. Strauss then confessed he had begun to cool on testing in Nevada with the arrival of the thermonuclear weapons. Considering the yields in the upcoming schedule, Strauss said, "If I were asked whether the two large shots should be made, and it were left to my sole decision, I would say load them on a ship and go out to Enewetak and put them on a raft and set them off."

However, he pointed out that such a step could entail a six-month delay, so the conversation returned to how to deal with growing concerns about fallout from testing. AEC Commissioner Libby hoped the furor would die down, but Strauss pointed out that while Moth "was a little one . . . they made as much fuss about it as if it had been a big one."

Libby, who had been concerned about delaying the weapons program, said, "People have to learn to live with the facts of life, and part of the facts of life are fallout." Strauss responded, "It is certainly all right to say [that] if you don't live next door to it." Commissioner Thomas Murray, who had argued publicly for halting tests over one hundred kilotons at the Nevada site, said during this closed meeting, "We must not let anything interfere with this series of tests, nothing." The top yield planned was to be a forty kilotons test. Previously, a shot of seventy-five kilotons had been the largest set off at the Nevada site.

After the Teapot tests, there would be none in Nevada for more than a year. The commission decided to write Anderson that a search would begin for another continental site—Point Barrow, Alaska, was one name circulated, but no new site was ever chosen.

32 · RETURN HOME TO RONGELAP

■ **IN JANUARY 1956,** Dr. Robert A. Conard, then a Navy captain, resigned from the service and moved to the AEC's Brookhaven National Laboratory on Long Island, New York, where he took charge of the annual Rongelap medical survey and supervised additional radiation research.

Work on government plans to return the Rongelap people to their home atoll was not halted by the AEC 1956 Rongelap survey, which had found that "radioactivity still was at levels at which permanent residence would have been of doubtful wisdom." Meanwhile, the original eighty-two people from Rongelap, who had been exposed to radioactive fallout, had increased with those who had married into the group, and more who had been born, so that by this time it was over two hundred men, women, and children.

Conard's second medical resurvey of the Rongelap people took place while they were living on Ejit, and he returned to Brookhaven on March 15, 1956. Two weeks later, he reported to the AEC that urine samples showed "some degree of internal absorption of radioactive materials, however, the total body burden was found to be below the tolerance levels that been established." He added that other than some skin lesions and hair loss, "there have been no illnesses or disease processes encountered which could be attributed to radiation effects." No deaths among the group had taken place.

However, Conard wrote that the skin lesions "require careful treatment," and "we cannot be certain that cancer will not develop at the site of skin lesions, but it does not seem too likely at this time." On the more positive side, "fertility did not appear to have been effected in view of the fact that about 10 sound babies have been born in the group since the exposure and new pregnancies are in evidence." He described the people appearing "to be comfortably quartered and well fed on Ejit," but wishing to return to their homes on Rongelap.

Two weeks later, Conard spelled out some problems with a move back to Rongelap Atoll. Writing to Dr. Dunham, who as the AEC's director of the Division of Biology and Medicine was now his boss, Conard explained that the returned Rongelap people "will be almost completely isolated and their only contact with the outside world will be the visiting Trust Territory field trip ship which will touch off there, at most, every three or four months. . . . This will mean that the people will have to depend almost entirely for medical care on their medical aide, a man . . . [whose] training is such that only the barest first aid care can be expected from him."

He pointed out, "Since these people have received significant amounts of radiation, the long-term effects of which are uncertain, and in view of the unique world-wide interest in these people, disproportionate radiological importance may be attached to any disease that may develop among them and any suggestion of negligence in medical attention may be cause of great embarrassment."

For the present, Conard suggested that it would be best for the Marshallese to remain on Ejit where day-to-day medical care would be immediately available, rather than hundreds of miles away attended only by a single medical aide. He recognized that going back to Rongelap was already a US commitment because of the "expressed wish" of the people to return to their homes, and it would soon take place. One proposal Conard made to mitigate the danger was to establish radio communications on Rongelap by either training a Rongelapese in receiving and broadcasting or placing an experienced radioman there. He also wanted air evacuation arrangements for ill persons put in place, through either the Trust Territory or the Navy, and monthly or semimonthly air visits by a physician.

Having made his point, Conard still worried about how returning to Rongelap might impact his study of the long-term effects of radioactivity on the people. Conard wrote, "The group of irradiated Marshallese people offers a most valuable source of data on human beings who have sustained injury from all possible modes of [radiation] exposure. . . . Even though . . . Rongelap Island is considered perfectly safe for human habitation, the levels of [radioactivity] are higher than those in other inhabited locations in the world. The habitation of these people on the island will afford most valuable ecological radiation data on human beings."

Within the AEC there was general recognition that a new phase of the Rongelapese exposure problem would begin once the islanders made it back to their atoll. AEC biologists had reported on the continued radiation in plants and marine life that would provide the basic food for all the returning people.

Some of the men, women, and children among the returnees had been exposed to the initial fallout; others had not. It was decided that the latter group would now be the control group.

The original control group selected in 1954 consisted of 115 Marshallese, principally from the village of Rita on Majuro Atoll. They had background and living conditions similar to those on Rongelap and Utirik, and were matched nearly as possible with each exposed person by age and sex. Blood samples from both groups became the primary basis for evaluation of any hematologic alterations in the group exposed to fallout.

Initial attrition and now relocation to Rongelap would force selection of a new control group from blood relatives who had been away at the time of the fallout, but again selected by size, age, and sex to those who had been exposed. They would be matched as nearly as possible by age and sex with each individual. A large number of returnees were to be children who would become part of the unexposed group that was regularly examined since they would be living in the same environment.

What would be the effect on these two groups after they ate radioactive foods? How could their condition be monitored?

An AEC memo, months later, showed that Conard's suggestions were taken seriously. Preparing for the Rongelap move, Dunham wrote to his boss that a plan for the Rongelapese people would see them "examined once a month and complete medical examination performed once a year by an American doctor. Arrangements have been made for urine collections and analyses every three months for the first year, and afterward on a yearly basis unless the findings indicate the necessity for more frequent analyses. These samples will be collected and the radiochemical analyses made in such a manner as to give an indication of body burden.

"A radio would be provided on Rongelap for communication with the Trust Territories Office on Ebeye (Kwajalein Atoll) where a plane would be available at all times for any emergency. A fully equipped dispensary would be provided on Rongelap and an experienced health aide (a Marshallese) would be present at all times. Before their return, the Marshallese would be given a complete medical examination, and immunized against smallpox, typhoid and tetanus."

Meanwhile, Trust Territory officials had decided that the copra produced by the returning Marshallese would have to be tested for radioactivity before being sold, and if the levels found were questionable, "arrangements would have to be made by the Atomic Energy Commission and the Trust Territory to continue underwriting support of the Rongelapese," according to a June 1, 1956, letter from Acting Trust Territory High Commissioner D. H. Nucker. He also noted the increase in population would require more housing than they had on Ejit, along with additional cisterns to provide adequate water supply.

On June 6, 1956, Holmes & Narver Co., the engineering company that worked for the AEC, was brought in to survey what existed on Rongelap and to prepare plans for rebuilding the village. In addition, a team was set up to do an additional radiological survey to guarantee the safety of the engineers and construction workers, who would be involved in the reconstruction effort and living on Rongelap Island for extended periods of time.

At the same time, the Rongelap returnees were questioned about what supplies, infrastructure, and amenities they wanted for their return home. When completed, in early June, the Majuro District Administrator provided Holmes & Narver with a preliminary statement of their requests "with respect to general criteria of living quarters, bath houses, eating houses, cisterns, storehouses, and other buildings." They also listed land in certain areas of the island for roads.

On July 11, 1956, a group of fifteen Holmes & Narver employees met on Rongelap with John Anjain and four other Rongelapese from the Native Council plus a Trust Territory representative, the latter having flown in from Majuro on a seaplane. Their purpose was to get the Rongelap returnees to establish locations of proposed buildings, so that the construction team could draw plot plans for three villages, one on Rongelap Island and two on nearby islands in the atoll.

It would be months later before revised drawings for the villages and cost estimates were sent to the AEC in Washington for approval and the necessary funding. In October, the Navy's Joint Task Force Seven was involved, since it would supply $1,300 a month to support the returnees for six months, after which that subsidy was to be phased out during the following year.

At that point, Holmes & Narver provided preliminary cost estimates for its activities, put at $276,190. It included an agreement that twenty-four Rongelap "natives" would be part of the labor force. At the same time, the Trust Territories agreed to supply "the necessary pens and cages for shipment of livestock, plus two 30-foot boats," which, at the right time, would be shipped from Majuro to Rongelap.

The initial estimate was that if plans and funding were approved by Washington by November 1, 1956, it would take another six months—from the time materials were first sent to Rongelap, through construction of the villages— for the project to be completed. The tentative plan was that ten days prior to completion of construction, the LST used by the contractors would return to Majuro, then pick up the Rongelap returnees at Ejit, collect their household goods and livestock, and take them to their home atoll for an arrival just after the last nail was hammered.

In early December 1956, another survey of Rongelap Atoll found houseflies and mosquitoes had become prevalent at the proposed construction sites. Spraying the atoll islands took time and delayed the first steps toward preparing the Rongelap Islands for the scheduled return.

On February 27, 1957, after a new radiobiological resurvey, the AEC informed the Navy that Rongelap Atoll was safe for the returnees, and the reconstruction operation was to begin. Internally that month, however, some AEC officials circulated a memo voicing concern that "since the Rongelapese are now subsidized by the United States Government, with little need or opportunity to actively engage in normal livelihood, there is the risk of an onset of indolence, to the detriment of the best interest of the Rongelapese."

It was April 28, 1957, when the LST arrived at the Rongelap Lagoon with construction materials and fifty-one workers. Even then, a demolition party had to dynamite the coral heads which obstructed the planned landing area. In addition, the beach sand proved very soft, requiring two bulldozers, one pulling, the other pushing, to move the loaded trailers ashore.

Three days later, five members of the Rongelap Native Council flew in, accompanied by the Holmes & Narver general construction superintendent, to inspect the newly begun work. They asked that old buildings still standing be demolished and three existing brackish water wells be rehabilitated. The Marshallese then requested the building of new burial vaults for an adult, two stillborn infants, and a four-month-old child.

Following the agreed-upon construction plan, the first structures completed were two warehouses and the Council House. At the end of May, the Interior Department in Washington issued a press release announcing, "Plans are being made for the return of the Rongelap people to their home atoll in about two months." It went on to say that the AEC "has carefully evaluated data from several radiological surveys made during the past two and one-half years. The results of the latest survey indicate the presence of residual radioactivity at a level that is acceptable from a health point of view, both as regards the potential external gamma radiation exposure and strontium-90 in the food supply, with the possible exception of land crabs."

By June 15, 1957, the new villages on Rongelap Atoll were all but completed. Wooden houses with corrugated metal roofs had replaced the old, thatched huts. A cement chapel was built to replace the burned-out church. Across from the new Council House was a new wooden school. Bulldozers had cut a new roadway to the coconut stands. At a final cost of more than $300,000, the US had constructed the finest villages in the Marshall Islands.

On Tuesday, June 25, 1957, an LST at Ejit began loading the returnees and their personal property, something that would take almost two full days. The number of persons had grown to 250, and the cargo included mats, galvanized wash tubs, and new airplane luggage, plus five coffins with the remains of those who had died. There also were thirty pigs, sixty chickens, a cat, a duck, a pet pigeon, and twelve outrigger canoes.

Another 150 others, who either were born on Rongelap or claimed land there, wanted to make the trip, but the LST was too small to take them all.

After their 6 a.m., June 27, departure from Ejit, and an absence of more than three years, the Rongelapese returned to their home atoll. "In spite of close confinement" on the voyage, according to a report written for Holmes & Narver, "the people appeared contented; they were going home. The trip was uneventful, and the vessel arrived at Rongelap at 9 a.m. on the morning of June 29, 1957."

Before disembarking, all the Rongelapese gathered beneath the deck awning. There they offered prayers and hymns of thanksgiving to God for their safe return to their native land.

On land was a sign in Marshallese strung between two palm trees: "GREETINGS, RONGELAP PEOPLE. WE HOPE THAT YOUR RETURN TO YOUR ATOLL IS A THING OF JOY AND YOUR HEARTS ARE HAPPY."

A Navy journalist, Joseph D. Harrington, who visited Rongelap six months after the resettlement, wrote, that "each house was placed in relation to the prevailing winds," and new "solidly packed roads . . . were meant to make travel between the two main stands of coconut, located on opposite ends of the island, much easier." The repaired water cisterns "more than doubled the storage capacity available . . . a great boon to a meticulously-clean people, whose predilection to frequent freshwater baths was curtailed in the past during dry seasons and when old cisterns leaked."

■   ■   ■

Planning for the future medical examination program for the Marshallese included equipment to measure radionuclides within the human body (whole-body counting).

In mid-March 1957, Conard had taken four of the exposed Rongelapese, including John Anjain, and two from Utirik, to the United States for tests that would determine the levels of gamma radiation already within their bodies. The level of beta radiation emitters such as strontium-90 had already been measured by urine tests. This was the first of such trips for exposed Marshallese.

In a pattern that would be repeated for decades, the Marshallese flew first to Hawaii where, in this case, winter clothes had to be purchased for them. They then flew to Chicago, where the Argonne National Laboratory had the sophisticated device—a whole-body counter—which measured gamma radiation in humans.

Normally, those being examined lay beneath a lead box containing a sodium iodide crystal. As the person passed under the lead box, the crystal reacted to the gamma rays in the body by giving off flashes of light that, translated into energy, were recorded to determine the type and amount of radioactivity being emitted.

At Argonne's Radiological Physics Division, specialists reported a series of problems they had encountered in dealing with the Marshallese. The two from Utirik were so overweight that they could not go through the normal machine. The four from Rongelap wanted to be measured sitting up rather than being measured in a horizontal position beneath the lead box.

As a result, they had to be given containers for submitting their excreta, which was measured for biological half-life of isotopes in the bodies. However, according to the Argonne report, "due to sightseeing trips, reporters, newsreel people and other extraneous activities, some samples were lost, and no samples could be regarded as representative of excretion rate over a definite period of time."

Using the data they had, the "results showed that low levels of radioactive elements could be easily identified and measured and were well within the permissible range," Conard later wrote. "The two men from Utirik had radioactive zinc in their bodies, which was later found to have come from eating contaminated fish."

Based on that experience, Conard decided to construct his own portable, whole-body counter and ship it halfway around the world to Kwajalein, where it was stored so that it could be taken to Rongelap for the yearly examinations.

At Brookhaven, Conard also had built a twenty-one-ton steel room, five-feet square and six-feet high, with walls four inches thick to keep out background radiation that naturally occurred from the sun and radioactive elements in the environment. It was mounted on a large trailer bed next to an air-conditioned room filled with the necessary monitoring and recording equipment.

To transport the whole-body counter and the steel examination room from Brookhaven to Rongelap, the Navy, in January 1958, supplied a floating dry-dock ship which sailed to the Marshall Islands by way of the Panama Canal and Hawaii. The body-counter and examination room were lashed down on the forward part of a Navy LST landing craft, and the landing craft was fixed in the drydock area of the larger ship.

When it reached Rongelap Lagoon, the drydock portion was flooded, and the landing craft, with the enormous steel room containing the body-counter on its deck, headed for the shore where it was beached.

Since the whole-body counter was new to the islanders, Conard called on John Anjain, as the island's magistrate, and several other village elders to be the first ones tested to show the others it had no ill effects. Before being tested, the Rongelapese were required to shower in the landing craft's crew quarters, then put on paper coveralls to reduce external contamination.

Each radioactivity test, for the eighty-two exposed Marshallese and an equal number in the control group, took fifteen minutes. Hawaiian music was piped in to relax them as they sat, one-by-one, alone in the body-counter's steel room on the landing craft.

The examinations included getting a medical history along with the taking of blood, x-rays, and, for some, urine and stool specimens. There were special exams given to children, such as hand and wrist x-rays to monitor growth and development. For some adults there was color photography of skin for selected lesions; and eye studies to check if those who saw the explosion directly had any long-term effects.

The examiners used tents ashore plus the school building, dispensary, and Council House. Doctors doing the examinations would also treat diseases that they came across. However, since not all the residents of Rongelap were being examined, there were a few early complaints from among the originally exposed examinees that they could be considered "guinea pigs."

While the medical exams were going on, biologists from the University of Washington collected examples of what was growing on the land, in the lagoon, and at sea—food that the Rongelapese normally would eat. The biologists also took soil and water samples, not only on Rongelap but also on several other islands in the atoll.

As expected, the bodies of the exposed Marshallese showed an increase of absorbed radionuclides since, as the earlier University of Washington report showed, their island had a persistent level of radioactive contamination, according to Conard's report. Overall, the level of cesium-137 rose by sixty times; the level of radioactive zinc by seven times. The strontium-90 levels, which had been slight a year earlier at their last exam on Ejit, were up significantly, but still below the maximum, safe, AEC level. Since local food made up only part of the islanders' diet, the radioactive burden was probably expected to rise even higher in coming years when the imported AEC food subsidy was scheduled to be halted.

For sixteen days, the doctors and biologists examined and sampled. Some Marshallese began to question why all this was being done if the exposed people were well and their atoll was safe.

Conard wrote in a June 1958 letter to Brookhaven colleagues, "I found that there was a certain feeling among the Rongelap people that we were doing too many examinations, blood tests, etc. which they did not feel necessary, particularly since we did not treat many of them." Conard continued, "I got the [Rongelap] people together and explained that we had to carry out all the examinations to be certain they were healthy, and only treated those we found something wrong with. . . . Perhaps next trip we should consider giving them more treatment or even placebos."

▪   ▪   ▪

Dr. Conard and the other scientists, nevertheless, believed the 1958 survey appeared to mark the successful beginning of a new phase of the Rongelap incident, one beyond just caring for the exposed people.

In his report that year to the AEC and passed on to Trust Territory officials, Conard wrote, "The habitation of these people on Rongelap Island affords the opportunity for a most valuable ecological radiation study on human beings. . . . The various radionuclides present on the island can be traced from the soil through the food and into the human being, where the tissue and organ distributions, biological half-times, and excretion rates can be studied."

In a later letter to a Brookhaven colleague, Conard wrote, "The levels are considered safe for habitation. However, the extent of the contamination is greater than found elsewhere in the world and, since there has been no previous experience with populations exposed to such levels, continued careful checks on the body burdens . . . are indicated to insure no unexpected increase."

Conard sent a series of other memos that he made sure were distributed not just to Brookhaven colleagues but to the Atomic Energy Commission and the State Department.

In one, dated June 3, 1959, he said of the Rongelap people, "Even at five years post-exposure it appears that the blood platelets have not returned to the levels of the unexposed comparison group, though there are no other evidences of their acute exposure other than atrophy and scarring of the skin from beta radiation in a few cases. These findings need continued observation. . . . Very little is known of the late effects of radiation in human beings. . . . At this stage of our knowledge, we must assume that any or all of these changes may occur in the exposed Marshallese. The seriousness of their exposure cannot be minimized."

Conard felt a responsibility to continue examining the exposed people, writing, "Any untoward effects that may develop may be diagnosed as soon as possible and the best medical therapy instituted. Any action short of this would compromise our responsibility and lay us open to criticism."

Conard's twin goals of caring for the exposed Rongelap people and gathering data to help understand the long-term effects on humans of low-level radiation would be continually misinterpreted for years to come.

A member of the University of Washington analytic team, Neal Hines, understood Conard's aims and wrote, "When the Rongelapese were once again on their native atoll, there existed both the impulse and optimum conditions for continuing medical studies. The Bikini accident of 1954 not only had created the most powerful obligation to do whatever had to be done to heal and care for the innocent Rongelap victims, but also had provided in its circumstances an unparalleled opportunity to continue the studies that might help avoid a tragedy in the future."

But Conard, from as early as the summer of 1959, had to deal with government colleagues who did not understand what he was trying to accomplish. He wrote to one of his AEC bosses in July of that year about a meeting he had with the then-Trust Territory High Commissioner Delmas H. Nucker, who apparently had no comprehension of the dual importance of the medical surveys.

"I was surprised that Mr. Nucker did not appreciate the fact that the body burdens of the Rongelap people had increased since their return to Rongelap," Conard wrote. Nucker seemed "to take the attitude that we are merely carrying out a scientific experiment using the Rongelap people as 'guinea pigs.' I tried to make our position in this matter clear and to emphasize the importance of continuity in our observations of the people for the early detection of significant changes that may be related to radiation so that they could be treated as soon as possible and that the scientific value of these studies was extremely important, but that this was secondary to the former objective."

To the present day, some Marshallese officials and others still refer to the US using the exposed people as "guinea pigs."

         ■   ■   ■

American government concern and interest over Rongelap began to cool with the results of the 1959 Conard medical survey.

Except for an early, speculative finding that children under six at the time of the fallout seemed to show some growth delays, the medical situation during the 1959 visit appeared to be stable. In fact, before examinations could be started, Conard and his colleagues had to hold a three-and-one-half-hour meeting because the Rongelap people had been told they were healthy and had many questions about why they should have to be reexamined.

Conard explained to them that although they had recovered from the acute effects of radiation, the examinations had to be continued to make certain they remained healthy, since there was little knowledge about the late effects of radiation on human beings. Their main complaint was the repeated blood sampling, which required Conard to explain that studying their blood was important in following their state of health.

They also objected to the continuing ban on eating coconut crabs, and Conard recorded that it was difficult to get them to understand that strontium-90 had built up in the crabs but not in the pigs in Rongelap that they were permitted to slaughter and eat. Some had to be talked out of their claims that recent cases of fish poisoning had been due to radiation.

In the end, 96 exposed people, including their recently born children, and 166 unexposed people were examined during the 1959 Rongelap medical survey.

The US compensation to the Marshallese as a result of the Bravo fallout also had become a major, continuing problem. It was generated in part by the Rongelapese lack of money due to overcrowding, since the atoll's population had continued to grow with the arrival of the exposed people's relatives. That created serious issues with regard to living space and copra production. Since copra was the prime source of outside income, the Rongelapese grew concerned about their inability to get back on their feet economically.

The US government had to expand its delivery of food beyond the time originally planned. One contributing element: Fishing was not being carried on as actively as it should have been.

In 1957, Amata Kabua had again hired the Hawaii-based, American attorney Earnest Wiles in an unsuccessful effort to get compensation for the Rongelap people, even though they were not US citizens and the fallout had occurred outside US territory. A new lawsuit, this time filed in US District Court in Hawaii, sought $8.5 million for "property damage, radiation sickness, burns, physical and mental agony, loss of consortium and medical expenses (past, present, future, and undetermined) by virtue of negligence on the part of the United States in the Bravo detonation."*

At the same time, Kabua got into politics as the US Interior Department began moving Micronesia toward self-government. In 1961, he was appointed to the Council of Micronesia. The Interior Department then established the Micronesian Congress, and Kabua successfully ran for a Senate seat in the first election in 1964.

Kabua and other Marshallese would continue to pressure the US to provide additional compensation for upsetting and possibly endangering their lives.

---

* After that suit was denied, it was filed in the High Court of the Trust Territory, which in January 1961 also denied it on grounds the Trust Territory High Court had no jurisdiction. Nonetheless, Amata Kabua continued to pursue the matter.

33 · MEDICAL ISSUES AND MONEY PAYOFF ON RONGELAP

■ **OVER THE YEARS,** Dr. Robert A. Conard and pediatricians he brought with him to Rongelap carefully watched the slow development of several children who had been exposed to the 1954 fallout. In the survey done in March 1963, the doctors' attention was initially focused on two boys who had been one-year-olds at the time of the fallout.

Both showed early signs of cretinism, a condition of stunted physical and mental growth owing to a deficiency of a thyroid hormone often related to iodine deficiency.

Also of particular interest was the development of a palpable nodule in the thyroid gland of thirteen-year-old Disi Tima, a fisherman's daughter, who had been exposed to the Bravo fallout when she was four years old.

Conard believed the findings about the three children possibly represented the first signs of long-term radiation effects, and he sent the thyroid nodule for laboratory examination. He also had called for an AEC restudy of the initial estimate of radiation dosage absorbed by children—an estimate that had been made from radiochemical analysis of a collective urine sample taken on Kwajalein in 1954, fifteen days after the fallout began.

That first study, nine years earlier, had detected low levels of radioactive iodine, but missed some shorter-lived iodine isotopes. Conard's restudy put the radiation dose to a child's thyroid gland at 1,050 rads, a level high enough to cause eventual trouble.

The thyroid is a butterfly-shaped gland, slightly less than an inch in adults and smaller by half in children. It rests on either side of the windpipe and is attached below the larynx by a thin tissue. The thyroid is one of several glands that regulate the chemical content of blood, its principal job being to pick up iodine that enters the bloodstream and use it in making the thyroid hormone called thyroxin.

An excess of thyroxin results in over-activity—fast heartbeat, high blood pressure, and rapid cell growth. A deficiency of thyroxin is marked by sluggishness, slow speech patterns, a thickening of the skin, and, in children, cretinism.

The two Rongelap boys both showed delayed growth patterns and an inability to concentrate on work, whether at school, as was the case with one boy, or at home, as was the situation with the other.

Early in 1963, Rep. Wayne Aspinall (D-CO), chairman of the House Interior Committee, had introduced a Kennedy administration compensation bill for the Rongelapese that called for the establishment of a trust fund of $1 million with each of the eighty-two exposed Rongelap people having an equal share. Under the legislation, an individual would receive each year the interest accrued on his or her share. It also said that a recipient could withdraw his or her share of their principal if needed upon special request to the secretary of Interior, at that time Stuart Udall, who under the draft legislation would control the overall trust fund.

In formulating that draft of the legislation, Interior Department officials considered the obvious problem of giving roughly $11,000 directly to men, women, and children who had no experience in handling that amount of money and whose society did not normally deal in cash. The approach followed what had been done six years earlier, in 1957, for the Bikini people. There, a $300,000 trust fund had been established six years earlier to pay the Bikinians for use of their atoll for nuclear tests. That trust fund's interest yielded an insignificant $15 per person, per year.

While the 1963 medical survey was still under way on Rongelap, Kennedy administration officials in Washington pressed for quick passage of the Rongelap compensation bill.

On March 14, 1963, Assistant Secretary of the Interior John A. Carver Jr. sent a statement to the House Interior Committee that said, "There is, to date, no evidence of leukemia or of radiation illness. Further, whether or not the radiation has any life-shortening effects is not yet apparent."

Carver did point out that bone development in some Rongelap children might have been restrained and there had been "somewhat greater incidences of miscarriages and stillbirths among the exposed women" than normal. Correctly, Carver noted, "We cannot say with any certainty that there will be not future illness or death and no diminution in life expectancy which can be attributed to the 1954 fallout."

The draft legislation also said, "A payment made under the provisions of this act shall be in full settlement and discharge of all claims against the United States arising out of the thermonuclear detonation on March 1, 1954." In the administration's presentation to the House Committee, the fallout from the Bravo test was always described as an accident which occurred when the winds changed. No negligence or error on the part of Joint Task Force Seven officials was ever hinted, and no legislator asked questions about the original cause of the radioactive fallout that reached the Marshallese on Rongelap.

A week after Carver's testimony, the bill passed the House Interior Committee and on March 21, 1963, it was unanimously approved in the House by a voice vote and sent to the Senate. The State Department wanted quick action before the subject of Rongelap was to be discussed by the UN Trusteeship Council during its May 1963 session.

The Senate, however, delayed taking up the Rongelap measure until 1964. Meanwhile, Amata Kabua did not like the provision that placed the money in a trust fund to be controlled by Interior Secretary Udall. He wanted the US payments made in a lump sum to the exposed people. Kabua conferred with Earnest Wiles, the Hawaiian lawyer, who then got in touch with Washington lobbyist Paul Aiken. Wiles and Aiken for several years had been working together on Marshallese land claims.

Aiken had been an assistant postmaster general in the Truman administration, and during the Eisenhower years he had built up a Washington law practice that was aided by his fundraising for Democratic candidates and their party.

Back in Rongelap, Conard's 1964 medical examinations had begun, and doctors discovered that the two boys, whose thyroids were less developed, continued to fall behind their peers. In fact, both boys were now shorter than their brothers with whom their mothers had been pregnant at the time of the fallout.

Three other children, who had been under ten at the time of the fallout, also showed measurable stunting in skeletal growth.

However, most attention during the 1964 examinations quickly focused on the thyroids of some other children. Nearly ten years after initial exposure to radiation, they were beginning to show problems. Disi's thyroid had enlarged slightly, and two other exposed girls, one 13 and the other 14, were found to have similar nodules.

Conard decided surgery would be the best method of discovering whether the growths were benign or malignant.

When the 1964 exams concluded, Conard, using a translator, told the three girls and their parents that their thyroid nodules should be removed and examined. He told them there was no way of knowing for sure that radiation had caused the growths, but an operation was necessary to find out. They agreed without really understanding the purpose, other than that the Americans wanted the surgery to happen.

It took months to arrange. Such special surgery for Trust Territory inhabitants was normally done at the US Navy hospital in Guam, some five hundred miles west of Rongelap. It was August before the first operation took place on thirteen-year-old Irmita. With Conard back in the US at Brookhaven National Laboratory, Navy surgeon, Capt. C. A. Broadus took responsibility and handled the operation on the assumption that a malignancy could be involved. Therefore, he removed Irmita's entire thyroid, and not just the nodule.

Tissue examined by a pathologist showed the nodule was benign, but because the entire thyroid had been removed, Irmita would have to take synthetic hormones her entire life to make up for those once produced by her thyroid and her parathyroid. The two other operations just dealt with the nodules, and both turned out to be benign.

Unaware of the new medical discoveries, the Senate Committee on Interior and Insular Affairs subcommittee, chaired by Sen. Frank Church (D-ID), held a hearing on the compensation bill in spring 1964. Kabua testified as a member of the Council of Micronesia, vice president of the Micronesian Congress, and also president of the Marshall Islands Import-Export Company. The latter was a business he had recently founded and an enterprise that was looking for additional capital. At the hearing, he was accompanied by John Anjain's brother, Jeton, a dentist, who spoke English more fluently than Kabua.

Kabua told the senators that before he and Jeton had come to Washington, the Navy had flown them to Rongelap for a meeting with the people there, and the group that they met supported passage of the House bill. Senator Church pointed out that the subcommittee had a substitute bill for the House-passed measure, which "would eliminate the trust fund provision and make direct payment to each individual entitled." Church then asked Kabua for his opinion of the two bills.

"Well, Mr. Chairman," Kabua responded, "I think the wishes of the people are that they want to receive the money directly, rather than have it in a trust fund." Church then asked if Congress "will be faced at some later date with further claims?" Kabua replied that if he were referring to the same people, "No sir; definitely not."

Church ended by saying he wanted to move the long-delayed legislation along and that Kabua's testimony "will be most helpful to us."

Later at the hearing, Sen. Gaylord Nelson (D-WI) asked, "Do the individual recipients understand the value, all of them, of that $11,000 [they are to get under the bill]?"

Kabua answered, "I'm sure they do."

In mid-August, the Senate approved authorization of the compensation bill with the $950,000 to be divided equally and paid in cash to the exposed Rongelapese after payout of "reasonable attorney fees for legal services . . . not to exceed five percent" of the overall appropriation. The measure also said the Interior secretary "shall give advice concerning prudent financial management to each person receiving a payment pursuant to this Act, to the end that each such person will have information as to methods of conserving his funds and as to suitable objects for which such funds may be expended."

The House quickly accepted the Senate version, and President Lyndon Johnson signed the bill into law on August 22, 1964, at a time when the three Marshallese girls were recovering from thyroid operations.

The 1965 medical survey substantiated Dr. Conard's earlier concern that thyroid growths represented the first long-term effect of radiation. Nodules of the thyroid gland were detected in two boys, twelve and seventeen years old, and in Mijjua, John Anjain's wife, who had been thirty years old at the time of the fallout and was forty-one when the nodules appeared. Examining her, the doctors had felt a number of small nodules in her thyroid that were not soft and malleable, but hard—a characteristic associated with malignancy.

On the trip home after the examinations, Dr. Conard consulted specialists about the thyroid issue and decided that surgery on the three new patients should be done by Dr. Bradley P. Colcock, one of the nation's leading thyroid surgeons, then on the staff of the Lahey Clinic in Boston. Colcock had dealt with a similar problem in the US where American adults, who as children had been given x-ray treatment for adenoids or scalp infections, had developed thyroid nodules.

In June, Mijjua, the two boys, and a translator travelled first to Brookhaven, where Dr. Conard examined them, and then to Boston. The operations on the two boys went easily. The nodules in both cases were limited to one lobe of the thyroid, and, once removed, the growths proved to be benign.

With Mijjua, however, while she was still on the operating table, the hardened nodule was quickly found to be cancerous. Dr. Colcock decided to take out almost her entire thyroid. Adjacent lymph nodes were dissected, and although one contained malignancy, it was determined the cancer had not moved to adjacent blood vessels.

Mijjua and the two boys convalesced in a Boston hospital without complications, and two weeks later they returned to Rongelap.

In September 1965, Dr. Conard made a special trip to Rongelap to explain the need for the more heavily exposed people to take a synthetic thyroid hormone pill. Under normal circumstances, the pills would have been given to some and not to others. Then observations would have been made to determine whether the treatment was working or not. That traditional course of random clinical trials was ruled out on ethical grounds—the doctors did not want to be accused later of experimenting with any of the Rongelap people.

During this trip, Dr. Conard examined Mijjua and the two boys and others. He found possible thyroid growths in two more teenage girls who had been one-year-olds at the time of the fallout. He tried to explain to the Rongelapese what was happening with these new thyroid cases and their relationship to the "poison" that the people now saw as the cause of all their ailments. The islanders listened to the translation of what Dr. Conard said and nodded their apparent understanding and agreement, he would later write.

Most, however, failed to take the thyroid pills he left with them.

■   ■   ■

For Amata Kabua, there was the matter of money. Besides being a politician, he was a businessman, always on the lookout for profit. One American government official called him "the merchant prince."

If the exposed Rongelap people got funds from Washington, Kabua's father, as iroij, would by tradition get a percentage of what any of his people received.[*]

On April 29, 1965, Congress finally approved an appropriations bill that contained $950,000 as compensation to the Rongelap people for the fallout from the Bravo test. The Trust Territory administration then decided to send a group to Rongelap in December 1965 to explain the payments and determine how the money was to be distributed.

On December 7 and 8, 1965, Robert C. Shoecraft, attorney general for the Trust Territory, met on Kwajalein with his Marshall Islands district adminis-trators and Al Yamamoto, of the Bank of Hawaii's Kwajalein branch, to work out details for the fund transfer. The $950,000, less than 5 percent for the Washington lobbyist, left $902,500 for distribution. Shoecraft advised that each exposed adult and minor would be entitled to about $10,500. He suggested Congress intended minors to receive their money only after reaching the age of twenty-one. Thus it was incumbent upon the Marshallese to devise a plan to accomplish that. The proposal called for each minor's parent to be given nearly $500 in cash at the time other payments were made; and that $10,000 would be put in a savings account in the Hawaii bank in the minor's name.

---

[*] The negotiations over Kwajalein took more than ten years, but eventually, in 1964, the Micronesian landowners, including Lejolan Kabua, settled with the US government for an annual $750,000 for a 99-year lease. That came down to $10-per-acre-per-year for a handful of Marshallese landowners. As iroij, Lejolan Kabua got a percentage of their shares.

The next day, Shoecraft and the group flew in a Trust Territory amphibious plane to Rongelap, accompanied by Iroij Lejolan Kabua and his son, Amata. Their purpose was to announce the compensation money and explain how it was to be distributed. Not surprisingly, they were greeted as they came ashore by village women attired in colorful new dresses, there to give flowered leis to the visitors.

At a Council House meeting, with the entire population present, Iroij Kabua announced the good fortune he was bringing them. Shoecraft then read the entire 1964 law as District Administrator Oscar DeBrum translated it into the Marshallese language. The people seemed to understand what was happening. Shoecraft said that adults were going to be interviewed to determine how each wanted his or her money paid—all in cash or part in cash and part in a bank deposit. Remember, the Rongelapese for the most part had lived in a barter economy. The idea of money—in fact American dollars—was totally new to most of them.

At the conferences, payment decisions would be made. At the end, each individual would be asked by Shoecraft their name, age, whether they had been living on Rongelap on March 1, 1954, and whether they had been injured by radiation. Finally, each would have to sign a statement as to the method of payment under Public Law 88–485. Some eighty claimants, including the heirs of thirteen people who had since died, were interviewed that day. Two others were interviewed later that week on Ebeye and Majuro.

On February 24, 1966, Shoecraft arrived by seaplane at Rongelap, again bringing with him Iroij Kabua, his son Amata, DeBrum, and Bank of Hawaii's Yamamoto. Washington lawyer/lobbyist Aiken and his wife were also part of the group. The visitors sat at a table at one end of the Council House while the people sat on benches facing the table. For two days the slow process of payment went on.

A name was called, and the individual came forward. Amata and DeBrum acted as translators. Each person would identify himself or herself and sign for the payment. Reflecting language in the authorization statute, the signers acknowledged that the money they received "shall be in full settlement and discharge of all claims against the United States arising out of the thermonuclear detonation on March 1, 1954." It is doubtful those who signed at that time understood what they technically were giving away.

Adults who wanted cash were told by Shoecraft and Aiken about the benefits of putting at least some of the money in the bank. Most took their cash. Many stopped by the iroij before they returned to their seats and gave him a $1,000 share of their money. After the payments were completed, Amata Kabua convinced several of the recipients who had taken cash to invest some in his trading company which imported products for Micronesians.

Less than a month later, Dr. Conard and a team of physicians arrived for the 1966 medical survey. It was completed in March, and Conard reported in an April 1, letter to Wilfred Goding, then-high commissioner of the Trust Territory, that there were five more exposed people with thyroid abnormalities believed to be due to radioactive fallout. That made the total eighteen altogether. The newest five included a mature woman of forty-six, another woman of twenty, and three teenage girls who were thirteen, or one-year-of-age at the time of the fallout.

They were later flown to the US and operated on in Boston in midsummer by Dr. Colcock. None of their nodules turned out to be malignant. Dr. Conard began to think that the malignant nodules found in Mijjua earlier had been an exception, and that early removal of the nodules would prevent their becoming malignant.

When Dr. Conard returned to Rongelap in March 1967 for that year's survey, he discovered only one new thyroid nodule case, this time in a seventeen-year-old boy who, because he had been away in an outlying atoll, had not been examined for several years. However, the new case became the sixteenth of nineteen exposed children who were ten years or less when the radioactive fallout arrived. Conard also reported that in two cases, where softness in nodules had earlier been found, there was regression in size apparently due to the children taking the thyroid pills.

Conard's 1967 visit also showed the impact of the year-earlier distribution of cash to the exposed people. Trading ships, which had previously avoided Rongelap because its copra production was low and the people had little money to spend, now made it the first stop. Newly purchased motorboats raced about in the lagoon, while no fewer than six small trucks ran along the island's six-mile road. Outside many houses stood new kerosene-run refrigerators and automatic washing machines, all of which required fuel that the people had to buy.

The Rongelapese were also buying more canned food—so much so that, because they had been eating less food from the irradiated local plant and marine life, it led to a drop in radionuclide levels in their bodies from the previous year.

Conard also saw that the people on Rongelap were paying little attention to producing copra, and, though they had motorboats, they hardly ever fished for food.

There did seem to be a positive side. Conard observed in a later report that newly purchased garbage disposals had reduced the Rongelap fly population, and children had begun using the outhouses rather than the beaches for defecation.

However, there was new disgruntlement among some islanders. Those who a year earlier had invested in Amata Kabua's trading company learned that their funds had been lost. They became jealous of those who still had money. The Rongelapese who had not been exposed to fallout, and who were regularly examined as part of Conard's medical survey, began to ask why they should not be paid.

This desire for money led some people to leave for Ebeye to seek jobs on the American missile-testing base on nearby Kwajalein,

One of those who had moved to Ebeye was Magistrate John Anjain. He and Mijjua had separated before her 1965 cancer operation. She had subsequently lived with different men, in the Marshallese fashion, and John had found a new wife. As a family, John and Mijjua and their three sons had received more than $50,000 from the compensation bill. The boys' money was left in the bank to be used for their schooling. John had been one of those to pay $1,000 to the Iroij, but he used some of his remaining money to purchase part-ownership of a store on Ebeye, which he then helped run.

In 1968, however, two new thyroid cases were found in children who were less than ten during the fallout. That meant eighteen of nineteen children under ten in 1954 when the fallout came had thyroid issues. In addition, a decision was made to send to the United States for evaluation three individuals whose thyroid nodules had not responded to use of pills, plus a twenty-nine-year-old woman with a tumor alongside her thyroid.

That pattern continued in 1969 when Conard reported a new thyroid case in an adult woman and three other cases in children who had not responded to thyroid pill treatment.

The September 1969 surgeries were conducted at Cleveland's General Metropolitan Hospital by Dr. Brown M. Dobyns, who had accompanied Conard during the examinations on Rongelap earlier that year. One of the four Marshallese was Morje, a twenty-one-year-old woman, who was six years old at exposure and at the time had received serious beta burns. She had a small thyroid nodule in 1965, which appeared to vanish, then reappeared. For a time, she said she had lived on a northern island of the Rongelap Atoll—where radiation had been higher—and had not been taking her thyroid pills.

When Dobyns operated on Morje, he found that cancerous cells had invaded two adjacent lymph nodes and several distant ones. Faced with this metastasizing, Dr. Dobyns performed a radical dissection on the left side of her neck in the most extensive surgery carried out on any of the fallout victims.

Morje became the first child to have cancer found in the thyroid nodules from the group of heavily exposed children. Surgery affecting the nodules of others up to then had been found benign. "These recent findings greatly increase the concern about radiation-induced neoplasms in this population," Conard wrote.

By the end of the 1969 examinations, Dr. Conard had seen enough. With colleagues, he put together a 135-page report covering the first fifteen years of examinations. In it he wrote, "The significance of radiation exposure of the thyroid glands in the Rongelap people had not been fully appreciated until the appearance of thyroid lesions. More careful review of the dose calculations indicated that considerable exposures from radioactive iodine absorption had probably occurred, particularly in the children."

Reconstructing what had occurred during the fallout, scientists decided the main source of iodine ingestion was water, and since it was rationed over the two days before the exposed Marshallese had left Rongelap, it was assumed both children and adults drank the same amounts. If both adults and children had the same amount of radioiodines, the smaller size of children's thyroids meant they had received a larger dose.

"The [overall radiation] dose to the thyroid glands was greater than that to other organs by a factor of two in adults and a factor of about seven in children," according to Conard's 1969 report. In adults, that meant a dose calculated at 160 rads. However, scientists estimated the dose "from the various iodine isotopes to the child's gland was about 1,000 rads, with a range of 700 to 1,400 rads." according to the report.

Another important conclusion was that the latent period for development of the Rongelap cancers was seven years in one case (John Anjain's former wife Mijjua) and fifteen years in the case of Morje. Conard warned, "It may be that we are just reaching the critical period in the post-radiation observations."

34 · FALLOUT AND TRUTH

■ **IN EARLY OCTOBER** 1967, a three-day meeting called the Second Interdisciplinary Conference on Selected Effects of General War was held in Princeton, New Jersey, sponsored in part by the Defense Atomic Support Agency. The first such conference, in January 1967, had gathered a distinguished group of current and former government scientists to talk about the effects of the atomic bombing of Hiroshima and Nagasaki.

This second, three-day conference, again with a similar group of scientists, discussed fallout from nuclear testing with a focus primarily on the Bravo test and resultant radioactive contamination of Marshall Islanders and Japanese fishermen.

Participating in the conference as discussion initiators for three of the five major subjects on the agenda were familiar names: Dr. Charles L. Dunham, to lead discussion of the 1954 Bravo thermonuclear test; Dr. Robert A. Conard to discuss the effects of fallout on the Marshallese populations; Dr. Lauren R. Donaldson to talk about ecological aspects of weapons testing; and Dr. Merril Eisenbud to discuss psychosocial reactions to nuclear weapons.

The discussions, as would be expected when scientists gather together, broadened out to include much more than fallout, plus the medical and psychosocial aspects of radiation exposure. Just one day's transcript ran 423 pages.

No surprise that the participants at times went beyond simply reviewing past scientific circumstances and events to include questioning why there was so much public misunderstanding about radiation, and whether it was caused by government officials, who had trouble telling the truth in difficult situations.

For example, some participants agreed that the government had not told the truth initially about what had happened to the Marshallese after Bravo, nor had they been honest about the impact of radioactivity in fallout.

One such exchange began when Conard complained, "When you talk to a [public] group, it's obvious that they just don't understand the simplest things about radiation."

Dr. Theodore (Ted) Taylor, a conference participant, responded, "I claim they haven't been helped by the official [government] spokesmen, at least in the United States." Taylor, back in 1967, was then a forty-year-old, theoretical physicist who had been a nuclear weapons designer at Los Alamos and deputy director of the Defense Atomic Support Agency. He was just beginning to become an antinuclear critic. The US government, according to Taylor, had always said, "Don't worry. We know what's being done." After the Bravo test, when Marshallese had been seriously irradiated, Taylor said government statements should have been, "Obviously we didn't know what the hell we were doing."

Taylor continued, "This has happened so many times. We deny the fact that we didn't know what we were doing, but there is no basis for confidence any more. I think that is central. I think that this central fact, that the public has, on the basis of the record, a positive lack of confidence in what they are told."

Dr. Wright H. Langham of the Los Alamos National Laboratory, referring to the fallout from Bravo, asked: "Why isn't it fashionable to admit a mistake when it involves radiation? Do you mean to tell me that the greatest nation in the world can't say, 'Okay, we made a mistake?'"

Dr. Frank Fremont-Smith, director of the New York Academy of Sciences Interdisciplinary Communications Program and co-chairman of the conference responded, "We can do so anywhere except in radiation. That is holy. That is part of our religion. We are the radiation people and we don't make mistakes in radiation."

Fremont-Smith added, "We're not the only government that didn't know how to handle a radioactive accident. . . . But the first thing to do on the government's part is to deny anything dangerous has happened, which is almost standard procedure, and then gradually it leaks out, whereas actually this is the way people lose faith in the government. The credibility gap gets bigger and bigger, and I think certainly this is true in this country."

Dunham, a radiologist who had been head of the AEC's Division of Biology and Medicine at the time of Bravo, joined in to discuss its impact: "People had begun to be aware that there is such a thing as fallout, but they didn't have any real feel for it, and I don't think the military did either. Certainly I didn't."

There was the following exchange about truth telling:

> **Taylor:** "It seems to me it's a very, very important fact of life that the worldwide public has lost confidence in the official spokesmen of the government of several nations as a result of consistent denial . . . "
> **Fremont-Smith:** "Of the truth."
> **Taylor:** " . . . of the truth by spokesmen for these governments . . . and that's the state of affairs that now exists."

Fremont-Smith came back saying: "Then we are also talking about the credibility gap between the younger generation and the adult generation in any country, which is part of the same thing. We have lied to the youngsters repeatedly, again and again, and the youngsters don't have any confidence in the adult world. I think it's a very broad problem we're talking about. This may be true in a good many other countries, too."

Eisenbud, the environmental scientist, complained about the government failure to investigate events surrounding the Bravo test. He said that after Bravo, "The next test shot was delayed 36 hours . . . [but] right up to the last minute, with the fallout . . . on the ground, the people just didn't go up to investigate." Eisenbud said he thought that was because Bravo "proved what a lot of people had suspected; you can have massive fallout following a surface detonation of a megaton bomb."

He pointed out that the AEC "had recommended an evacuation capability up there [in the Bravo test site area] and this was denied [by the military-dominated, Task Force commander] on the basis that it wasn't necessary, that there would not be any fallout; that there just couldn't be enough fallout to warrant keeping ships on station so that they could evacuate natives on short notice."

Eisenbud went on, "There was just a complete breakdown as far as information was concerned, in taking the steps that were necessary in order to evaluate the situation, and to take the necessary palliative measures." He noted that even at the time of this discussion, twenty-three years after the event, there had been no full-blown investigation of Bravo and its aftermath. At one point he noted as "surprising" that there had not been an official US government report on the *Lucky Dragon* incident. Referring back to the time in 1954 when he had been sent to Japan, he said, "I never wrote a report on my own experience in Japan beyond the first two weeks because I just waited and waited, presuming I was going to be able to fit it into some sort of overall report."

He specifically focused on the pre-Bravo meteorology report, which he said described the development of wind patterns starting a day or two before the test and running right up to shot time, which was not available. "I asked for it before I came down here and it's still classified," he said.

Talking about fallout patterns, Eisenbud said those available were based on early, very sketchy data collected by two or three individuals, and that certain isodose curves (lines on a map showing different radiation levels from the detonation point) were drawn which were, "at best, approximations."

Eisenbud continued, "Those of us who have had the experience of actually measuring these fallout patterns from smaller weapons find that they are not quite so uniform, that they tend to be amoeba-like and are harder to find. There arose out of this experience the need for an experiment which would make it possible to get better approximations of the total amount of debris that falls out . . . physical and chemical characteristics. This wasn't done, and as far as I know hasn't been done in any other subsequent explosions during the period when they were still testing in the Pacific. I think that, from the point of view of national security, we are without information which is badly needed. . . . It may be totally useless in the sense that there may not be, even with the present information, a satisfactory answer to all the complications of mass fallout and the way it would interact with blast."

Having made those points, Eisenbud said he worried about "underestimation or miscalculation of the magnitude of the nuclear event and its psychosocial consequences. What motivates so many people to deny facts, when they are so readily available? . . . One wonders if there isn't something in our national culture which makes us prefer getting on and moving, rather than waiting and listening and finding out. . . . Clearly evident facts will be denied up and down and proved not to be so by other authority. . . . There is a lack of a logical approach to the realities of the problem that can be counted upon no matter where we stand."

Eisenbud was not finished. During another discussion, he put forward his own view about thermonuclear weapons: "I think that this event [the Bravo test] is one of the really few historical events in all of history. We woke up one morning and found that we had bombs that could be exploded if we knew how to use them. It threw our government into a turmoil. . . . They knew they had to say something, but couldn't decide what to say."

Referring to the AEC's February 15, 1955, statement on fallout, Eisenbud continued, "It took a year [after Bravo in March 1954] for your government to formulate a position. This wasn't because they were dismissing it, or that this wasn't important, but it was because they couldn't agree on what their actual position was."

After a pause, Eisenbud said, "But, you see, there's one element that hasn't been brought out. That is that anyone could take that diagram [of fallout created by Bravo] and lay it on a map of Europe, let's say, by putting Bikini near some important Soviet airbase, and point the wind anywhere you choose to, and get 800 rads per hour running through friendly nations. This is why I say we have bombs which we are probably no longer in a position to use; imagine the impact of this possibility militarily."

Dr. John A. P. Millet, a psychiatrist and psychoanalyst at New York City's Columbia University's College of Physicians and Surgeons, and consultant at the New York Psychoanalytic Clinic at Columbia, followed Eisenbud, saying, "We've been talking about our dissatisfaction with leaders for not giving us the information that we ought to have. . . . I think we're getting into the area of the mystique of the leader in this country, and perhaps one of the great problems hasn't been touched upon sufficiently yet, which is that our leaders are not sufficiently well educated to know what to think, and therefore, how to act or what to say. They are constantly changing their minds from one position to another, which is one of the problems . . . due to their political needs and their careers.

"It seems to me we have two ends to work on here," Millet continued. "How to get correct information that is capable of solving problems to our leaders, and how to educate the public. Now, if the general public doesn't want to be educated, this is something we've got to know. . . . We can only do a limited amount in getting them interested in the world in which they live.

"On the other hand," Millet said further, "the leaders are certainly very interested in the world in which they live. Perhaps this is the primary goal for our efforts, to try to get the proper knowledge to our leaders."

Dunham asked, "Isn't it one of the fundamental problems that leaders, almost by definition, are amateurs? They've never faced a particular crisis until they face it?"

Fremont-Smith responded, "That's right."

Dunham went on, "This is a dilemma that the world has been facing for a good many years, and I don't know how you can just suddenly say that these people are more stupid than somebody else. It's a personal problem, as you hinted at."

To which Fremont-Smith said, "And the thing is partly compounded by the election every two or four years, which means leaderships change, or there are desperate efforts to maintain leadership at any cost, because that's the time you'll be able to really show your responsibility, after you've been re-elected."

When Dr. Conard's conference session arrived, he turned the discussion to the medical condition of the Marshallese, but also talked about how their situation—living on a coral atoll when the fallout arrived—could differ from other areas subject to a nuclear attack.

"Characteristics of a particular fallout situation depend on many factors," he said, "such as whether the bomb is detonated over water, underwater, over land, the geography of the terrain, the populations exposed, time of fallout arrival, length of fallout, etc."

He explained, "Fallout effects are somewhat different from those produced by direct effect of the bombs. In Japan, for instance, [where the nuclear explosions were at an altitude high enough so the fireball did not reach the ground] the major casualties came from blast and heat, with fewer casualties from radiation exposure. Whereas with fallout, [as with the Marshallese] it was a purely radiation exposure situation."

"In Japan there were psychic trauma, physical trauma, starvation, disease and many complications," Conard said. "In the Marshall Islands, the Marshallese people had a minimum of these factors involved. In addition, the fallout produced a more complicated type of radiation exposure, in that you had not only whole-body exposure, but also the exposure of the skin and internal deposition of radioactive materials."

Two years later, Dr. Conard spelled out in detail his 1969 medical findings about the Marshallese in a publication entitled "Medical Survey of the People of Rongelap and Utirik, Thirteen, Fourteen, and Fifteen Years after Exposure to Fallout Radiation."

"By far the most significant of the late effects of fallout exposure noted in these people is a high incidence of thyroid abnormalities," Conard wrote. It was 1963, nine years after exposure to fallout, that a thyroid nodule was first detected in a twelve-year-old girl. By 1966, eighteen cases had been detected of whom fifteen were in children exposed to the radioactive fallout when they were less than ten years of age. The others were adults when the fallout happened. Of the total eighteen, sixteen involved nodular glands, one of which proved cancerous. The two other cases, the two developmentally delayed boys mentioned earlier, resulted in complete atrophy of their thyroids. No such thyroid abnormalities were found in the unexposed Rongelap children, or the slightly exposed Utirik children, and only a low percentage of nodules were found in the older, unexposed population.

Three more thyroid nodule cases were found between 1966 and 1969. One, discovered during the 1968 survey, was Lekoj Anjain, son of John Anjain and his wife Mijjua, who was a one-year-old at the time of the fallout. He had his benign nodule removed surgically in Boston later in 1968.

Overall, Conard's fifteen-year report said of sixty-six exposed Rongelap Marshallese living at that time (from the original eighty-two initially exposed), three had developed malignant thyroid lesions, and sixteen benign thyroid nodules. The three malignant lesions of the thyroid in the heavily exposed Rongelap people appear to be the first such cases clearly associated with radio-iodine exposure.

Until development of many of these abnormalities in 1965, the Marshallese people were considered to have no thyroid gland issues. After 1965, the remaining exposed Rongelap population was placed on thyroid hormone treatment.

Dr. Conard concluded his fifteen-year report, "These findings indicate the seriousness of the exposure to radioiodines in fallout." This specific radioactive fallout threat particularly affected young children, as shown by the nodules found in 89.5 percent of the Rongelap children who were under ten years of age at the time of the 1954 fallout. The incidence of thyroid lesions in those exposed as adults was lower than the children but higher than the adults in Utirik or Marshallese not exposed.

Thyroidectomies, partial to complete, by 1969 had been carried out on eighteen Marshallese.

For Rongelap's exposed adults and children, the main source of radioactive iodine ingestion was considered to be through drinking water. Since it was being rationed at the time of fallout, it was assumed that the children drank the same amount as adults and had the same thyroid burden of radioiodines.

Years later, a Brookhaven Laboratory study determined that another pathway contributed to the intake—"fallout debris falling directly on food prepared and consumed outdoors during passage of the fallout cloud."

The small size of the children's thyroid resulted in a substantially larger dose. The total estimated dose from the various iodine isotopes to the child's gland was about 1,000 rads, with a range of 700 to 1,400 rads. The glands received an extra 175 rads from external gamma radiation.

Over the fifteen years of the study, sixteen deaths had occurred among the exposed Rongelap people. Those 13.0 deaths per 1,000 persons per annum compared with 8.3 per 1,000 for the Marshall Islands as a whole (1960).

The incidence of miscarriages and stillbirths in the exposed women was about twice that in the unexposed women during the first four years after exposure. From 1955 to 1958, among the twenty-two exposed Rongelap women, there were thirty pregnancies, twelve miscarriages, and eighteen births. No such difference has been seen since then, and records "show no impairment of fertility in the exposed women," according to the report. In addition, the report said, "No radiation-induced sex ratio alteration has been seen."

While genetic effects had not been specifically studied because of the small number of people involved, no apparent radiation-induced genetic changes had been found on routine physical examinations in the first-generation children of exposed parents. The possible exception was those earlier increased miscarriages and stillbirths among the pregnant, exposed women.

The general health status and incidence of physical abnormalities of the exposed Rongelap people was about the same as of the unexposed people on the island except for the thyroid abnormalities. More than one hundred Rongelap people, who were relatives of the exposed people but had been away from the island at the time of the accident, had moved back to Rongelap Atoll. They served as "an ideal comparison population for the studies," according to Conard's fifteen-year report. The overall population of Rongelap Atoll, which had risen and fallen over the years, at 1969 was just short of two hundred people.

Dr. Conard reported, after the 1968 survey, that the thyroid tablets were not being regularly taken, and thereafter a weekly dose of the seven tablets would be done "in the presence of a responsible individual who will use a check-off list." This step was taken because "based on preliminary evidence, some of the children who have been taking their medication [showed] an improvement in growth," Conard wrote.

Conard and his medical colleagues were learning about dealing with the long-term effects of the radioactive fallout on the exposed Marshallese. But how much of what was going on did the Marshallese understand?

Eight days after the test, Walmer Strope was part of a group that took sand samples on one of Rongelap's southern islands to test for radiation. The next day, Strope took radiation readings on a northern Rongelap island and had to quickly return to the boat because levels became too high.

On Kwajalein Naval Station, four days after the fallout, Task Force Commander Maj. Gen. Percy W. Clarkson (far left) and Naval Station Commander Adm. Ralph S. Clarke (far right) meet with Kabdo of Utirik (left) and John Anjain, magistrate of Rongelap (right).

Magistrate John Anjain during the first weeks that the Rongelap people were cooped up in a secluded section of the Kwajalein Naval Station, being examined medically almost every day as the American doctors tried to figure out what exposure to radioactive fallout had done to their health systems.

Joint Task Force Commander Maj. Gen. Percy W. Clarkson visited with Marshallese Islanders who had been brought to the Kwajalein Naval Base after their home atoll had been subject to radioactive fallout from the Bravo explosion on Bikini more than a hundred miles away.

Dr. Robert A. Conard, who first went to the Marshall Islands during World War II as a navy medical officer, returned as a radiological safety officer for Operation Crossroads, and was among doctors in 1954 who initially examined the Bravo Rongelap fallout victims. From 1956 to 1979, when he retired, Dr. Conard organized and led the annual medical examinations of the exposed Marshallese, along with their followup treatment.

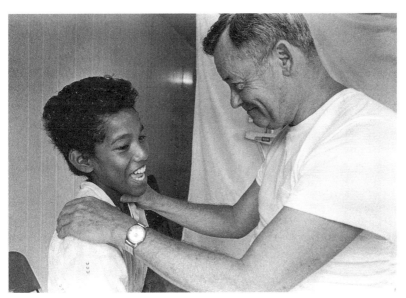

Dr. Conard's examinations focused on the thyroid glands of exposed Rongelap children when nodules began to appear in 1963, nine years after they were victims of Bravo's radioactive fallout. Overall, sixteen of nineteen children under ten at the time of the fallout turned up with nodules, leading Conard to conclude exposure to radioactive iodine had a greater effect on a child's smaller and more active thyroid glands than it did on those of adults.

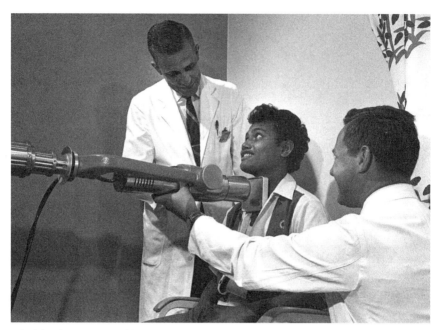

Lekoj Anjain, a one-year-old when the radioactive fallout came, was a fourteen-year-old in 1968 when he arrived at Brookhaven National Laboratory for examination and later had a thyroid nodule removed. At the time, no one knew that four years later he would become the first cancer death caused by the Bravo fallout.

The gravestone of nineteen-year-old Lekoj Anjain buried in a spot overlooking Rongelap Lagoon where, as columnist Stewart Alsop wrote in 1972, "Maybe Lekoj will see the waves of the green Pacific lapping at the shore of Rongelap again, and maybe not."
CREDIT: David Robie Publishing Limited

The Runit Dome on Enewetak Atoll was part of the cleanup of radioactive matter that began in 1977. To deal with soil and other matter contaminated by plutonium-239, whose half-life is 24,100 years, a crater on Runit Island was chosen as a burial pit. In 1979, filled with plutonium-contaminated soil mixed with cement and other radioactive waste, it was capped by an eighteen-inch-thick cement dome that was more than one hundred yards in diameter.

The rebuilt Rongelap village area with the pier, church, and school in the distance, and end of the aircraft landing strip on the left. Some of the fifty rebuilt homes are in the foreground, only a few of which are occupied by some workers employed by either the Marshall Islands or U.S. governments and their families who keep things repaired.

CREDIT (*top and bottom*): Terry Hamilton, Scientific Director, Marshall Islands Dose Assessment and Radioecology Program, Lawrence Livermore National Laboratory

John Anjain with a nephew in 1985, the year the Marshallese on Rongelap left their home atoll for a second time. Anjain, himself, had a thyroid nodule removed in 1973 and became a spokesman for his people. He died in 2004 at the age of 81.

CREDIT: David Robie Publishing Limited

35 · DEATH COMES HOME TO RONGELAP

■ IN THE SUMMER of 1971, legislators in the Micronesian Congress began to publicize to the rest of the world what they said was the plight of the exposed people of Rongelap and Utirik. It came at a time when Marshallese politicians had begun to chafe at their status as a US Trust Territory, and had become more active in seeking their long-promised independence.

Marshallese Congressman Ataji Balos, whose district included Rongelap, was the most outspoken. A gregarious, former school teacher, Balos had for years been an activist in controversies between the United States government and the Marshall Islanders, including payments to the landowners for use of their property that made up the Kwajalein military base. In fact, he once led a sit-in near the target area that delayed testing of a US anti-ballistic missile element at a time when Kwajalein was key to the American offensive and defensive missile complex.

One immediate result of calling international attention to the radioactive fallout on the Rongelap and Utirik Marshall Islanders had been an invitation in August 1971, for Balos and Magistrate John Anjain to attend a symposium in Japan on nuclear bomb testing. They went as guests of the organizers, a Japanese group called the Congress Against Atomic and Hydrogen Bombs (Gensuikyo). After visits to Hiroshima and Nagasaki, they were interviewed on Japanese radio and television. In Tokyo, Balos invited the Japanese antibomb group to send its own medical team to examine the exposed Marshallese on Rongelap.

Upon his return to the Marshall Islands, Balos asked US Trust Territory officials at their Saipan headquarters to provide visas for his Japanese invitees, but he got no answer. When, in December 1971, the Japanese group, including two doctors and six journalists, arrived in Majuro, the capital and largest city in the Marshall Islands, they were denied visas to visit Rongelap. Nevertheless, before they returned to Japan, they did examine eight exposed Rongelap Marshallese who were in Majuro.

The visa denial decision had been made by Trust Territory Assistant Attorney General Robert Bowles after consulting both the State Department and US embassy in Tokyo. Bowles wrote in a cable to Balos that he was denying the visas out of a necessity "to protect the immigration security of the Territory," according to Micronesian Congress records.

A subsequent AEC report on the matter referred to the Japanese group's sponsors as "a radical left socialist party" and labeled one of the two Japanese physicians, Dr. Yoshima Hondo, as "an activist." The other, Dr. Haruo Ezaki, was recognized as an expert, since he was a consultant on thyroid disease for the Atomic Bomb Casualty Commission and the new chair of surgery at Hiroshima University.

Balos cabled back to Bowles that if the Japanese were not allowed to visit the islands, "he would encourage the exclusion of Dr. Robert Conard and the AEC [from Rongelap] until the matter of the Japanese was cleared up," according to the Micronesian Congress records.

In response to the treatment in 1971 of the Japanese team, newspapers in Japan and Micronesia criticized the Trust Territory leadership and the AEC, accusing them of trying to suppress facts about the Marshallese radiation exposure.

Bowles's denial of visas effectively gave Balos and Amata Kabua—by then president of the Micronesian Congress—the opportunity to step up their anti-American campaign. On January 25, 1972, Balos spoke on the floor of the Micronesian House of Representatives about the visas being turned down for "people who were interested in helping the people of the Marshall Islands."

He then said, "While I am not in possession of conclusive evidence to prove it to this House, I am now convinced that the United States knowingly and consciously allowed the people of Rongelap and Utirik to be exposed to the 1954 fallout. This was done to the Rongelapese and Utirikese so that the United States could use them as human guinea pigs in the development of its radical capabilities to treat its citizens who might be exposed to radiation in the event of war with an enemy country. This is a crime unmatched in peacetime."

Balos introduced a resolution, co-sponsored by all members of the Micronesian House of Representatives, which was then approved unanimously. The Balos resolution established a "Special Joint Committee of Congress Concerning Rongelap and Utirik Atolls." The legislation authorized that committee to "investigate the results of irradiation of people of Rongelap and Utirik Atolls, to secure medical assistance and aid, to obtain compensation for injuries and, having power to conduct hearings and investigations, to issue subpoenas for witnesses and to bring suits in any court of the Trust Territory."

To calm the situation, Dr. Conard recommended that Dr. Toshiyuki Kumatori, whom he had recently met, join the planned March 1972 medical team's visit to Rongelap. As one of those doctors who had dealt with the *Lucky Dragon* seamen, he was acceptable to the new Micronesian Congress Joint Committee. At the same time, Balos asked that Dr. Ezaki also join the group, and Conard agreed.

Then, on the eve of the Japanese doctors' departure to Kwajalein, a Marshallese activist talked Dr. Ezaki out of going. As a result, Dr. Kumatori also backed out. When Conard arrived at Rongelap, the promised two Japanese doctors were not with him.

At a village meeting the evening after the team arrived, Conard tried to address Balos's criticism directly. "We have to carry out many examinations on you to find these sicknesses and test them. This is not experimenting on you, or using you as guinea pigs, as some people have said. We are doctors and interested only in helping people."

The Rongelap people sat and listened as Conard spoke, his words immediately translated, but no one replied. At last, in the question period, Tima, the fisherman, said, "You draw blood from us, which makes us weak. We need more food. Why don't you give us any food for the blood you draw?" Conard responded that the amount of blood taken was not enough to make a person weak, and that it was quickly replaced by the body's systems. Tima then mumbled that if he did not get food, he would not let his family be examined.

Another person asked why the Japanese had received compensation for fish that might have been irradiated while the Rongelap people had been paid nothing for coconut crabs, which they were not allowed to eat. Conard replied that fish were a staple for the Japanese, but that the coconut crab made up only a small part of the Rongelap diet. He added that he hoped to permit the Rongelapese to resume eating the coconut crab soon. Although one older person rose to thank Dr. Conard and the other doctors for their examinations, most of the group at the meeting remained unhappy.

The next morning, Congressman Balos sent word to the Rongelapese that he did not want them to be examined because Dr. Ezaki had not arrived as an observer. As Conard and other doctors waited, no one came to be examined.

After two days, Dr. Conard decided to concede defeat. The plan had been to then go to Utirik before returning to Majuro. But Conard learned via radio that the Utirik people, being told what had happened on Rongelap, would also refuse examinations. Conard later wrote in his 1992 memory of the situation, "The main problem centered around the fact that they had been told not to submit to the examinations. The team held sick call and a few people came in to be examined, in spite of admonitions from some of their elders. The refusal of examinations was a discouraging and frustrating experience for our team, who were prepared for a detailed survey."

Dr. Conard returned to the United States disturbed that for the first time in eighteen years the examination process had been interrupted. Perhaps most frustrating for Dr. Conard was that over the years the Rongelapese might have been telling him one thing, while telling their own congressman another. Balos claimed he had been told by the Rongelap people that Dr. Conard was cold and uncommunicative. The examinations required people to spend the day going from one line to another in military fashion. Often, they had to wait in line for hours. When people were told that they were well, they wondered why the examinations continued. But they never asked. As food became short and money limited, the people wanted to be paid for their time spent on examinations.

Senator Olympio T. Borja, the Northern Mariannas legislator who had been named chairman of the Special Joint Committee, met in April 1972 with President Nixon's surgeon general, Dr. Jesse Steinfeld, who was visiting Micronesia to look into the allegations of poor treatment of the Rongelap and Utirik people. Steinfeld offered one of his Public Health Service team who could do an "objective job" of examining the people, but Borja turned the offer down.

Instead, Borja wanted to bring in Dr. Ezaki and Dr. Kumatori, plus doctors from the World Health Organization—Dr. E. Eric Puchin of the British Medical Council and Dr. William S. Cole, assistant director of the US Public Health Service—to join Conard's regular Brookhaven team on the next visit to Rongelap.

The four additional doctors would be independent observers and report back to Borja's Special Committee. They would assess "the [Brookhaven] medical team's methods of examination, an evaluation of the medical treatment given to the Rongelapese in the past, and that given to them in the proposed survey," according to Borja's July 7, 1972, letter to then-Trust Territory High Commissioner Edward E. Johnston.

This expanded medical group arrived on September 10, 1972, aboard the Trust Territory ship, *M. V. Militobi*, which anchored in Rongelap Lagoon. The Conard medical team worked, ate, and slept in temporary quarters and facilities on Rongelap. The Micronesian Special Committee's observers, which included Chairman Borja and the four medical consultants, were housed aboard the *Militobi*. They disembarked daily to Rongelap Island to meet with the people and observe the medical team at work.

At a meeting with the people the first evening, a colleague of Senator Borja on the Joint Special Committee explained the presence of the consultants and asked that the people cooperate with the examinations. Dr. Conard gave his standard talk while introducing the doctors and technicians. He made a special announcement: coconut crabs would no longer be forbidden to eat, but only one crab per person per day could be consumed. During the question period one islander asked that since only one crab a day could be eaten, were they still radioactive? "Yes, but only slightly," Dr. Conard replied. He added that some crabs would be taken back from the trip for study in Brookhaven's laboratory to see if they could be considered safe for unlimited eating.

During the examinations, Dr. Conard and his colleagues found two new cases of thyroid nodules, both in women. Dr. Dobyns would later in November perform surgery on the eighteen- and twenty-two-year-old women at Cleveland Metropolitan General Hospital and find the nodules were not malignant.

However, the most important discovery during examinations involved nineteen-year-old Lekoj Anjain, son of John and Mijjua Anjain. Back in 1954, Lekoj had just celebrated his first birthday, ten days before the fallout, and was the youngest person exposed. During the two days before the Americans evacuated the Rongelap people, Lekoj drank water and ate food that delivered radioactive elements to his tiny body and organs. In 1968, he successfully had thyroid nodules removed and seemed well when examined in 1971.

In June 1972, Lekoj had graduated from Marshalls Christian High School in Majuro. He was handsome with bright eyes and hopes for his future. However, according to his father, Lekoj first started having bruises appear on his body in 1970. He was once hospitalized for more than two weeks because of them. They became worse in 1972. Lekoj was not examined when Dr. Conard came to Majuro in spring 1972. But when the September examinations were set, he was homesick for Rongelap, so he accompanied Dr. Conard and his team on the *M. V. Militobi* to Rongelap.

On the boat, Lekoj complained that he felt sick and sleepy, but there had been flu at his school, and so no one took his complaints seriously. During the second day of examinations on Rongelap, Sebio Shoniber, a Marshallese lab technician, drew the blood sample from Lekoj and passed it to a Brookhaven analyst, Michael Makar, for testing. Later, when Sebio was cleaning up for the day, Makar asked him how Lekoj's blood sample had been taken, noting that the white count was very low. Because of the potential health danger it signaled, they called Lekoj back and repeated the test, and again the count was low.

Lekoj's blood test showed his white blood cell count was two thousand per cubic millimeter of blood; the normal human white blood cell count is about eleven thousand per cubic millimeter of blood. That low, white-cell count meant that Lekoj's body was less prepared to fight off bacteria and infection, and it also indicated trouble in either the bone marrow or lymph glands where the cells are produced.

That evening, the doctors discussed the case. The next day, after talking with Lekoj, Dr. Conard and his team decided to take him to Hawaii, where, at Tripler Army Hospital, he could be given more sophisticated tests. Lekoj was on board when the *Militobi* left Rongelap on September 13, 1972, for examinations on Utirik and then to Kwajalein. On the flight from Kwajalein to Hawaii, Lekoj was in good spirits, looking forward to possibly having an interview at the University of Hawaii, where he wanted to go to college.

At Tripler Army Hospital, Dr. Conard had a bone marrow test attempted, but the doctors there were unable to do it successfully. As a result, Conard decided that further testing needed to be done at Brookhaven. The examination at Brookhaven revealed that Lekoj had chronic myeloid leukemia (CML), a type of cancer that starts in certain blood-forming cells of the bone marrow. The leukemia cells grow and divide, building up in the bone marrow and spilling over into the blood, replacing the normal white and red blood cells. CML is a fairly slow-growing leukemia, but it can change into a fast-growing leukemia that is hard to treat.

The Brookhaven doctors found that normal bone-marrow elements in Lekoj had decreased and been replaced by cancerous ones. As a result, his white-cell count had dropped down to seven hundred with almost half the cells being abnormal. His temperature rose to 101 degrees, and he was put into isolation, since infection with such a low white-cell count could be fatal.

For John and Mijjua Anjain, this was another terrible blow. In 1963, their youngest son, born after the fallout, had been stricken with poliomyelitis when an epidemic of the disease hit the Marshall Islands. As a result, he had a paralyzed leg and used a crutch to get around.

On October 2, Lekoj's temperature returned to normal, and he was flown to Washington, DC, for admittance to the National Institutes of Health (NIH) at Bethesda, Maryland. His care would be supervised by the National Cancer Institute's doctors. Dr. Conard arranged for John Anjain to be flown to Washington from his home in Ebeye, accompanied by Sebio Shoniber, the Marshallese lab technician, as translator. They were put up in a motel near to Lekoj's NIH hospital building, and they visited daily.

Then occurred a great coincidence. Lekoj was placed in an NIH hospital room that was also occupied by Stewart Alsop, the *Newsweek* columnist, who, eighteen years earlier—with his brother Joe—had written two columns about the military implications of the Bravo test, as well as the fallout. Alsop, who himself was fighting an unusual form of leukemia, was there with a then-case of lobal pneumonia, which his cancer doctor wanted monitored.

"I had the privileged bed beside the window, and the bed beside the door was occupied by a muscular young man with brown skin, curly black hair and a huge grin," Alsop wrote in his book, *Stay of Execution.*

Alsop went on, "I am still haunted by a mental image of Lekoj as a cheerful brown baby playing in the sand under the palm trees of Rongelap, as the sky lit up above him from the great explosion on Bikini, and playing still, feeling no harm, as the dust of the fallout settled around him. The cheerful brown baby was now my roommate . . . [with] a particularly vicious variety of acute myelogenous leukemia. There was no doubt at all that the bomb and the leukemia were cause and effect. . . . Lekoj was the first case of leukemia from the fallout of a hydrogen bomb test."

For ten days, Lekoj was given chemotherapy and was starting to feel better. But he later developed a fever and blood transfusions were needed. His twenty-one-year-old brother George, who was attending Maui Community College in Hawaii, was tested for the special type of blood that Lekoj needed, but his blood was not the right type. Other relatives were tested, but none were a match.

Alsop described John Anjain who "for hours at a time he would sit by Lekoj's bedside, saying nothing at all. Once in a long while he would reach out and touch Lekoj's hand, and sometimes Lekoj would mutter something, in Marshallese, and grin." Since Lekoj could speak almost no English, there was little said between him and Alsop, other than smiles back and forth and an occasional "How you feel?"

Toward the end of Alsop's twelve days in the room, Lekoj more often responded with a grin, but occasionally said, "No good, feel deezy." Alsop wondered if Lekoj really knew how sick he was. Both agreed the bone marrow tests "hurt bad," and "blood make you feel good," when both got transfusions on the same day.

By October 27, 1972, Alsop was well enough to go home, but before leaving he got Lekoj's permission to write a *Newsweek* column about the younger man's situation. LEKOJ AND THE UNUSABLE WEAPON, was Alsop's column in *Newsweek*'s

October 30, 1972, edition. In it he described the young Marshallese as "the first, and so far only leukemia victim of an H-bomb." He also wrote, "The Bikini bomb showed for the first time that fallout could be even more lethal to human life than the great fireball itself."

Alsop recalled in the column that Lekoj "spends a good deal of his time curled up in a ball on his bed, wrapped in a blanket, only his curly black hair and his brown skin showing above the blanket. He speaks very little English, but he grins broadly, showing strong white teeth when anybody smiles at him."

Later, with Alsop gone, Dr. Conard arranged for Lekoj's mother, Mijjua, to be brought to Washington. But on November 8, Lekoj came down with a cough. His sputum was tinged with blood. In the next two days he had increased trouble breathing. On November 12, Lekoj was put on a respirator. He died on November 15, 1972. His immediate cause of death was described as pneumonia, but the real cause was the leukemia.

At that moment, Dr. Conard was in Cleveland, attending the two Marshallese girls who were undergoing thyroid surgery. He quickly arranged for his chief administration aide, William Scott, to go to Washington to supervise arrangements for the Anjain family, and to have Lekoj's body returned to Rongelap. Mijjua and John Anjain, Sebio the translator, and Scott boarded an airplane for Los Angeles the day after the death, with Lekoj's coffin in the baggage compartment.

The death of Lekoj had medical and political implications, but the political implications had to be faced first by Dr. Conard and the AEC. When the plane arrived in Los Angeles, Scott and the Anjains remained on board trying to decide what to say to news reporters said to be waiting for them. By the time they disembarked, the reporters were gone.

But troubles were far from over. When they arrived in Hawaii on the morning of November 17, and moved to an Air Micronesia flight to Kwajalein, it was discovered that Lekoj's coffin had been left in Los Angeles. The Anjains went ahead to Kwajalein with Scott staying behind in Hawaii to await the coffin and then accompany it to Kwajalein the next day.

On November 19, the church on Ebeye was filled for Lekoj's funeral service. When the coffin was set before the altar and opened, the whole congregation walked by. They were amazed at how healthy Lekoj looked as he lay there. No one, however, during the service publicly mentioned that radiation was the cause of his death.

The next morning, the Trust Territory leadership had arranged for a Navy S-16 amphibious airplane to take the Anjains and the coffin to Rongelap for the internment on the island. The flight was then canceled after officials discovered the aircraft had ripped its pontoons on a previous trip.

Trust Territory officials quickly arranged for the family and the coffin to be put aboard the *M. V. Militobi*, along with 120 people from Ebeye who wanted to attend the burial on Rongelap. As a final irritation, the *Militobi*, in a fog, ran aground on Rongelap's coral reef at 2 a.m., and passengers had to wait until morning to be pulled off.

Once on Rongelap, people gathered in the church and listened as Peter Coleman, deputy high commissioner, spoke for the Trust Territory and the American government. Brookhaven's Scott talked briefly of the grief of Dr. Conard and the doctors and nurses at Brookhaven and NIH. Finally, John Anjain, who had been silent in his grief, stood up. District Administrator Oscar DeBrum translated his words into English. Scott later recalled that at one point John said, "The people in the United States did all they could." He would be much harsher toward the US as years went on.

The next morning, Lekoj Anjain was buried at a spot overlooking Rongelap Lagoon. In his *Newsweek* column before the young Marshallese died, Alsop had written, "Maybe Lekoj will see the waves of the green Pacific lapping at the shore of Rongelap again, and maybe not."

The *New York Times* reported Lekoj's death in its November 21 edition in a news story buried on the middle of page twenty-eight. The headline was MARSHALL ISLANDER'S DEATH TIED TO FALLOUT. Written by science reporter Walter Sullivan, it began, "The first known death from a disorder typical of radiation exposure has occurred among those subjected to heavy fallout from nuclear weapons tests. The victim was a 19-year-old Rongelap islander named Lekoj Anjain, who died last Wednesday of leukemia at the National Institutes of Health in Bethesda, Md., after an intensive effort to stem rapid progress of the disease with chemicals."

On November 22, Dr. Conard wrote Trust Territory High Commissioner Johnston about his sorrow at Lekoj's death, adding that at NIH "everything possible was done for him. However, in spite of all this, pneumonia, complicating the leukemia, proved to be too much for him resulting in his death." As a result of Lekoj's leukemia, Dr. Conard said that when examinations resumed, "The usual careful hematological examinations will include bone marrow study if indicated, and we also plan to carry out additional examinations of the blood on the Rongelap exposed people at six-month intervals."

Conard added that reports from the four outside medical observers on the September Rongelap examinations had been sent to Senator Borja. Furthermore, Conard had learned the observers' reports "were favorable," and so "this hopefully should clear the atmosphere" for future visits, assuming that they would continue. Dr. Conard closed by saying, "I am greatly indebted to many people for their sympathetic assistance during these trying times . . . [and] I sincerely hope that things will calm down and we will have smoother sailing from now on."

Stewart Alsop, who was deeply affected by Lekoj's death, worried that it was not only catching his pneumonia that ended the young man's life but also the Bravo bomb.

In his book, Alsop wrote, "There was the further feeling, as hard to shake off, that we Americans were responsible for his death—that we had killed him with our bomb. His was the world's first death from a hydrogen bomb, and the bomb was ours. . . . Before Lekoj died, I had long believed in my mind that the nuclear weapon, in its indiscriminate, unimaginable brutality, was an insane weapon, suicidal, and inherently unusable. Now I knew it in my heart."

36 · RONGELAP DEPARTURE

- **THE DEATH OF** Lekoj stirred anger among the Rongelapese toward not only Dr. Conard and the Brookhaven-sponsored doctors but also US government officials. The late 1960s and early 1970s had marked a period of social and political change in the United States, topped by the Watergate scandal and the August 1974 resignation of President Richard Nixon. Less apparent were the political stirrings among the previously docile, but at-that-time, increasingly less-isolated, people in the Marshall Islands.

The children of hereditary chieftains had moved into politics and dominated the Micronesian Congress. Some activists formed alliances with Japanese groups and, with some American Peace Corps alumni, pushed the United States for additional aid to nuclear victims. In this younger, Marshallese leadership view, "Bravo doctors were AEC doctors. Bravo doctors supported AEC policies. Their attachment to science, based on these policies, interfered with their ability to practice medicine. In a nutshell, trust was the issue of contention," Dr. Laura J. Harkewicz wrote in here 2010 doctoral thesis "The Ghost of the Bomb: The Bravo Medical Program."

Senator Borja's Special Joint Committee of the Micronesian Congress on compensation for the people of Rongelap and Utirik issued a preliminary report in 1973, "dedicated to the memory of a young Marshallese man, Lekoj Anjain, who was one year old when the world's greatest nuclear explosion was detonated one hundred miles from his home on March 1, 1954, and who was nineteen years old when he died during treatment for acute myelogenous leukemia."

The panel's report read, "By the very nature of the unique set of circumstances surrounding the daily lives of the two groups, the people of Rongelap and Utirik are 'guinea pigs.'" The report explained further, "No other group of people . . . has been exposed to the same amounts and differing kinds of radioactivity, and no other group in the world has been so carefully studied for the results of such effects." The report raised the question of whether Dr. Conard and the AEC were carrying out examinations of the Marshallese where "their [individual medical] treatment is only of secondary [import], or whether they are, by virtue of their experience as a group being examined and studied for the best of humanitarian aims," i.e. guinea pigs. This preliminary report left the answer "up to the reader."

Changes began after Dr. Conard's 1973 examinations discovered that fifty-year-old, shrunken, white-haired, John Anjain had a thyroid nodule. Borja got US funding for a doctor to give regular medical care to the Rongelapese. He also arranged for a former Navy landing craft to regularly serve the island, plus the promise of a new radiological survey of both Rongelap and Utirik. A problem emerged for US bureaucrats because while the AEC and its successor agencies could pay for the medical examinations related to the Bravo bomb, AEC officials insisted that the Trust Territory or its parent, the Interior Department, had to cover any primary care. It was a task that neither entity had the personnel or equipment to handle.

On February 28, 1974, Borja delivered his final report, telling the Micronesian Congress, "It is a sad but true fact that compensation for these people can never make them 'whole' again. What wrong has been committed can never be erased by better medical treatment, or even the payment of money. However, compensation can attempt to do this and the gesture, if nothing else, may have some helpful effect upon these people." He added, "We are recommending what is probably much less than the cost of building the bomb, which has caused so much misery and suffering to the people."

He cautiously hoped, "These funds will help compensate the people for what damage has been done. That it will not create dependency, which compensation tends to do, and which is the antithesis of the intent of compensation; and that it will encourage those who have left these islands to return."

The committee recommended that the AEC pay the family of Lekoj Anjain $50,000 to be used as it pleased; the people who have had thyroid operations because of the radiation were to get $2,500; the Utirik people displaced by the fallout were to receive $1,000 each, plus the $116 already offered by the AEC; and finally, "There will be no release of liability for the AEC or the US Government if the people take this money."

Borja's report was just the first of decades of governmental studies of Rongelap, its radiation contamination, and the health of exposed people, including what should be done about them, or to them. Micronesian, United Nations, Japanese, and US governmental bodies, including congressional committees, even Greenpeace, the environmental activist organization, all would play roles.

There were also outsiders, such as the Japanese antinuclear group, Gensuikyo, that used the Marshallese for their own purposes and to expand existing differences between the Rongelapese and Americans.

For example, on April 9, 1975, Nelson Anjain, uncle of Lekoj, who had replaced his brother John Anjain as chief magistrate of Rongelap, wrote to Dr. Conard about having "learned a great deal" after spending time with the Gensuikyo. As a result Anjain wrote, "I've made some decisions that I want you to know about. The main decision is that we do not want to see you again. We want medical care from doctors who care about us, not about collecting information for the US government's war makers. . . . We've never really trusted you. So we're going to invite doctors from hospitals in Hiroshima to examine us in a caring way."

Four months later, at a Gensuikyo conference in Hiroshima's Labour Hall on August 3, 1975, Nelson Anjain read excerpts from his Conard letter to some seventy delegates from two dozen countries. The British magazine *New Scientist* described him as "one of the most impressive delegates at the conference" who was "also probably the least obtrusive."

Meanwhile, Dr. Conard and the Brookhaven staff had grown concerned by the unexpected development of thyroid nodules appearing among fifteen Utirik people in the mid-to-late 1970s. Of that group, there were three cases of malignant thyroid nodules whereas for Rongelap, of the nineteen thyroid nodules, four turned out to be cancerous. The surprising number of Utirik malignancies were definitely a higher incidence than would be expected, based on the original estimated exposure dose that was one-tenth that of the Rongelapese.

In an October 25, 1976, statement prepared for delivery to the Marshallese, the AEC explained, "The development of thyroid cancer in the Utirik people within the past few years was unexpected. Statistical comparison of thyroid cancer incidence at Utirik with the larger experience of the United States indicates that radiation very likely was involved. Accordingly, it was recommended to the Department of Interior that all Utirik people who have thyroid operations be considered for compensation similarly to the Rongelap people."

That year, an arrangement was reached under which a resident Brookhaven physician was to provide medical care for Rongelap and Utirik patients at hospitals either on Ebeye Island in the Kwajalein Atoll or on Majuro.

On October 15, 1977, President Carter signed into law legislation that provided $25,000 to the exposed Rongelap people who had their thyroid glands removed, and $1,000 to those on Utirik. "Not more than $100,000" was to be paid to the heirs of anyone whose death "is related to the thermonuclear detonation," i.e. Lekoj Anjain. The measure also required the interior secretary to report to Congress by the end of 1980, if any "additional compassionate compensation may be justified" for individuals related to radiation injuries or illnesses related to the Bravo test.

In addition, Rongelap and Utirik were to receive $100,000 each, the funds to be provided by the Interior Department for "community purposes." This section of the legislation carried a provision which said these payments would be "in full settlement and discharge of all claims against the United States arising out of the thermonuclear detonation on March 1, 1954."

In December 1977, Brookhaven released the results of three radioactivity surveys it had taken over the previous fifteen months, each covering the same five atolls in the Northern Marshall Islands including Rongelap, Utirik, and Rongerik. The report of the surveys found that the exposure rate on Rongelap Island, where all the people lived, was "about twice the background radiation level of uncontaminated atolls in the Marshall Islands." At the same time, the report said exposure rates were varied, and greater on Rongelap Atoll's northern islands, where the people sometimes visited.

The study's major and most controversial finding, which took into consideration only the external radioactive elements and not those that had been ingested, was that the people living on Rongelap Atoll may be close to exceeding the limitation standard for thirty-year exposure set by the International Commission on Radiological Protection (ICRP), although it was within the level set for individuals within a single year.

Despite those troubling results, the Brookhaven group wrote, "At this time, we do not recommend any remedial action until a complete dose commitment can be determined by means of examining the external, dietary and whole body counting data available to date."

In 1978, administrative responsibility for Rongelap and Utirik was shifted to the Marshall Islands government from US officials at the Trust Territory, as a step toward eventual independence. At the same time, the US had Lawrence Livermore National Laboratory conduct new surveys of the Northern Marshall Islands, based in part on the increased thyroid nodules that developed on Utirik. There had been recognition of the lack of knowledge about radioactivity on other islands and atolls in the area on which people had been living during the Pacific nuclear tests. It would be four years before results of that group of surveys would be shared with the Marshallese.

Dr. Conard retired in 1979, and a year later he warned, "The findings in the Marshallese emphasize the importance of thyroid exposure to radioiodines that may result from warfare or accidents in which radioiodines are released. Exposure to penetrating gammas or neutrons is a more serious hazard, not only because of their acute effects, but also because of the fatal nature of malignancies such as leukemia which may develop."

Conard acknowledged that there had been an underestimation of the original exposures, and "it seems likely that in the Marshallese the exposure to the short-lived isotopes of iodine-132, iodine-133 and iodine-135, which have more energetic betas and deliver a faster dose rate than iodine-131, might account for the high incidence."

The Livermore radioactivity surveys and other data brought a new realization—the Rongelap people's consumption of locally grown food, such as breadfruit and coconuts, along with fish, were all contaminated by radioactive substances taken up through the soil and water. That meant ingested radiation, of which cesium-137 was more than 90 percent, with small added amounts of plutonium-239 and plutonium-240.

Cesium-137, with a half-life of thirty-three years, is a toxic substance if inhaled or ingested. As it decays it produces short-lived, radioactive barium whose gamma radiation can pass through the human body and organs, destroying cells and causing malignant tumors. The new, potential danger of cesium-137 ingestion to Rongelapese and others in the Northern Marshalls who had been consuming locally grown food was substantial. But, results of the Livermore surveys and others done by Brookhaven and contractors, again, were not made available to the Marshallese until 1982.

Those results were a shock, not just to the Marshallese but also to American officials. During the period from July 1981 to June 1982, the average Rongelap male's body burden for cesium-137 rose 56 percent, while the average female level increased by 11 percent. Including children, the overall population showed a 1.8 percent monthly rise in cesium-137, after showing a constant level of cesium-137 in the previous two years. The latest Brookhaven study claimed the rapid rise "may have resulted from the relaxing of restrictions to the northern islands of Rongelap Atoll as a source of coconuts and coconut crabs."

Relaxation of restrictions on eating local food was justified during a December 1982 presentation by a Department of Energy (DOE) Defense Program officer. He told the Rongelap people on Majuro that they could "make their own judgments on radiation protection, after being told they could eat food that had been restricted for many years," according to a July 22, 1985, internal DOE memo from T. F. McCraw, an AEC operational safety official.

McCraw's memo also explained that the Rongelap people saw a conflict of their interests: The DOE's Defense Office was in charge of their health and safety, rather than the AEC; but it was Defense Department personnel that had detonated the bomb in the first place.

Further confusion came to the Rongelapese, according to the McCraw memo, with the distribution of a booklet at that time, written in both Marshallese and English, which misstated whole-body radiation standards. The booklet predicted a high exposure for Rongelap Island residents of 400 mrem/ yr which, McCraw wrote, "would not be acceptable." The acceptable whole body level was less than 100 mrem/yr, with the Rongelap people obeying locally-grown food restrictions, he wrote, adding "To my knowledge, this error [in the booklet] has never been corrected. . . . Right or wrong, I have argued that exposures not found acceptable for the US population are also not acceptable in the Marshalls."

That 1982 advice, which suggested the Rongelap people could make their own decisions about the local food they ate, resulted in even higher body burdens showing up in 1983. Nonetheless, while some Rongelap leaders began to call their atoll unsafe, senior Energy Department officials maintained it was safe, but only if consumption of local foods were limited.

In 1983, John Anjain's younger brother, Senator Jeton Anjain, who then represented Rongelap in the Micronesian Congress, joined with Marshall Island's president, Amata Kabua, in presenting that legislative body with a resolution calling for evacuation of the Rongelap people from their atoll. It passed unanimously. Jeton Anjain would later tell the UN Trusteeship Council that the resolution was drafted "after we were told that one-half of our atoll was under quarantine, and after we saw in the Department of Energy's own words that many islands on Rongelap were as contaminated as islands at Bikini."

In 1984, after the US took no action on the Micronesian Congress' Rongelap evacuation resolution, Senator Anjain, sought assistance from Greenpeace, the global, nonviolent, proenvironmental group that was using peaceful protests to oppose nuclear testing. The Greenpeace organization recognized that aiding the evacuation of Rongelap would be a departure from its past activities, since there would be no crowds to gather, banners to hang, or inflatable boats to place in nuclear testing waters.

Instead, what it named Operation Exodus involved using its refitted, 131-foot, British-built trawler, named the *Rainbow Warrior*, beginning on May 17, 1985, to ferry the Rongelapese and one hundred tons of their personal belongings and stripped-down housing to Majetto Island 112 miles away from Rongelap in the northwest corner of the Kwajalein Atoll.

Operation Exodus took eleven days and four trips, roughly thirteen hours each way, to move some three hundred Rongelapese, ranging from eighty-year-olds to newborns, and their belongings. The *New York Times*, on May 21, 1985, carried a short, four-paragraph story on page seven, based on a Greenpeace press release after the first trip of seventy-five people took place. Under the headline, 70 ARE MOVED FROM ATOLL USED IN '54 TEST, the article reported that DOE's Roger Ray, identified as the Energy Department's "project officer for the radiation program in the Marshall Islands," described the atolls, including Rongelap, as having average radiation levels lower than many areas of the US. "I feel confident there is no major justification in disrupting the lives of the islanders," Mr. Ray said, echoing the US government's view that the move was unnecessary.

On the other hand, the *Rainbow Warrior*'s crew was emotionally affected by exposure to victims of radioactive fallout from the 1954 nuclear test. "These fourteen days had a lasting impression on the crew, motivating almost all of them to continue their activism," according to New Zealand journalist David Robie, who witnessed and later described Operation Exodus. "Their experience at Rongelap exposed to them the consequences of nuclear testing on these isolated South Pacific communities, stirring up powerful emotions," Robie wrote.

One irony: the *Rainbow Warrior*, once the evacuation was completed, left Majetto on May 31, 1985, and headed toward its new home in Auckland, New Zealand. There it was to prepare for a July voyage to Mururoa Atoll in French Polynesia, where the group would seek to halt nuclear tests planned by the French government. The ship never left Auckland. On July 10, 1985, two bombs, planted by French commandos, ripped open the trawler's hull, and the ship went to the bottom of the harbor. Moving the Rongelap people was the *Rainbow Warrior*'s final operation.

37 · CLEAN UP OF BIKINI AND ENEWETAK

■ **THE EXPOSED RONGELAP** and Utirik people have provided a look at what radioactive fallout from use of nuclear weapons would do to the long-term health of humans. What's happened to the Bikini and Enewetak Atolls has shown the long-term, destructive effects any use of nuclear weapons would have on future target areas and their ecosystems.

The US used Bikini and Enewetak Atolls from 1946 to 1958 as bases for sixty-six US nuclear tests, with detonations carried out on or above islands in their atolls. Of those tests, seventeen were thermonuclear devices. The total yield of all tests has been estimated at 107 megatons, or the equivalent of over 7,000 Hiroshima bombs.

Washington's continuing connection with Bikini and Enewetak has been through the Defense Department, which since 1946 has been primary user of those atolls for atomic and more recently anti-ballistic missile tests. For some Marshallese, the Pentagon became a separate source of income.

In 1956, the US drew up an agreement, which was signed by Paramount Chief Juda, the traditional leader of the Bikini people then living on Kili. It gave rights to use Bikini Atoll for future testing in exchange for $325,000. Testing on Bikini ended in 1958.

The US signed a similar deal in 1956 with the Enewetak people then living on Ujelang. The offer was $25,000 in cash and a $150,000 trust fund in exchange for full use rights to their home atoll. Eventually, after a vote by the Enewetak people, the deal was signed on November 19, 1956, by the two iroijs, Joanej and Ebream, on behalf of the atoll's two tribal groups.

The agreement was to run "until such time as it will not be necessary to occupy and use Enewetak Atoll in the interest of the maintenance of international peace and security." Much later, as Kwajalein Atoll became a key part of the Pacific range testing of US ICBMs and anti-missile systems, Enewetak was utilized as a "target and impact area" for tests of ICBMs launched from Vandenberg Air Force Base in California.

■   ■   ■

It was not until December 1966, eight years after testing ended, that the AEC, at the request of the Interior Department, agreed to determine whether Bikini Atoll and its lagoon were safe enough for its people to return. The AEC then conducted an extensive radioactivity survey of the atoll in April and May 1967. However, it was not until August 12, 1968, that President Lyndon Johnson announced plans for the return of Bikinians to their home atoll within the next two years. At that time, there were 300 on Kili, compared to the 167 that had left Bikini more than 20 years earlier.

Plans were prepared, but it was understood that living would be restricted to the islands of Bikini and Eneu, and then only after a layer of coral rock had been spread over the village sites. In addition, local food diets would be restricted, and returnees would face continued monitoring "to assure that they do not, in some unforeseen way, accumulate a greater radiation dose than predicted," according to an Interior Department memo at that time.

In June 1969, despite some lingering radiation, the first forty Bikinians returned to their home atoll. In 1974, greater radiation was discovered in the interior part of Bikini Island, and in June 1975, another Energy Department survey showed "that food plants were recycling radionuclides from the island's soil." Adding to the data was an AEC 1975 urine analyses of one hundred of the returned Bikinians, which showed cesium concentration had quadrupled and strontium doubled since similar tests that had been done in 1970.

After the cesium findings were released, Congress in June 1975, approved a $3 million "ex gratia" payment to the Bikinians "in recognition of the hardship suffered by the people . . . due to displacement from their atoll since 1946." The Trust Territory decided to add to the shipment of food to Bikini, but there were not enough ships for monthly service, so during this interim period, the returnees depended almost entirely on contaminated food grown on the islands.

In April 1978, Brookhaven researchers did whole-body counting measurements on the 139 returnees then living on Bikini Atoll and confirmed that many had acquired cesium-137 body burdens well in excess of the maximum permissible levels.

With the danger recognized, the Bikinians back on the Kili Council voted against the rest of the Bikini community returning to their atoll, and the Interior Department sought $15 million from Congress to prepare to move people off Bikini and back to Kili.

On August 18, 1978, Congress approved the $15 million, and added $3 million to a $3 million trust fund that had been established in 1975. At $6 million, the trust fund would produce roughly $14 per person per month. By September 5, 1978, three ships had taken the Bikini people and their possessions back to Kili.

Months later, following a new radiation study of all the Northern Marshall Islands, the US government informed the Bikinian Kili Council that Bikini Atoll was to be considered off-limits for at least the next thirty to sixty years.

■   ■   ■

The Bikini experience had an impact on US handling of the resettlement of Enewetak Atoll, made up of only 2.75 square miles of dry land.

In the four months before the US halted nuclear testing in August 1958, the AEC and Defense Department had detonated twenty-two near-surface tests at Enewetak Atoll in anticipation of Soviet Union acceptance of President Eisenhower's call for suspension of all atmospheric testing. The American flurry of tests took place on platforms, barges, or underwater, and thus created extensive fallout. Even after tests were halted, Enewetak continued to be used for years by the Pentagon as a target and impact area for the US anti-missile and ICBM test programs. Meanwhile, living conditions for the Enewetak Marshallese who had been moved in 1947 to Ujelang Island had become difficult. With food and water scarce, field trip vessels came at most three times a year. In March 1967, a UN mission reported housing built in 1947 by Navy Seabees had become dilapidated. The original population of 147 had increased to over 230, creating a housing shortage. A Trust Territory official visited in August 1967, and described Ujelang Village as presenting "an ugly, pathetic, and altogether depressing picture."

In May 1972, a group of Enewetak leaders, and their lawyer, visited their home atoll and made clear their desire was to return to their ancestral properties. "For Marshall Islanders in general, and Enewetak people in particular, land is a part of one's person and one's entire identity. . . . One's sense of self, both personal and cultural, is deeply embedded in a particular parcel of land on a particular atoll," was how Montana State Anthropology Professor Laurence Carucci once described land relationship before the Micronesian Claims Tribunal

In June 1972, faced with a US government public commitment to return Enewetak to the Trust Territory by 1973, the Air Force and AEC began to consider a cleanup program, as had been done in Bikini. The total of forty-three tests detonations at Enewetak created radiological conditions "too hazardous for unrestricted use of the atoll for future residents," according to an Energy Department report at that time.

Initial radiological studies, finished in April 1973, estimated the Enewetak cleanup would cost over $28 million. Another AEC study, finished in early 1974, dealt with whether or not all or any parts of the atoll could be safely reinhabited. An interagency task force report was released in June 1974 and presented to the Enewetak people in September of that year. Based on the study, AEC officials said that it could be five years before they could return to resettle Enjebi Island, second largest in the atoll and home to one of the atoll's two tribes. In addition, islands in the northwest portion of the atoll, where most of the tests had been detonated, would be permanently off-limits because of radioactive contamination.

An AEC master plan for Enewetak was presented to the US Congress in spring 1975, but not funded until June 1976, when the first $20 million was approved. Even then, the money was withheld until the secretary of defense certified that the Enewetak landowners agreed that those funds would be the "total commitment of the Government of the United States for the cleanup of Enewetak Atoll."

Initial cleanup operations began March 14, 1977. The first unexpected obstacle appeared in June 1977, when questions arose about what to do with soil and other matter contaminated with plutonium-239, whose half-life is 24,100 years. By August 1977, it was agreed that plutonium-contaminated materials from all Enewetak's other islands would be buried on Runit Island in a crater left from the Cactus test of an eighteen-kiloton thermonuclear device.

Plutonium-contaminated soil was mixed with cement, turned into slurry, and dumped into the crater. With a diameter slightly larger than a one-hundred-yard football field and a depth of roughly thirty feet, it would eventually hold about 105,000 cubic yards of highly radioactive waste. In 1979, the crater, filled with highly radioactive plutonium and other radioactive waste materials, was encapsulated under an eighteen-inch-thick, cement dome.

By 1980, 19,600 new coconut trees had been planted on the islands of Enewetak, Japtan, and Medren. Enjebi was to come later. Planting some 13,000 coconut seedlings on the northern Enewetak islands was originally delayed when the soil was found to have cesium-137 and plutonium levels equal to those in Bikini, which made the resultant copra unacceptable for commercial purposes. An AEC group determined in October 1978 that coconuts from those trees would also not be suitable for human consumption. Nevertheless, 10,700 seedlings were planted in February 1980, under the theory that "in the 8-to-10-year interim, the trees could harm no one, but contribute substantially to the ecological restoration of the islands," according to a DOE study.

As part of the cleanup, roughly six inches of surface soil was removed on those islands with some contamination, and where deeper excavation was needed, clean soil was brought in. Work went on ten hours a day, Monday through Saturday, with at least two hours each day taken up with people moving by boat from island work camps to island work sites. Overall, some four thousand personnel were involved.

The safety steps taken to protect the cleanup workers were extensive, according to a fact sheet put out in 1980 by the Defense Nuclear Agency

(DNA): "When earth-moving operations were conducted in contaminated areas, water sprinklers were normally rigged to minimize re-suspension of the soil. Personnel were trained (and directed on-site) habitually to remain upwind of any dust-producing operation. Air samplers were set up downwind to monitor for any airborne hazard. These power-operated samplers forced air through filters, which were monitored and changed at two-hour intervals, and subjected to laboratory analysis."

The 1980 DNA fact sheet claimed it was "the largest radiological cleanup operation ever conducted." Imagine trying to do such a complex cleanup after nuclear weapons were used if a country were in the midst of a war, or even after peace had taken place.

And then there would be the cost. By mid-1979, the director of the Defense Nuclear Agency estimated, "the total cost for cleaning up, rehabilitating, and resettling Enewetak Atoll could amount to about $100 million to $105 million," according to a GAO report at the time. The final figure was $104 million by the time the Enewetak people returned in 1980. That cost included not only the military personnel employed on the project and support activities but also Interior's rehabilitation costs and Department of Energy's technical assistance.

Pause for a moment and think what cleanup would be after a nuclear weapon fireball hit the ground in cities such as Washington, DC, Chicago, or New York. Or consider what it would be like trying to clean up the farmlands after fallout came down in Iowa, Nebraska, or Kansas.

Forty years ago, cleaning up 2.75 square miles (1,760 acres) of Enewetak Atoll took more than two years, some 1,000 people at a time, and over $100 million.

Just think what would be needed for decontamination if nuclear weapons were used against the US. For a hint, Chernobyl, so far, has cost roughly $235 billion for cleanup, relocation, and medical expenses.

We would be looking at trillions of dollars.*

---

* There would be other costs—the health of those working on the cleanup. In January and February 2017, legislation was introduced in the House and Senate to help veterans of the Enewetak cleanup get recognition which, under past laws, gave coverage to those service personnel who participated in the Nevada and Pacific atomic tests. Those Enewetak bills never got out of committee, and new ones were introduced in 2019.

In 1980, the Interior Department constructed 116 homes for the return of the Enewetak people on the atoll's three southern islands of Enewetak, Medren, and Japtan, along with community centers, piers, and other life support elements. Enjebi, which had been flooded by a one-hundred-foot wall of radioactive water generated by the 1952 Mike hydrogen blast, was at that time not considered safe for occupancy for another thirty years.

Beginning in early April 1980, 450 Enewetak people moved back to the three habitable islands. The homecoming ceremonies, on April 8, 1980, included presentation of two inscribed bells donated by the Cleanup Project and Rehabilitation Program personnel for the church towers at Medren and Enewetak. When it became clear that those hoping to return to their properties on Enjebi could not do so, a number of original returnees within a few days went back to Ujelang. Even more left Enewetak within the month and returned to Ujelang, leaving no one living on Japtan.

Those who remained, about 175 men, women, and children, have had regular radiological monitoring, the need for a supplemental feeding program, observance of special living patterns, and loss of cash crops. In addition, there is Runit Island and a warning, contained in a 1982 National Research Council task force report, that Runit's cement dome could be breached by a severe typhoon.

38 · MARSHALL ISLANDS INDEPENDENCE

■ **DURING THE CARTER** administration in the late 1970s and into the 1980s under President Reagan, US government officials dealing with Rongelap, Utirik, Bikini, and Enewetak negotiated a broad independence agreement with the Marshall Islands government and other Micronesian entities.

Called the Compact of Free Association, it gave the Marshallese and Micronesians self-government and the ability to conduct their own foreign affairs—as long as they were carried out after consultation with Washington. In return, the US government retained authority and responsibility for security and defense of these emerging, tiny, Pacific Ocean island nations.

The US agreed to end nuclear testing in the Pacific, and the Compact's Section 177 called for Washington to provide "just and adequate settlement" of all claims arising from past nuclear tests. At that time, the Department of Justice faced the task of defending pending court claims from Marshallese entities amounting to over $7 billion. Under the Compact, all those testing claims would be considered settled once the agreement came into effect.

Section 177 required the US to create a $150 million Nuclear Claims Trust Fund to be distributed by the Marshallese government. A new, Marshallese-run, Nuclear Claims Tribunal would deal with claims by individual Marshallese, replacing all previously filed federal court litigation.

The trust fund was supposed to yield $18 million a year during the first 15 years of the agreement. During those 15 years, the expected total income of $183.7 million would be divided among the people of four atolls affected by nuclear testing. Bikini was to get $75 million, Enewetak $48.7 million, Rongelap $37.5 million, and Utirik $22.5 million. After 15 years, 75 percent of annual earnings would go for uncompensated claims and other nuclear-related issues.

Special medical care was to continue for the then remaining 174 Rongelap and Utirik people exposed in the 1954 fallout, with $22.5 million promised via the Compact to support that effort for the next 11 years. In addition, US food programs for the Bikini and Enewetak people would continue, as would planting and other agriculture programs on Enewetak, the latter for only five years.

The US would also pay for a complete new survey of Rongelap, with the goal of restoring the habitability of the atoll. In addition, the Compact authorized, but did not appropriate, whatever US funds would be necessary "to restore the habitability of Rongelap Island and return the Rongelap people to their homeland. The latter step is to be taken in consultation with the Marshall Islands government and the Rongelap local government council." There were also provisions calling for a new US effort to clean up Bikini and Enjebi Island on Enewetak.

One of the most important long-term provisions granted all Marshallese citizens the right to reside and work in the United States without requiring a visa or labor certification, and with no limitation of how long they could stay.

After fourteen years of negotiations, Amata Kabua, then-Marshall Islands president, signed the Compact on June 25, 1983. US Ambassador Fred M. Zeder II signed for the Reagan administration.

It was approved by the Marshallese in a plebiscite on September 7, 1983, but with support from only 58 percent of the voters. Opponents, led by the Kwajalein Landowners Association, wanted a higher rent for their properties being used by the Navy, more than the $9 million annual payment contained in the Compact.

Passage of the Compact through the US Congress took longer than expected because of its costs, but it was finally approved and signed by President Reagan on January 14, 1986. However, it was not until early October 1986 that the US was able to inform the UN secretary general that the Republic of the Marshall Islands would assume its new, independent status on October 21, 1986.

On October 30, 1986, the new Marshall Islands government set up a Nuclear Claims Trust Fund account with the promised $150 million from the US. It took another five years to establish the three-judge, Nuclear Claims Tribunal along with procedures for handling individual claims and their distribution. The Tribunal established payments, from $12,500 to $125,000, based on twenty-seven medical conditions caused by exposure to radiation for all those who were physically present in the Marshall Islands during the testing program, not just survivors of fallout on Rongelap and Utirik.

By the year 2000, the Marshall Islands government saw that the Compact's $150 million would not be enough, and in September that year it submitted to the US Congress a "Changed Circumstances Petition," seeking additional compensation. That petition first noted that the fund had lost 15 percent of its value in the 1987 stock market collapse, and that it suffered additional later investment losses. Meanwhile, claims that came in were greater than expected, so that funds by the end of 2004 would be all but depleted.

One reason for the large number of claims was that the Nuclear Claims Tribunal had decided the payment pool would be all Marshallese who were on the islands at some time between July 1, 1946, and September 30, 1958, including those in utero at the ending date of American testing. That created a number far greater than those exposed to the Bravo fallout who were on Rongelap and Utirik.*

The Tribunal's argument was that the US had failed to monitor all Marshallese who may have been exposed during that twelve-year-period when the sixty-seven American tests had a combined explosive yield—and resultant fallout—that was one hundred times greater than all atmospheric tests conducted at the Nevada Test Site.

This scattering of the exposed Rongelap and Utirik people, along with the Marshallese alive during the years of testing, complicated health care for

---

* At the time of the Bravo fallout, the population on Rongelap and Utirik totaled 253 people. The Bikini and Enewetak evacuees had numbered 302. Nonetheless, the Nuclear Claims Tribunal awarded damages "to [all] persons throughout the Marshall Islands" at the time of the testing, which by 1996 numbered 10,919.

the fallout survivors and their families. Beginning in 1998, the Department of Energy contract for Rongelap and Utirik survivors was taken over by Hawaii-based Pacific Health Research Institute. Care was promised not only to people with radiological ailments but also comprehensive health care to all considered eligible. Pacific Health would have doctor visits four times a year at Mejato, rather than twice a year, along with year-round access to medical facilities on Majuro and Ebeye. Annual examinations were done on survivors living in Hawaii and the US mainland.

In 2011, the contract was taken over by International Outreach Services, a Hawaii nonprofit organization. In February 2014, Dr. Ashok N. Vaswani, who had worked at Brookhaven Laboratory on Marshall Islands issues, told a workshop in Japan that none of the Rongelap patients who were in utero at the time of the fallout developed thyroid cancers, although some developed nodules. Examinations carried out by International Outreach showed "their most prevalent disease is type 2 diabetes mellitus." Vaswani added, "The ubiquitous scourge of obesity, diabetes, and the metabolic syndrome as part of the lifestyle change is perhaps of greater clinical and medical significance than the potential for thyroid cancer."

A 2017 article on "Radiation Effects in the Marshall Islands," published under the auspices of the National Institutes of Health, reported, "Even though many of the Rongelap people, and almost all of the children, have already had thyroid surgery, most of them have residual thyroid tissue that remains vulnerable to neoplastic change. Despite the apparent decrease in the rate of occurrence of thyroid nodules, continued observation of the exposed Marshallese people must be pursued."

When the Compact ends in 2023, medical examinations for the fallout survivors will continue, as will the Section 177 health care for the four affected atolls including Rongelap.

■   ■   ■

The Compact's provision allowing any Marshallese to freely seek employment in the US, whether in Hawaii or the mainland, has led to migration from the home atolls. In 2010, according to census figures, there were just over 4,300 Marshallese in the US mainland. As of February 2018, some 12,000 Marshallese were living in Northwest Arkansas and in nearby towns in Oklahoma, Kansas, and Missouri. They had been drawn by work in local poultry plants. While they paid US income taxes, the Marshallese in the US, as Compact authorized migrants, were barred from receiving most forms of federal assistance including Medicare, Medicaid, and Social Security.

The long-term effect of radiation directly affected the bodies of those who had lived for some time on Rongelap, Bikini, and Enewetak. But also having a negative health effect were the years of US government-provided, processed food that had to be eaten to limit their eating radioactive, locally-grown foods.

As outlined in a 2005 hearing before Congress, the Western diet made available through the US Agriculture Department for free food distribution to the Marshallese was high in fat, high in carbohydrates, low in fiber, and lacked Vitamin A and iron. In addition, reduced physical activities, such as food gathering and fishing, had led to a more sedentary lifestyle for the islanders, which in turn led to diabetes, atherosclerotic diseases, and hypertension.

Under the Compact's Section 177, the Department of Interior has managed what's called the Four Atoll Health Care Program for Marshallese with roots in Rongelap, Utirik, Enewetak, and Bikini. The successor to Dr. Conard's annual examinations, it provides comprehensive health care to some eighteen thousand Marshallese.

But the focus of the medical care has changed because so many Marshallese have turned up with diabetes. In September 2018, the Interior Department had to add $188,000 in supplemental funds because, "The number of diabetic patients this year has reportedly increased by 86 percent over prior year cases," according to a department press release.

A study by the director of the Office of Community Health and Research of the University of Arkansas, published in the March 2016 *American Journal of Health Behavior*, reported that 46.5 percent of the roughly 12,000 Marshallese adults in Arkansas had diabetes, which was traced to the radioactive contamination from nuclear tests that "altered . . . traditional, subsistence lifestyle of the Marshallese."

■   ■   ■

Another new issue for the Marshallese relates to climate change and the environment. The Marshall Islands are among the most threatened land masses in the world. Rising seas and dwindling rainfall will likely make the atolls uninhabitable as early as midcentury, according to various studies. The more than one thousand atoll islands that make up the Marshalls sit an average of less than six feet, or two meters, above sea level. Some places are only three feet above sea level.

A draft National Climate Assessment, released in January 2013, noted the drop in rainfall in the Marshall Islands over the period of 1950–2010. On Kwajalein there were 0.3 inches less of monthly rainfall per decade and 0.4 inches less of monthly rainfall per decade in Majuro. Freshwater supplies obtained from aquifers and catchment, while already constrained, will decline even further as a consequence of warmer, drier weather and saltwater intrusion from rising sea levels.

"Nowhere else in the world is the threat posed by climate change more immediate and real than it is here," then-Marshall Islands President Christopher Loeak told a United Nations meeting in Majuro in April 2014. "The rising oceans around us, our disappearing coastlines, the increasing salt in our fresh water, and the corroding coral beneath our feet tell us loud and clear that climate change is here," he said.

■   ■   ■

Also looming is the end of most US financial assistance to the Marshalls, now scheduled to end by 2023, based on provisions in the 2003 amended Compact. Under those provisions, the US government has made contributions to a trust fund for the Marshall Islands government that was to provide the future base for its financial self-sufficiency. The trust fund, to which the United States, Taiwan, and the Marshall Islands governments have annually contributed, totaled $290 million in 2016.

The United States will provide the Marshall Islands with approximately $70 million annually through 2023, according to the State Department. This payment has included contributions to a jointly managed trust fund and financial assistance from other US federal grants.

US Compact payments to the Republic of the Marshall Islands (RMI) government in 2017 totaled $31 million and provided approximately 60 percent of the RMI annual budget for health, education, and infrastructure development and maintenance. The United States will continue to lease the US base on Kwajalein Atoll, so national security ties will remain strong after 2023.

However, the Kwajalein base rental money, which under a 2011 lease agreement now amounts to $20 million a year, has been going to private Kwajalein landowners. Family members of three current and former iroijs (Marshallese chiefs), including those of former President Amata Kabua, own one-third of the property. Kabua, who died in September 2019, had managed the rental negotiations, which went on for eight years. When agreement was reached in 2012, and runs through 2066, Kabua and the other landowners took in $32 million in back rental payments that had been built up in an escrowed account in Marshall Islands banks. That distribution from the escrowed account alone reduced total bank deposits in the Marshall Islands banking system by 13 percent in 2012, as money was transferred to Kwajalein landowners living in Hawaii.

At a March 2018 meeting, the Trump administration's Department of Interior assistant secretary for Insular and International Affairs, Doug Domenech, told representatives of the Republic of the Marshall Islands, "I recognize there are both technical and administrative issues that the United States and the Republic of the Marshall Islands need to address before we approach 2023 and the transition to trust fund distributions."

One of them was the legal status, post-2023, of the Marshallese already living in the United States under the Compact but who were not American citizens. Their future status has yet to be determined.

39 · THE ATOLLS UPDATED

- **THE FOUR ATOLLS** impacted by US nuclear tests, Rongelap, Utirik, Bikini, and Enewetak, have seen their leaders take different approaches over the past years.

The Bikini Atoll Rehabilitation Committee had been placed in charge of its trust fund, and a tourism program was begun in the 1990s. Visitor quarters were established for sixty to one hundred people, a power plant and electric distribution system built, along with a dock and small boat landing. By 1996 there was an airport system with a lighted runway, renovated airport building, and refueling station so that a tourist underwater divers' program could be established.

That year, the Bikini Atoll Council approved a tourist diving operation that allowed limited numbers of experienced divers to explore sunken warships, such as the carrier *USS Saratoga*, that rest at the bottom of Bikini Lagoon as relics of the 1946 Crossroads nuclear tests. Charging up to $5,000, the funds raised provided additional income for those on Kili.

In 1998, when the first clearing of some three hundred acres on the lagoon side of Bikini Island had been completed in preparation for radiological cleanup, the project was put on hold after the Nuclear Claims Tribunal adopted the fifteen millirem/per year standard. For the next eight years, the

Bikinians upgraded the dive program, doubling the capacity to produce oxygen for divers' tanks and built four-plex living quarters for tourists.

On April 11, 2006, the people of Bikini, along with those of Enewetak, made a new effort to get the $563 million awarded by the Nuclear Claims Tribunal by suing the US government for the money in the US Court of Federal Claims. That court turned down the Bikini claim in August 2008, and that decision was affirmed by the US Court of Appeals for the Federal Circuit on January 29, 2009, saying it was without jurisdiction given Section 177 of the Compact Agreement, which withdrew jurisdiction for all US courts. On April 5, 2010, the Supreme Court declined to take case, ending that legal avenue for additional compensation.

Work on Bikini's resettlement program ceased in 2007, and in the years since only a small team remained on the island to maintain facilities. Tourism continued, although airline service ended in 2008. The resort aspect has since been served only through ships that have brought divers to search sunken warships, resting at the bottom of the Bikini Lagoon.

On July 31, 2010, the United Nations Educational, Scientific and Cultural Organization (UNESCO) World Heritage Committee designated Bikini Atoll as a World Heritage site, citing how Bikinians had "conserved direct tangible evidence that is highly significant in conveying the power of the nuclear tests, i.e. the sunken ships sent to the bottom of the lagoon by the tests in 1946 and the gigantic Bravo crater." In seeking the UNESCO recognition, Bikini's Senator Tomaki Juda wrote that he wished "for the world to remember the role of our tiny atoll in the global politics of the 20th Century—for the role of the Bikini tests in the start of the Cold War and the nuclear arms race."

The Bikinians, during the Obama administration, wanted terms of their trust fund changed so the money could be used to relocate them elsewhere, including the United States. While the island may have been considered habitable at that time, virtually all of the Bikinians alive in 2013, about 4,880 at that time, had never lived there.

In 2015, Congress allowed the Bikini people to spend 90 percent of the trust fund's earnings to support those on Kili and Ejit. Both over-populated, low-lying islands had been hit by climate change and faced more and more flooding from rainfall and rising sea levels. A Senate report that year said, "The people on these islands have limited living space, lack suitable sustainable resources to provide water and food for their population, and they are exposed to tidal flooding on an increasingly frequent basis."

Four times a year, the Bikini Claims Trust Fund, based on very strict rules, has paid about $120 per each qualified Bikinian person, or about $480 annually, according to a former official of the Kili/Bikini/Ejit local government.

In August 2017, after President Trump's election, the Kili/Bikini/Ejit Local Government Council (KBE Council), the elected governing body of the Bikini people, voted to take total control over the Resettlement Trust Fund, ignoring US congressional language that created the fund. By statute, the Interior Department had to approve expenses other than for resettlement. This newly-elected KBE Council wanted to use the funds to provide for current needs of the Bikini people and to invest in "income-generating projects" that would provide long-term support for Bikinians.

In response, the Trump administration agreed to give up approval of the trust fund's disbursements. On November 16, 2017, Interior Assistant Secretary for Insular Areas Douglas W. Domenech gave the KBE Council total authority and direction over the current Trustee of the then roughly $70 million fund. However, Sen. Lisa Murkowski (R-AL), chairman of the Senate Energy and Natural Resources Committee, with congressional authority over the matter, questioned the legality of Interior's ending its authority over use of the trust fund's money.

At a February 2018 hearing before her committee, Murkowski said, "I am very sensitive to the notion that Washington, DC, should not dictate local government decisions. I am also mindful, however, that the trust fund was established for the people of Bikini, with its statutory purpose being the 'rehabilitation and resettlement of Bikini Atoll' as a result of the United States' nuclear testing." The Alaska senator noted that Bikini Mayor Anderson Jibas had approved a resolution that resulted in $11 million being withdrawn from the trust fund. She added, "We've heard reports of large sums being spent on things such as an airplane, two landing craft, an elaborate function in Hawaii, and cash payments to households on Ejit and Kili in the name of disaster relief, but without any damage assessment being conducted."

At that hearing, Mayor Jibas said some of the recent $11 million withdrawal from the trust fund was spent on renovating and building new houses damaged by high tides, but that he and his colleagues wanted to fund projects that would generate future income.

Marshall Islands ambassador to the United States, Gerald M. Zakios, told the committee his government "was quite surprised" at the Trump administration's "relinquishing its responsibility of administration and oversight of these funds." He added, "We honestly believe that had some prior consultations taken place before this action, arrangements could have been made in the RMI to provide a proper framework to assure accountability of the Bikini Resettlement Fund."

Murkowski introduced a bill that would limit annual withdrawal from the fund to about $2 million annually, an amount that was 5 percent of its market value. Her legislation would also give the Interior secretary specific authority to disapprove withdrawals until such time that a new resettlement plan is submitted to Congress.

Overall, as of 2021, the population of Bikini Atoll descendants has exploded to 6,600, although only 11 had survived from the 167 that lived on the atoll at the time of the 1946 testing. Some 2,550 lived on Majuro; 800 on Kili; 300 on Ejit; 350 on other Marshall Islands; and 1,400 in Hawaii. Others live in communities across the US mainland, including 250 in Springdale, Arkansas.

■   ■   ■

The Enewetak Atoll resettlement program focused living quarters and village life on the two southern islands of Enewetak and Medren, which before cleanup had mostly been used for various housing and other facilities required by the nuclear testing program. Many of the remaining islands of the atoll that had been used for the forty-three nuclear tests were devastated and nearly all vegetation destroyed. Thus, resettlement had to include revegetating those portions of the atoll that could sustain food growing.

In 2000, the Enewetak/Ujelang local government signed an agreement with the Energy Department to establish the Enewetak Radiological Laboratory, which was completed in May 2001. The laboratory included a permanent whole-body counting system, to assess radiation doses from internally deposited cesium-137, and laboratory areas for collecting bioassay samples. Scientists from the Lawrence Livermore National Laboratory have continued to support the operation of the facility. They have monitored the level of "fission emitting radionuclides" in the bodies of the resettled islanders, ingested through locally grown foods or inhalation of contaminated dust particles in the air.

In a 2008 article, Montana State University professor of anthropology, Dr. Lawrence M. Carucci, who had spent time in Enewetak, wrote that the resettled Marshallese "cannot be at home in the very land that is their home, since the contours of the land are no longer the same. Its productive capacity is lacking, and, without those products, the wide array of day-to-day activities that allowed people to make local products into canoes, and sleeping mats, and foods, have lost their meaning. For nearly twenty years, people have not been able to make themselves into 'real Enewetak people,' since the materials required for this self-fashioning are not available to them. This is the grand contradiction of life on Enewetak."

Lawrence Livermore scientists regularly examine the Runit cement dome and the leakage of radioactive waste into the adjacent Enewetak Lagoon. In a 2010 report, the Livermore team said, "Extensive environmental studies conducted over the past two decades continue to show that the radionuclide concentrations in lagoon water and sediment, and fish collected in close proximity to the Runit Dome are very similar in range to that observed in samples collected from other parts of the lagoon. Based on these data there appears to be no evidence that Runit Dome is significantly impacting the environment." The report added that as a safety matter, the scientists are using the whole-body counter on Enewetak regularly to monitor the resettled Marshallese and have seen no unusual increase in their radiation levels.

In 2016, the Defense Threat Reduction Agency (DTRA), which regularly has reported on monitoring of Enewetak, released a fact sheet that read, "Today, all of the atoll islands and the lagoon are accessible with the exception of Runit Island." Runit Island, site of the buried highly radioactive materials from the previous testing, although quarantined, regularly has been visited by researchers and tourists.

The present residents of Enewetak live in cement, two-story houses mostly built before 1980 and survive primarily on locally-caught fish and US-supplied supplementary food, such as canned tuna and white rice, which has made up over half the average diet. The atoll's outer islands are not considered safe for the harvesting of bird eggs, turtles, shellfish, and other food resources, which once made up their diets.

Assisted by the University of South Florida, the Enewetak community is involved in an effort "to provide greater amounts of locally produced food and to better integrate necessary imported food into the local diets," according to the Interior Department. "The replanted vegetation is producing at pre-nuclear testing period levels, when the population was about 150 people, but is not sufficient for the current population of about 800 people," according to the department.

Many residents use their share of the Enewetak and Enjebi Trust Funds, which has provided each with approximately $98 quarterly, for loans to buy goods. Others paddle or sail to the northern islands to gather copper wire and pipe from still existing abandoned test sites or waste dumps for sale to visiting tradesmen. It is a hard daily life, made tougher given the occasional winds, rainstorms, and droughts.

■   ■   ■

Utirik in 2018 had a population of 435, occupying a land area of less than one square mile.

Of the 167 residents who were on the atoll when the fallout came after the 1954 Bravo test, which included eight in utero, ninety-eight were still alive in 1994, and only forty-two in 2014. In 2012, only six of those survivors actually lived on Utirik.

Most Utirik fallout survivors have resided in Majuro or Ebeye, with a handful in Hawaii or the United States. For those living on Utirik, a 1999 radioactivity study showed that at a safe four millirem per year, no impact on health was expected from the external exposure and ingestion of cesium-137, plus some strontium-90, in local food crops such as coconut, pandanus, and breadfruit.

Nevertheless, under a 2002 agreement between the Utirik Atoll Local Government and the Department of Energy, a whole-body counting facility was established in 2003 in Majuro to serve the Utirik survivors and general Marshallese public. It has been supervised by scientists from Lawrence Livermore National Laboratory but maintained and operated by Marshallese technicians.

Utirik Atoll residents have received whole-body counts during their scheduled visits to Majuro as part of their routine Department of Interior-supported medical surveillance program. Under US legislation, those exams will continue through 2023.

Utirik's Claims Trust Fund stood at $13.6 million in 2010, according to an audit of the Utirik Atoll Local Government, completed in 2015 by Deloitte & Touche LLC under authority of the Marshall Islands auditor. The fund in 2010 paid out $1.5 million in claims arising out of the US nuclear testing program for health issues or loss or damage to property to the people of Utirik.

As of 2014, nineteen Utirik women had turned up with thyroid nodules. Of those women, three later developed thyroid cancer. Six Utirik men developed thyroid nodules, and one turned cancerous. Remembering how small the exposed dose was on Utirik, 301 miles from Bikini, the nodules on average showed up between twenty-five years and thirty years after exposure.

Under the DOE health program, those exposed persons are examined once or twice a year, although care is primarily limited to radiogenic issues. Examinations do not include individuals who subsequent to the 1954 fallout have lived in Utirik's radioactive areas.

In August 2018, Charge d'Affaires of the Japan embassy, Hiroshi Watanabe, and Utirik Mayor John Kaiko signed an $83,000 grant contract for construction of a base for fishing at Utirik Atoll. The Utirik Atoll Local Government will construct a fish base equipped with three freezers and one ice machine. A solar system will be installed on the fish base funded by other donors. Once built, Marshall Islands Marine Resources Agency will start trips to Utirik to collect fish.

40 · RONGELAP TODAY

• **THE DEPARTMENT OF** Energy, under the 1986 Compact of Free Association, directed a resurvey of Rongelap that in 1989 found the atoll was "safe for habitation by adults, provided the diet is equivalent to that formerly used." However, the plutonium and radiation doses that would affect infants and small children needed further study. Those conclusions left the Rongelap people needing greater assurance before they would return to their atoll.

An earlier Departments of Energy and Interior memorandum of understanding with the Rongelap government set a "primary condition" for initiating resettlement of Rongelap Atoll as one where "the calculated maximum whole-body radiation dose equivalent to the maximally exposed resident shall not exceed 100 millirem/year above natural background, based upon a local food only diet." That "local food only diet" meant "local food taken, grown and/or gathered from the southern islands of Rongelap Atoll and the immediately surrounding waters," according to the memo.

At a 1991 Senate hearing, Peter Oliver, undersecretary of Commerce for the Republic of the Marshall Islands pointed out, "Unfortunately, the people of Rongelap never found the practical advice and administrative support that became available to the other two Marshallese communities [Bikini and Enewetak] that were gravely impacted by the US nuclear testing program. Nor were the people of Rongelap granted the large sums of money for decontamination and resettlement given to these other communities." He made clear the Marshall Islands government expected the US eventually to make funds available for the cleanup of Rongelap, as had been done for Bikini and Enewetak, because of Defense Department participation.

Congress in 1992 provided $2 million for further government studies, including one by the National Academy of Sciences, to answer the question of whether Rongelap was safe for rehabilitation. Meanwhile, a Rongelap Habitability Work Plan was initiated.

The National Academy of Sciences report, released in 1994, entitled, "Radiological Assessments for Resettlement of Rongelap," came up with a troubling conclusion: "Some people returning to Rongelap, and subsisting on a local-food-only diet, might receive an annual dose in excess of 100 millirem above background, if there is no remedial action."

The report recommended costly cleanup actions similar to those carried out by the Defense Department on Enewtak to reduce the dangerous cesium-137 in the soil. It recommended removal of "the upper layer of surface soil from the village and on each house plot," adding, as also needed, "a layer of crushed coral around the houses and to common areas of congregation throughout the village. Both these actions are recommended as they could be accomplished as housing is rebuilt on Rongelap Island before resettlement with little impact on resettlement activities."

A second recommendation called for application of a potassium chloride fertilizer in Rongelap agricultural area to "reduce the uptake of cesium-137 by coconut, breadfruit, and pandanus fruit." That fertilizer had already been proven successful in reducing cesium-137 in coconuts grown on Bikini and in increasing plant growth in the relatively potassium-deficient tropical soils of the Northern Marshall Islands, according to the report.

On September 19, 1996, an agreement was signed by US Interior Secretary Bruce Babbitt on behalf of President Bill Clinton to assist the Rongelap people with $45 million to fund resettlement of their atoll. As described in a later House of Representatives report, the money was "in lieu of appropriation of the full amount of $90 million . . . [but] at a lesser cost to taxpayers by contributing to the trust fund lesser amounts and encouraging methodical and cost-effective resettlement."

In 1998, the Rongelap government contracted for the first phase of the resettlement program, which included establishment of a base camp, the construction of essential infrastructure, and implementation of cleanup measures proposed by a team from Lawrence Livermore National Laboratory. That team had recommended removal of several inches of topsoil in the housing and communal areas and replacement with clean crush coral sand to remove most external fission products, and particularly cesium-137.

To eliminate the cesium-137 from the food chain, the agriculture areas were seeded with potassium chloride fertilizer, which American researchers in the 1980s had discovered would block 95 percent of cesium plant uptake while enhancing growth. That combination would allow consumption of traditionally grown foods, including coconuts.

On Rongelap between 2001 and 2010, local government officials supervised multimillion-dollar construction projects that were to provide for future resettlement of their home atoll. Included were a 3,950-foot runway and small airport terminal, a dock and large concrete pier, a power plant, a desalinization water-making facility, a school, reconstructed church, paved roads, and nine of a planned fifty homes.

Following advice of US government scientists, areas where community facilities and homes were located had fifteen inches of topsoil scraped off and replaced by crush coral rocks. The ground with food crops such as coconut trees was spread with potassium fertilizer to block their roots taking up radioactive cesium-137.

The Energy Department in 2010 had done replanting in Rongelap to help identify any issues associated with using groundwater as irrigation for growing vegetables, rather than reverse osmosis filtered water or cistern water.

The general aim of the Rongelap garden projects was "to develop data on the uptake of the cesium-137 and strontium-90 in leafy vegetables, and other root and grain crops," Glenn Podonsky, a health and safety officer of the Energy Department told the House Subcommittee on Asia in May 2010. "Such activities will allow the DOE to address future concerns about the potential impacts of changes in diet on exposure conditions in the Marshall Islands," he added.

In short, along with the Marshallese themselves, their exposed land continued to be used for experiments that, in addition, allowed the US to gather information on the long-term effect of low-level radiation on the environment.

At the same time, the Department of Energy developed with the Rongelap Local Government Authority a radiological surveillance program for any return-ees to the atoll. That included providing on Rongelap a whole-body counting facility and adjoining health physics laboratory. During reconstruction, the 115 workers on the island were monitored, and none showed elevated levels of plutonium in their urine, according to Lawrence Livermore scientists who carried out studies in 2010.

In early 2010, with tens of millions of dollars already invested in the cleanup, members of Congress and Obama administration Interior Department officials publicly announced they wanted to see Rongelap resettled by the end of the fiscal year 2011. In response, the Rongelap local government council, on March 30, 2010, passed a resolution that allocated $7 million more in resettlement trust funds to be used for moving those willing from Mejato to Rongelap by October 1, 2011.

Late in 2011, a number of US Interior Department officials and local Mar-shallese leaders, including Rongelap Mayor James Matayoshi, went to the atoll to inspect progress on housing construction. Matayoshi, who lived and worked in Majuro, was a Japanese American whose mother had been one of the Rongelap fallout survivors. After the Rongelap visit, he said that all fifty houses on the atoll "should be completed over the next six to nine months."

They were, but the major resettlement never took place.

In 2012, Mayor Matayoshi said the local government was not forcing anyone to return to Rongelap, but said the US had changed its position. "It's up to the people to make their own decision about whether to resettle," he said, adding that older Rongelapese were concerned about the safety of other islands in the atoll which had not been cleaned up.

The Rongelap Resettlement Trust Fund had been spent down to about $11 million, forcing the local government to take a conservative approach to spending. "If the Trust Fund falls under $10 million, we can only use 50 percent of the income and we'll have to reinvest the rest," Matayoshi said.

In 2016, the Marshall Islands government released an environmental report which said only "about one-third" of Rongelap Island had been cleaned up, and that US authorities had instructed that returnees were "not to venture into the other parts of the island." The Rongelap people, as a result, held back and began to "insist that the island should be remediated fully so people are not exposed to potential harm."

The 2016 Marshall Islands government report's main conclusion vis-à-vis local food growing was not promising. It said, "The tests of the 1940s and 1950s have forever changed the terrestrial environment and human interaction with the environment by reducing the already scarce area available to grow annual and perennial crops, by rendering crops such as coconuts unusable, by removing thousands of metric tons of topsoil, and by the vaporization of several islets."

Deloitte & Touche LLC audited the Rongelap Local Government's finances for the year ending September 30, 2015. At that time, the holdings of the Rongelap Resettlement Trust Fund stood at $6.2 million, according to the audit which was posted in July 2018 on the website of the Marshall Islands Office of the Auditor General. During 2015, some $1.5 million went for resettlement expenditures.

The audit showed that the Resettlement Trust Fund had been used for investments that appeared promising but had yet to pay off. The major investment went to Aquaculture Technologies of Marshall Islands Inc. (ATMI). Founded in 2013, ATMI was a Marshall Islands start-up company that raised Pacific Threadfin, commonly known as "Moi," for export to Hawaiian and Taiwanese wholesale markets where this fish is considered a high-grade, sushi-quality product.

The Rongelap Local Government invested $300,000 in ATMI, owns 51 percent of its stock, and has loaned the company more than $500,000. A US AID $1.7 million grant in 2015 allowed the company to purchase a feed mill machine to manufacture its own fish feed. ATMI began supplying surplus feed to other fish farmers in the region. Although not yet profitable, ATMI expanded.

Mayor Matayoshi announced in 2016, "The company has exceeded $100,000 in sales as a start-up company." The hatchery produced fingerlings in numbers that exceeded the volume available in sea-cages, which had been placed in lagoons at Majuro and Rongelap. In those floating cages, the fish grew into larger fish. As part of ATMI growth, an additional six sea-cages were established, four in Rongelap, as part of expansion into full commercial phases. The 2016 goal was to harvest, process, and sell fifty thousand pounds of Moi.

Not all of Mayor Matayoshi plans have succeeded. The Deloitte audit showed a 2010 equity investment of $5,000 in a Nevada communications firm named AcionMobile Inc., along with a $400,000 loan to that company, both from Rongelap Local Government funds. By 2015, both investments had to be written off with news from Nevada that AcionMobile was no longer an operating company.

In April 2018, at the Asia World Expo, Matayoshi announced a radical plan. He proposed turning Rongelap into a "Special Administrative Region," much like Hong Kong. The atoll would be a site for foreign investment with relaxed visa requirements, a new tax-free shipping port, and services for offshore companies registered in Rongelap. A Chinese businessman with Marshallese citizenship was behind the project, which raised concerns from the US embassy.

Matayoshi's plan had not gotten support in the Marshall Islands Nitijela legislature, which would have to pass a law to permit the project to happen. With Rongelap's people scattered over the Marshall Islands and the US, the future of the rebuilt but unpopulated atoll remains uncertain as of this writing.

With their home atoll unavailable, the Rongelap community has become more dispersed. In 2002, there were 360 Rongelapese on Mejato; ten years later, in 2012, there were only 150. The others had settled in Majuro, Hawaii, and mainland United States. Of the eighty-six Rongelap fallout survivors from 1954, twenty-five were alive in 2014, with only six of them reported residing on Mejato.

In 2018, only several dozen people lived on Rongelap, primarily workmen and employees of the Marshall Islands or US governments.

Descriptions of contemporary Rongelap were captured by vacationing sailors who passed through the atoll in recent years. In January 2013, Pete and Daria Friday sailed their boat, *Downtime*, into the lagoon off Rongelap Island. On their website they recorded, "We anchored next to a small resort with six bungalows on the beach. The resort was just a year old and sat empty."

The next day they went ashore "to meet the locals, all 50 of them! We approached the newly constructed concrete wharf that most likely cost more than the whole island and tied up," they wrote. "We walked ashore into what looked like a ghost town! No less than 50 new homes with shiny tin roofs and gleaming white block walls stood all around us with only three people in sight, sitting under a big fruitless mango tree next to the generator station."

The workers on the island were living in trailer-like structures that "all had a wall unit pumping cold air . . . and were sitting the hottest part of the day out in refrigerated comfort. As we walked around wondering who was going to live in all these houses we saw a 30-foot-wide, freshly paved road heading south into the jungle towards the beach. As we walked down the middle of the road we had no fear of traffic because there were only four cars that we counted in the entire settlement. The road continued on along the shore and then turned right to a newly built airport terminal that was surrounded with thousands of square feet of asphalt paving. Was I missing something? There are all of 50 people on the island and they have to pave a road to and from the airport that is 500 yards away from town center? Will the homes be given to people that were relocated 60 years ago? Will the country buy a new plane so the airport will be of any use? Who knows?

"The generators chugged on 24 hours a day consuming more diesel [fuel] per person per day than the average person would use in a year if he had to buy it. . . . You have to wonder where it will all end. . . . People will move back out here after having lost all their skills as islanders from the hustle and bustle of Majuro and then what? Learn how to fish? How to farm radioactive soil? I just don't know. . . . It sadly reminded me of what we did to the Indians back in the states."

■   ■   ■

On June 15, 1977, the former Rongelap Magistrate John Anjain and a small delegation of fellow Marshall Islanders appeared at an unusual 7 a.m. meeting of the Senate Energy and Natural Resources Committee to press their claim for further compensation.

I covered and wrote about that session, having been to Rongelap with Dr. Conard three years earlier, and aware of what had happened to the

Rongelap people when the radioactive fallout came down like snow for hours.

In the intervening years, Anjain's wife had contracted cancer. His son, just a year old the day the ash fell out of the sky, had died of leukemia. Anjain and another son had their thyroids removed to avoid cancer.

Sen. Howard Metzenbaum (D-Ohio), who chaired that morning hearing, apologized for "the exigencies of time," and allotted 15 minutes for Anjain's testimony, while saying his written statement was "very, very, moving," and would be included in the record.

Anjain said he had nothing to add to his statement, speaking through another islander with better command of English.

His statement had ended:

"Now, it is 23 years after the bomb . . . I know that money cannot bring back my son. It cannot give me back twenty-three years of my life. It cannot take the poison from the coconut crabs. It cannot make us stop being afraid."

41 · EPILOGUE

■ **PEOPLE TODAY HAVE** forgotten, if they ever knew, what a single nuclear weapon can do.

That has been an obsession of mine for the fifty-plus years that I have been writing about national security affairs.

When the first two atomic bombs were dropped over Japan in August 1945, I was twelve years old, spending the summer swimming and playing baseball at Schroon Lake Camp for Boys in New York's Adirondack Mountains.

While newsreel and newspaper pictures of mushroom clouds became fixed in my mind, the actual devastation never was real to me. All I knew was that the Second World War would soon be over and that was enough.

In the following years, as nuclear testing began, I remember sitting in the Fantasy Theater in Rockville Centre, my suburban hometown of twenty-eight thousand on Long Island, New York, as the "News of the Day" newsreels at our Saturday afternoon double feature showed the various explosions out there in the South Pacific. In the 1950s, as testing moved to Nevada and then back to the Pacific, there was little talk of radioactive fallout in the eastern part of the United States. But we followed stories about fallout as it drifted over Europe and Asia.

By the early 1960s, I was working in Washington and well aware, through newspaper and television coverage, that radioactive fallout from Pacific and Nevada test shots had resulted in cows in Denmark eating grass and scientists measuring strontium-90 levels that had turned up in the milk produced for Europeans. I drank a lot of milk then and still do.

It was in February 1966, well after the 1963 atmospheric test ban treaty, that I first wrote about the impact of nuclear weapons. It was a rather flip, three-paragraph note in the *Reporter Magazine*, which no longer exists. The story concerned a law that had passed Congress the previous month, a measure which required the US government to pay $11,000 to each inhabitant, or their survivors, who had been on Rongelap the day the atoll had been dosed with radioactive ash from Bravo.

The April 26, 1986, accident at the Chernobyl nuclear power plant near Kiev, where radioactive fallout from a reactor's explosion and fire spread as far north as Sweden, reopened my interest in what had happened to the Marshallese. Two weeks later, on May 13, 1986, I wrote an article in the *Washington Post* that compared what happened after the Bravo test to the situation following Chernobyl, at one point noting that, like the Soviet Union, the US government initially had delayed admitting what had occurred.

I wrote at that time, "The question of public disclosure was eventually overshadowed by other serious issues—again, perhaps foreshadowing Chernobyl—including the question of how the released radiation would affect the health of the exposed individuals and the land."

However, it was not until I began research for this book in 2013 that I realized the initial worldwide knowledge of Bravo's radioactive fallout had little to do with the Marshallese on Rongelap. The first international attention to fallout came from disclosure of what had happened to the twenty-three Japanese crew members of the *Lucky Dragon*. Although some of the crew saw a reddish flash west-southwest of the boat, and heard an explosion some seven minutes later, it was another two or three hours before a fine white dust began to come down on the boat. With a light rain, the radioactive dust continued to settle on crewmen and the fish on the deck as they worked for more than five hours to bring in their lines.

There was no such fallout from Hiroshima or Nagasaki, where both bombs exploded 1,800 to 2,000 feet high and their fireballs never reached the ground.

Radioactive fallout and its long-term effects—few people today realize—would be the result from any future nuclear weapons explosion that touched the earth's surface.

Fallout, when it occurs, does not just affect the target but also the surrounding areas which could be as far as hundreds of miles away. And the effects could last for years, if not decades thereafter.

It should be noted that remaining concentrations of radioactive materials from fallout created by US nuclear tests in the Northern Marshall Islands in the 1950s still far exceed those resulting from the July 1986 nuclear power accident at Chernobyl, according to a study published in July 2019 in the Proceedings of the National Academy of Sciences.

"Background gamma radiation and soil activity measurements in the northern Marshall Islands" from plutonium and cesium in the soil were "significantly higher" on Rongelap and other atolls than levels that resulted from fallout from the Chernobyl accident, according to the study done by researchers from Columbia University.

Through this book I want to remind people of the long-term health and environmental damage these weapons could cause if ever used again in war. The Rongelap Marshallese and Japanese seamen who were exposed to fallout on March 1, 1954, can be seen as surrogates for anyone caught in a future nuclear war. The tiny islands of the atolls of Bikini, Enewetak, and Rongelap, for the most part, still cannot be inhabited, despite attempts to decontaminate them, more than sixty-five years later.

In telling of these people, I hoped to show how much is owed to Marshall Islanders who were living simple, isolated lives far away in the South Pacific but who, like the handful surviving Japanese seamen, are symbols of what would be the unthinkable short- and long-term medical results should nuclear weapons ever again be used.

Bravo was America's deadliest nuclear test, but deaths did not occur immediately.

In the first years after the fallout, there were additional miscarriages and still births among exposed Rongelap women. The Marshallese blamed the fallout while Conard and other doctors found no direct link.

Marshallese still refer to fallout as "the ashes of death."

Death from cancer probably would have accompanied the thyroid growths that began appearing nine years after exposure to almost all Rongelap's children under twenty years of age, had their thyroids not been surgically removed. Those surviving children, now adults, still must take thyroid pills.

Eighteen years later, fallout-created cancer did kill nineteen-year-old Lekoj Anjain, as well as his mother, perhaps his father, and others.

But Bravo's fallout in a broader sense also killed the Rongelapese way of life by making their home islands radioactive and forcing them to live elsewhere. Dependent on US government-supplied surplus food, much of which was high-caloried, their eating habits changed, which years later has led to diabetes, atherosclerotic diseases, and hypertension—other paths to an early death.

Seeking a more powerful weapon for warfare, the US unleashed death in several forms on peaceful Marshall Island people and their tiny, coral atolls, leaving some of their islands still too radioactive for human life.

Bravo's hellish results must serve as a cautionary tale to all people should nuclear weapons ever strike today's cities, towns, agricultural areas, or even harbors.

ACKNOWLEDGMENTS

■ **THIS BOOK HAS** a long history and therefore there are many people to whom I owe thanks.

First and foremost is the late Elisabeth Sifton, my long-time friend and editor, who more than forty years ago encouraged me to go to Rongelap and write about the Marshallese. In succeeding years, she repeatedly told me to write this book and when I actually started six years ago, she read draft chapters as they were finished up until December 2019, when she died.

I also must mention Mr. William Shawn, who as editor of the *New Yorker* took a chance and paid for my 1974 trip to the Marshall Islands and the sixty-thousand-word article I wrote the next year. It was put into galleys, but never published and became the starting point for this volume.

Tribute is also due to Dr. Robert Conard, who allowed me to accompany his medical team not only to Rongelap but also to Bikini, leaving me with a permanent appreciation for what he was trying to do, and my long-term concern for the exposed Marshall Islanders.

Starting with Ben Bradlee, a legion of *Washington Post* editors and reporters have supported my decades of writing about nuclear weapons. It started with Meg Greenfield, a wonderful friend and colleague who actually edited my 1966 note in the *Reporter Magazine* while she worked there, and before she joined the

Post. Larry Stern, Bob Kaiser, Karen DeYoung, Dan Morgan, Kathy Sawyer, Rick Atkinson, David Broder, Dan Balz, Don Oberdorfer, Maralee Schwartz, Jonathan Randal, Mike Getler, Peter Milius, Jim Hoagland, Haynes Johnson and dozens of other *Post* veterans, as well as the entire Graham family made those years for me a golden age at the newspaper.

I want to also thank Keith Urbahn and Matt Carlini of Javelin, for believing in the book, and Keith Wallman of Diversion Books for his marvelous editing of the manuscript.

Finally, I want to pay tribute to two close personal friends who were great writers in their own right—Ward Just and John Newhouse. Their work in both fiction and nonfiction inspired me to undertake and finish this book with memories of them in my mind.

BIBLIOGRAPHY

Abella, Maveric K. I. L. et al., *Background Gamma Radiation and Soil Activity Measurements in the Northern Marshall Islands*, PNAS, July 30, 2019, Vol. 116, No. 31.

Advisory Committee on Human Radiation Experiments, *ACHRE Report, What We Now Know*, Part II, Chapter 11, US Government Printing Office, Washington, DC, 1995.

Advisory Committee on Human Radiation Experiments, *ACHRE Report, The Marshallese*, Part II, Chapter 12, US Government Printing Office, Washington, DC, 1995.

Advisory Committee on Human Radiation Experiments, *Documentary Update on Project Sunshine "Body Snatching,"* Memorandum, Advisory Staff Committee, Washington, DC, June 9, 1995.

Advisory Committee on Human Radiation Experiments, Final Report, US Government Printing Office, Washington, DC, October 1995.

Agnew, Harold, *Harold Agnew's Interview (1992)*, Voices of the Manhattan Project, National Museum of Nuclear Science & History, Los Alamos Historical Society.

Ahlgreen, Ingrid MS, Seiji Yamada, MD, and Allen Wong, MD, "Rising Oceans, Climate Change, Food Aid, and Human Rights in the Marshall Islands," *Health and Human Rights Journal*, Vol. 16, No. 1, June 2014.

Air Force Special Weapons Center, *History of Task Force 7.4 Participation in Operation Castle, 1 January 1953 to 26 June 1954*, Historical Division, Air Research and Development Command, Kirkland Air Force Base, New Mexico, November 1954.

Alford, Lt. Cmdr. L. H., *Report of Evacuation of Natives, Utirik Atoll, 4 March 1954*, To: Commander Task Group 7.3.

Allison, Ambassador John M., American Embassy Tokyo to Department of State, Embassy Dispatch 1428, Secret, April 16, 1954, *Public and Private Official Papers Relating to the case of "Fukuryu Maru No. 5" [Lucky Dragon]*, Documentation March 17 to April 23, 1954.

Allison, Ambassador John M., to the Department of State, Top Secret, Tokyo, May 20, 1954—2 p.m., Telegram 2853. *Subject: Fukuryu Maru*, Foreign Relations of the United States, 1952–1954, China and Japan, Vol. XIV, Part 2.

Alsop, Joseph and Stewart Alsop, "Atom Bomb Tests Must First Overcome Inter-Service Rivalry," *Washington Post*, Feb. 10, 1946.

Alsop, Stewart, "Lekoj and the Unusable Weapon," *Newsweek,* October 30, 1972.

Alsop, Stewart, *Stay of Execution, a Sort of a Memoir*, J. B. Lippincott, New York, 1973.

American Heritage Foundation, Marshall Islands.

Anjain, Senator Jeton, *Statement on Behalf of the Rongelap Local Government and the Rongelap People Living in Exile at Mejato*, House Subcommittee on Insular and International Affairs, Washington, DC, November 16, 1989.

Anjain, John, *The Voices of Rongelap: Cautionary Tales From a Nuclear War Zone*, Barbara Rose Johnston and Holly M. Barker, posted November 22, 2008.

Anjain, Nelson, *Letter to Dr. Robert Conard*, Rongelap Island, April 9, 1975.

Armed Forces Special Weapons Project, *Evaluation of Radioactive Fall-Out*, AFSWP-978, September 15, 1955.

Armed Forces Special Weapons Project, *Operation Castle: Summary Report of the Commander*, Task Unit 13, Military Effects, Albuquerque, New Mexico, January 30,1959.

Associated Press, "Forrestal After Survey Said Navies Will Go On," *New York Times*, July 2, 1946.

Atomic Energy Commission, Division of Biology and Medicine, *Conference on Long Term Surveys and Studies of the Marshall Islands*, Washington, DC, July 12–13, 1954.

Atomic Energy Commission, *Eighteenth Semiannual Report of the Atomic Energy Commission*, July 1955.

Atomic Energy Commission, *Minutes, Thirty-sixth Meeting of the General Advisory Committee to the US Atomic Energy Commission*, Washington, DC, August 1953.

Atomic Heritage Foundation, *William Penney, Mathematician and Physicist*, 2019.

Bainbridge, K. T., *Trinity*, LA-6300-H History Special Distribution, Los Alamos National Laboratory, Los Alamos, New Mexico, Issued: May 1976.

Baker, Nicole, *Bikini Atoll Nomination by the Republic of the Marshall Islands for Inscription on The World Heritage List 2010*, January 2009.

Barker, Holly M., *Bravo for the Marshallese: Regaining Control in a Post-Nuclear, Post-Colonial World*, Wadsworth, Belmont, California, 2004, 2013.

Bethe, Hans, *Letter to Dr. W. F. Libby*, December 17, 1954.

Bikini Atoll, *Bikinian Demographics*, As of March 2016.

Bikini Atoll, *US Reparations for Damages: People of Bikini vs. US Lawsuit*, Court Filings and Updates 2006–2010.

Bikini Atoll Rehabilitation Committee, *Report No. 4: Status March 31, 1986*, Department of Interior, Washington DC, No. TT-158X08.

Bordner, Autumn S. et al., *Measurement of background gamma radiation in the northern Marshall Islands*, Proceedings of the National Academy of Sciences, June 21, 2016.

Bradley, Omar N., *A General's Life: An Autobiography by General of the Army Omar N. Bradley*, Simon & Schuster, New York, 1983.

Brown, George P., "The Atomic Bomb and Japan's Surrender," *Education About Asia*, Vol. 11, No. 1, Spring 2006.

Chernobyl Forum, *Chernobyl's Legacy: Health, Environmental and Socio-Economic Impacts*, IAEA Division of Public Information, Vienna, Austria, April 2006.

Chrestensen, Capt. Louis B., *SUBJECT: Island Evacuation, To Whom It May Concern*, Pages 11–15 of online document.

Clark, Dr. John C., "We Were Trapped by Radioactive Fallout," *Saturday Evening Post*, July 20, 1957.

Clarkson, Major General P. W., *History of Operation Castle*, Defense Nuclear Agency, Washington, DC, June 1983.

Clarkson, Major General P. W., *Reports on Evacuation of Natives and Surveys of Several Marshall Island Atolls*, Headquarters, Joint Task Force Seven, San Francisco, California, April 9, 1954.

Compton, Arthur H., to Frank E. Jewett, *Report of National Academy of Sciences Committee on Atomic Fission*, May 17, 1941.

Conant, James B., Chairman, A. H. Compton, and H. C. Urey, comprising a Subcommittee of the S-1 Executive Committee, *Memorandum written to Brig. Gen. L, R, Groves, Manhattan Project*, October 30, 1943.

Conard Robert A., Draft Statement, *To The Chiefs and All The People in Utirik Atoll*, 401326, October 25, 1976, Page 3 (Page 3 in the online version).

Conard, Robert A. et al., *March 1957 Medical Survey of Rongelap and Utirik People Three Years After Exposure to Radioactive Fallout,* Brookhaven National Laboratory, Upton, NY, 1958.

Conard, Robert A. MD, *An outline of Some of the Highlights of the Medical Survey of the Marshallese Carried Out in February–March 1959, 5 Years After the Fallout Accident*, 1959.

Conard, Robert A. MD et al., *Medical Survey of the Rongelap People, March 1958, Four Years After Exposure to Fallout*, Brookhaven National Laboratory, Upton, NY, May 1959.

Conard, Robert A. MD et al., *Medical Survey of Rongelap People Five and Six Years After Exposure to Fallout*, Brookhaven National Laboratory, Upton, NY, September 1960.

Conard, Robert A. MD et al., *Medical Survey of the People of Rongelap and Utirik Islands Nine and Ten years after Exposure to Fallout Radiation (March 1963 and March 1964),* Brookhaven National Laboratory, Upton, NY, September 1960.

Conard, Robert A. MD et al., *Medical Survey of the People of Rongelap and Utirik Islands Eleven and Twelve Years after Exposure to Fallout Radiation (March 1965 and March 1966)*, Brookhaven National Laboratory, Upton, NY, April 1967.

Conard, Robert A., *The 1968 Annual Medical Survey of the Rongelap People Exposed to Fallout in 1954: interim Report*, Brookhaven National Laboratory, Upton, NY.

Conard, Robert A. MD et al., *Medical Survey of the People of Rongelap and Utirik Islands Thirteen, Fourteen, and Fifteen Years After Exposure to Fallout Radiation (March 1967, March 1968, March 1969)*, Brookhaven National Laboratory, Upton, NY, June 1970.

Conard, Robert A. et al*., Twenty-year Review of Medical Findings in a Marshallese Population Accidentally Exposed to Radioactive Fallout*, Brookhaven National Laboratory, Upton, NY, January 1975.

Conard, Robert A. MD et al., *Review of Medical Findings in a Marshallese Population Twenty-Six Years After Accidental Exposure to Radioactive Fallout*, Brookhaven National Laboratory, Upton, NY, January 1980.

Conard, Robert A., "The 1954 Bikini Atoll Incident: An Update of the Findings in the Marshallese People," *Medical Basis for Radiation Accident Preparedness*, Elsiver North Holland, Inc., 1980.

Conard, Robert A., *Fallout*, Upton, NY, Associated Universities Inc., 1992.

Conard, Robert A., *Return of the Rongelapese to their Home Island*, AEC 125/30, material in proposed press release, Note by the Secretary, February 6, 1957.

Conard, Robert A., *Letter to Senator Olympio T. Borja, Chairman Special Joint Committee Concerning Rongelap and Utirik Atolls*, September 17, 1975.

Conard, Robert A., *Letter to Dr, Charles Dunham, Director of Biology and Medicine, Atomic Energy Commission*, March 28, 1956.

Conard, Robert A., *Letter to Mr. N. J. Emerson*, October 27, 1972.

Conard, Robert A., *Letter to The Honorable Edward Johnston*, November 22, 1972.

Conard, Robert A., *Letter and 1974 medical report to Dr. James L. Liverman, Director, Division of Biomedical and Environmental Research*, AEC, May 8, 1974.

Conard, Rear Admiral Robert, *Oral History: Operation Crossroads, Nuclear Weapons Test at Bikini Atoll*, 1946, Naval History and Heritage Command, November 9, 1993.

Congress, *Public Laws Authorizing Payments to Rongelap in 1964 and 1977*.

Congressional Record—House, *The Three Surprises*, March 17, 2004, Page H1139.

Congressional Record, June 16, 1966, Senate, Page 13,501.

Congressional Research Service, *Republic of the Marshall Islands Changed Circumstances Petition to Congress*, RL 32811, Updated May 16, 2005.

Cowan, George, *George Cowan's Interview (1993)*, Voices of the Manhattan Project, National Museum of Nuclear Science & History, Richard Rhodes Collection.

Cronkite, E. P. et al., *Study of Response of Human Beings Accidentally Exposed to Significant Fallout Radiation, Operation Castle Final Report Project 4.1*, Naval Medical Research Institute, Bethesda, Maryland, October 1954.

Defense Atomic Support Agency, *Fallout Phenomena Forum, Proceedings Part I*, Naval Postgraduate School, Monterey, California, April 12–14, 1966.

Defense Atomic Support Agency, *Proceedings of the Second Interdisciplinary Conference on Selected Effects of a General War*, DASA 2019–2, DASIAC Special Report 95, July 1969

Defense Nuclear Agency, *Castle Series 1954*, DNA 6035F, Washington, DC, 1982.

Defense Nuclear Agency, *Operation Castle, Radiological Safety Final Report*, Vol. I, Headquarters, Joint Task Force Seven, Technical Branch, J-3 Division, Washington, DC, Spring 1954.

Defense Nuclear Agency, *Cleanup, Rehabilitation, Resettlement of Enewetak Atoll—Marshall Islands*, Washington, DC, April 1975.

Defense Nuclear Agency, *Operation Dominic I—1962*, DNA 6040F, Washington, DC, February 1983.

Defense Nuclear Agency, *Fact Sheet—Enewetak Operation*, Washington D.C., April 1980.

Defense Nuclear Agency, *Evaluation of Enewetak Radioactivity Containment*, National Research Council, National Academy Press, Washington, DC, March 1982.

Defense Nuclear Agency, *For the Record—A History of the Nuclear Test Personnel Review Program, 1978–1993*, DNA 001-91-C-0022, March 1996.

Defense Nuclear Agency, *Operation Castle-Shot Bravo, Analysis of Radiation Exposure—Service Personnel on Rongerik Atoll*, DNA-TR-86-120, Science Applications International Corp., McLean, Virginia, July 9, 1987.

Defense Nuclear Agency, *Operations Crossroads 1946, United States Atmospheric Nuclear Weapons Tests Nuclear Test Personnel Review*, DNA 6032F, May 1, 1984.

Defense Nuclear Agency, *Operation Ivy: 1952*, DNA 6036F, December 1982.

Defense Nuclear Agency, *Operation Sandstone: 1948*, DNA 6033F, December 1983.

Defense Nuclear Agency, *Operation Teapot: 1955*, DNA 6009F, Washington, DC, November 1981.

Defense Threat Reduction Agency, *Enewetak Atoll Cleanup Documents*, 1973–2010.

Defense Threat Reduction Agency, *History of the Defense Threat Reduction Agency*, and other DTRA reports.

Defense Threat Reduction Agency, *Operation Crossroads, Fact Sheet*, 2015.

Deines, Ann C. et al., *Marshall Islands Chronology 1944 to 1990*, History Associates Incorporated, Rockville, Maryland, 1991.

Delgado, James P., and Larry E. Murphy, *The Archeology of the Atomic Bomb, Chapter Two: Operation Crossroads*, National Park Service.

Delgado, James P. et al., *The Archeology of the Atomic Bomb: A Submerged Cultural Resources Assessment of the Sunken Fleet of Operation Crossroads at Bikini and Kwajalein Lagoons*, National Park Service, Southwest Cultural Resources Center Professional Papers Number 37, Santa Fe, New Mexico, 1991.

Denyer, Susan, *Cultural Landscapes of the Pacific Islands*, ICOMOS, December 2007.

Department of Energy, Office of History and Heritage Resources, *Debate Over How To Use the Bomb, Late Spring 1945*, The Manhattan Project An Interactive History.

Department of Energy, *Enewetak Radiological Support Project, Final Report*, NVO-213, Las Vegas, Nevada, September 1982.

Department of Energy, *Involvement in the Evacuation of Rongelap Atoll*, Memo to Edward J. Vallario, July 22, 1985.

Department of Energy, Human Radiation Studies, "Oral History of Merril Eisenbud," *Remembering the Early Years*, Office of Human Radiation Experiments, May 1995.

Department of Energy, *Report on the Status of the Runit Dome in the Marshall Islands*, Report to Congress, June 2020.

Department of Energy, *Special Medical Care and Logistics Program*, International Outreach Program, Final Technical Report, October 2011–October 2012.

Department of Interior, *Interior Authorizes Full Decision-Making Power to Bikini Leaders over Annual Budget of the Bikini Resettlement Trust Fund*, Press release, November 28, 2017.

Department of Interior, *Interior Provides Complete FY2018 Funding for Bikini, Enewetak, Rongelap, and Utrik Healthcare Programs*, Press release, May 21, 2018.

Department of Interior, *Marshall Islands Nuclear Testing Compensation*, Statement of Thomas Bussanich, Senate Subcommittee on Insular Affairs, September 25, 2007.

Department of Interior, *Second Five-Year Review of the Compact of Free Association, as Amended, Between the Government of the United States and the Republic of the Marshall Islands*, Washington, DC.

Department of the Interior, *Interior Authorizes Full Decision-Making Power to Bikini Leaders over Annual Budget of the Bikini Resettlement Trust Fund*, Assistant Secretary for Insular Areas Doug Domenech Statement, November 28, 2017.

Department of Interior, *Interior Provides Funding to Ensure Continued Services under Four Atoll Healthcare Program in the Marshall Islands*, December 21, 2018.

Department of the Interior, *Budget Justifications and Performance Information Fiscal Year 2020: Office of Insular Affairs*.

Department of the Interior, *A Report on the State of the Islands 1999*, Office of Insular Affairs, Washington, DC.

Department of the Interior, *US and Marshall Islands Hold Mid-Year Meetings on Compact Funding and Trust Fund*, March 26, 2018.

Department of State, *United States Relations with and Policies Toward Japan*, a variety of documents dealing with the *Fukuryu Maru* including Numbers 844,832, 819,817, 809.

Department of State, *The Ambassador in Japan (Allison) to the Department of State, Subject: Fukuryu Maru*, Tokyo, May 20, 1954.

Department of State, *Report Evaluating the Request of the Government of the Republic of the Marshall Islands Presented to the Congress of the United States of America*, Bureau of East Asian and Pacific Affairs, Jan. 4, 2005.

Director of Ship Material, Joint Task Force One, *Historical Report, Atomic Bomb Tests Able and Baker*, Washington, DC, Vol. 2 of 3.

Director of Ship Material Technical Inspection Report, *Radiological Decontamination of Target and Non-Target Vessels*, Vol. 3, Operation Crossroads Joint Task Force One, AD473908, 1946.

Donaldson, Lauren R., *Radiobiological Studies at Enewetak Following the Mike Shot of November 1952*, University of Washington, Seattle, Washington, June 1953.

Donaldson, Lauren R., *A Radiological Study of Rongelap Atoll, Marshall Islands 1954-1955*, UWFL-42, University of Washington, Seattle, Washington, August 1955.

Doulatram, Desmond N., *The Lolelaplap (Marshall Islands) in Us*, University of San Francisco, Master's Projects and Capstones, 2018.

Dunning, Gordon M. (editor), *Radioactive Contamination of Certain Areas of the Pacific Ocean From Nuclear Tests*, US Atomic Energy Agency, Washington, DC, August 1957.

Eisenbud, Merril, *Minutes of A.E.C. meeting*. AEC Health and Safety Laboratory Advisory Committee on Biology & Medicine, January 13–14, 1956, Page 232.

Eisenbud, Merril, *An Environmental Odyssey*, Seattle, Washington, University of Washington Press, 1990.

Eisenbud, Merril, *Oral History*, Conducted January 26, 1995, Human Radiation Studies: Remembering the Early Years, Department of Energy, May 1995.

Eisenhower, Dwight D., *Dwight D. Eisenhower: 1954: The Public Messages, Speeches, and Statements of the President, January 1 to December 31, 1954,* Government Printing Office, Washington, DC, 1960.

Fehner, Terrence R. and F. G. Gosling, *Battlefield of the Cold War: The Nevada Test Site*, Vol. I, Washington, Office of History and Heritage Resources, Department of Energy, 2006.

Fehner, Terrence R., and F. G. Gosling, *Battlefield of the Cold War: The Nevada Test Site, Atmospheric Nuclear Weapons Testing1951–1963, Atmospheric Nuclear Weapons Testing 1951–1963*, Vol. I, National Nuclear Security Administration, Washington.

Fehner, Terrence R. and F. G. Gosling, *Origins of the Nevada Test Site*, History Division, Department of Energy, DOE/MA-0518, Washington, DC, December 2000.

Forrestal, James V. (author), Walter Millis (editor), *The Forrestal Diaries*, Viking Press, New York, 1951.

Garcia-Gomez, Ruth, "A Marshall Islands' Successful Aquaculture Venture," *SPC Fisheries Newsletter* #153, May–August 2017.

Glasstone, Samuel, *Proof that fallout was clearly visible where there was a short term hazard at the Mike and Bravo H-bomb tests*, April 3, 2011.

Glasstone, Samuel, *The Effects of Nuclear Weapons*, Washington, DC, Atomic Energy Commission, 1962.

Glasstone, Samuel and Philip J. Dolan, *The Effects of Nuclear Weapons*, Third Edition, US Department of Defense, 1977.

Gleason, S. Everett, *Memorandum of Discussion at the 235th Meeting of the National Security Council*, Washington, February 3, 1955, Document 5, Foreign Relations of the United States, 1955–1957, Regulation of Armaments; Atomic Energy, Vol. XX.

Gleason, S. Everett, *Memorandum of Discussion at the 236th Meeting of the National Security Council*, Washington, February 10, 1955, Document 7, Foreign Relations of the United States, 1955–1957, Regulation of Armaments; Atomic Energy, Vol. XX.

Gosling, F. G., *The Manhattan Project, Making the Atomic Bomb*, Washington, DC, Office of History and Heritage Resources, Department of Energy, 2010.

Government Accountability Office, *Actions Needed to Prepare for the Transition of Micronesia and the Marshall Islands to Trust Fund Income*, GAO-18-415, May 17, 2018.

Government Accountability Office, *An Assessment of the Amended Compacts and Related Agreements*, Statement of Susan S. Westin before the House Committee on Resources, July 10, 2003.

Government Accountability Office, *Enewetak Atoll—Cleaning Up Nuclear Contamination*, PSAD-79-54, Washington, DC, May 8, 1979.

Government Accountability Office, *Kwajalein Atoll Is the Key US Defense Interest in Two Micronesian Nations*, GAO-02-119, Washington, DC, January 2002.

Government Accountability Office, *Marshall Islands Status of the Nuclear Claims Trust Fund*, GAO/NSIAD-92-229, Washington DC, September 1992.

Government Accountability Office, *Operation Crossroads: Personnel Radiation Exposure Records Should Be Improved*, RCED-86-15, November 8, 1985.

Government of the Marshall Islands, *State of the Environment Report 2016*, SPREP 2016, Apia, Samoa.

Graybar, Lloyd J., *The 1946 Atomic Bomb Tests: Atomic Diplomacy or Bureaucratic Infighting?*, The Journal of American History 72, No. 4, 1986.

Gross, Gerald G., "A-Bomb Sinks 2 Ships and Damages 17; Capital Vessels Escape Extensive Harm; Destroyer Capsizes; Experiment 'Success,'" *Washington Post*, July 1, 1946.

Groves, General Leslie R., *Now It Can Be Told: The Story Of The Manhattan Project*, (paperback) De Capo Press Inc., 1983.

Gudiksen, P. H. et al., *External Dose Estimates for Future Bikini Atoll Inhabitants*, Lawrence Livermore Laboratory UCRL-51879, Livermore, California, March 1976.

Guskova, A. K. et al., *Acute Radiation Effects in Victims of the Chernobyl Accident*, UNSCEAR 1988 Report, Appendix to Annex G, United Nations, New York, 1988.

Hacker, Brian C., *The Dragon's Tail*, Berkley, Calif., University of California Press, 1987.

Hacker, Brian C., Elements of Controversy: *The Atomic Energy Commission and Radiation Safety in Nuclear Weapons Testing, 1947–1974*, University of California Press, Jan 1, 1994.

Hagerty, James, *Diary Entry by the President's Press Secretary (Hagerty)*, Washington, January 4, 1955, Document 3, Foreign Relations of the United States, 1955–1957, National Security Policy, Vol. XIX.

Hamblin, Jacob Darwin and Linda M. Richards, "Beyond the Lucky Dragon: Japanese Scientists and Fallout Discourse in the 1950s," *International Journal of the History of Science Society of Japan,* January 2015.

Hamilton, Terry, "Consequences of the Nuclear Weapons Testing Program in the Northern Marshall Islands," *Marshall Islands Monitor,* Lawrence Livermore National Laboratory, Vol. 2, Numbers 1 & 2, July 2010.

Hamilton, T. F. and W. L. Robison, *Overview of Radiological Conditions on Bikini Atoll,* Lawrence Livermore National Laboratory, UCRL-MI-208228, November 29, 2004.

Hamilton, Terry, "Return to Rongelap," *Science and Technology Review,* Lawrence Livermore National Laboratory, July/August 2010.

Hanna, John, *Treatment on Rongelap, Utirik: Callous, Condescending. Marshalls Medical Program . . . Disregarding Potential Danger, Food Declared Safe,* Micronesia Support Committee Bulletin, Vol. 4, Number 1, Honolulu, Hawaii, March 1979.

Hansard 1803–2995, *Summary Information for Mr. Winston Churchill,* 1954.

Hara, Kimi, *Cold War Frontiers in the Asia-Pacific: Divided Territories in the San Francisco System,* Nissan Institute/Routledge Japanese Studies, Routledge, New York, 2007.

Harkewicz, Laura J., *The Ghost of the Bomb: The Bravo Medical Program, Scientific Uncertainty, and the Legacy of US Cold War Science, 1954–2005,* University of California, San Diego, California, 2010.

Harvard Law Student Advocates for Human Rights, *Keeping the Promise: An Evaluation of the Continuing Obligations Arising Out of the US Nuclear Testing Program in the Marshall Islands,* Harvard Law School, April 2006.

Herken, Gregg, *The Winning Weapon,* New York, Alfred A. Knopf, 1981.

Hewlett, Richard G. and Oscar E. Anderson Jr., *The New World, 1939–1946,* Pennsylvania University Press, University Park, PA., 1962.

Hewlett, Richard G. and Francis Duncan, *Atomic Shield, 1947–1952,* Pennsylvania State University Press, 1969.

Hewlett, Richard G. and Jack M. Holl, *Atoms for Peace and War 1953–1961,* University of California Press, Berkley, Calif., 1969.

Higginbotham, Adam, *Midnight in Chernobyl,* New York, Simon & Schuster, 2019.

Hirschfelder, J. O. and J. Magee to K. Bainbridge, "Danger from Active Material Falling from Cloud Desirability of Bonding Soil Near Zero with Concrete and Oil," June 16, 1945, DOE/NV Nuclear Testing Archive, Las Vegas, Nevada.

Hollaway, David, *Stalin and the Bomb,* New Haven, Conn., Yale University Press, 1994.

Holmes & Narver Inc., *Report of Repatriation of the Rongelap People,* November 1957.

Homei, Aya, "The contentious death of Mr Kuboyama: science as politics in the 1954 Lucky Dragon incident," *Japan Forum,* 25:2, December 2012.

Hoshina, Zenshiro, *Daitoa Senso Hishi: Hoshina Zenshiro Kaiso-roku* [Secret History of the Greater East Asia War: Memoir of Zenshiro Hoshina] (Tokyo, Japan: Hara-Shobo, 1975), excerpts from Section 5, "The Emperor made *go-seidan* [= the sacred decision] – the decision to terminate the war" [translation by Hikaru Tajima].

House Committee on Resources, *Rongelap Resettlement Act of 1999,* House Report 106–404, Government Printing Office, October 20, 1999.

House Committee on Resources, *The Status of Nuclear Claims, Relocation and Resettlement Efforts in the Marshall Islands*, House Hearing 106–26, May 11, 1999, Government Printing Office, 1999.

House Subcommittee on Oversight and Investigations, *Radiation Exposure From Nuclear Tests in the Pacific*, Hearing February 24, 1994, Washington, DC, Committee on Natural Resources, Serial No. 103—68.

Ishibashi, M. et al., *Radioactive Analysis of the Bikini Ashes (The Radioactive Dust from the Nuclear Detonation)*, Bulletin of the Institute for Chemical Research, Kyoto University, 1954.

Japan Times The, "Lucky Dragon's Lethal Catch," March 18, 2012.

Jessee, Emory Jerry, *Radiation Ecologies: Bombs, Bodies, and Environment During the Atmospheric Nuclear Weapons Testing Period, 1942–1965*, Montana State University, Bozeman, Montana, January 2013.

Johnson, J. Christopher et al., *Mortality of Veteran Participants in the Crossroads Nuclear Test*, National Academies Press, Washington, DC, 1996.

Johnson, Giff, *Rongelap Housing Project Completed in Marshalls*, Pacific Islands Report, March 27, 2012.

Johnston, Barbara Rose and Holly M. Barker, *Consequential Damages of Nuclear War: The Rongelap Report*, Routledge, London and New York, 2017.

Johnston, Barbara Rose and Brooke Takala, "Environmental Disaster and Resilience: The Marshall Islands Experience Continues to Unfold," *Cultural Survival Quarterly Magazine*, September 2016.

Joint Chiefs of Staff, *Joint Chiefs of Staff Decision Amending J C S 1552 74 Test C Operation Crossroads*, JCS 1901 10, 24 August 1946, Top Secret, Appendix C.

Joint Committee on Atomic Energy, *The Nature of Radioactive Fallout and Its Effects on Man*, Special Subcommittee on Radiation, Part I, Washington, DC, May 27, 28, 29, June 3, 1957.

Joint Task Force One, *Commander, Report on Atomic Bomb Tests Able and Baker* (Operation Crossroads, Vol. I, Defense Nuclear Agency, December 5, 1979.

Joint Task Force One, The Office of the Historian, *Operation Crossroads, The Official Pictorial Record*, Wm. H. Wise & Co. Inc., New York, 1946, Page 89 (Page 88 in the online version).

Joint Task Force One, Commander, *Historical Report, Atomic Bomb Tests Able and Baker*, Washington, DC, Vol. 1, 1946.

Joint Task Force One, *Public Information Estimate No 1*, March 7, 1946.

Joint Task Force One, Director of Ship Material Technical Report, *Radiological Decontamination of Target and Non-Target Vessels*, Vol. 2 of 3, Defense Nuclear Agency, Washington, DC.

Joint Task Force One, Director of Ship Material Technical Inspection Report, *Radiological Decontamination of Target and Non-Target Vessels*, Vol. 3, Operation Crossroads, AD473908, 1946.

Joint Task Force Seven, Commander Group 7.1, *Operation Castle Pacific Proving Grounds, March– May 1954*, Los Alamos Scientific Laboratory, Los Alamos, New Mexico, June 1954.

Joint Task Force Seven, Technical Branch, *Operation Castle: Radiological Safety*, Final Report, Vol. 1, (Extract) Defense Nuclear Agency, Washington, DC, September 1985, Page C-3.

Joint Task Force 132, *Radiobiological Studies at Eniwetok before and after Mike Shot*, Project 11.5, Operation Ivy, Pacific Proving Grounds, November 1952.

Jones, Matthew, *After Hiroshima: The United States, Race and Nuclear Weapons in Asia*, 1945–1965, Cambridge University Press, New York, 2010.

Jones, Vincent C., *Manhattan: The Army and the Atomic Bomb*, Center of Military History, United States Army, Washington, DC, 1985.

Kazel, Robert, *Ex-Chief of Nuclear Forces General Lee Butler Still Dismayed by Deterrence Theory and Missiles on Hair-Trigger Alert*, Nuclear Age Peace Foundation, May 27, 2015.

Kohn, Dr. Henry I., *Report [of the] Rongelap Reassessment Project, Corrected Edition*, Berkeley, Calif., March 1, 1989.

Kovaleski, Serge F., "The Most Dangerous Game," *Washington Post Magazine*, January 15, 2006.

Kunkle, Thomas and Ristvet, Bryon, *Castle Bravo: Fifty Years of Legend and Lore*, Fort Belvoir, Va., Defense Threat Reduction Agency, 2013.

Kuznick, Peter, "Japan's nuclear history in perspective: Eisenhower and atoms for war and peace," *Bulletin of the Atomic Scientists*, April 13, 2011.

Lamm, Joanne, *The Island is Missing!*, US Army Heritage and Education Center, October 28, 2010.

Lapp, Ralph E., *The Voyage of the Lucky Dragon*, New York, Harper & Brothers, 1957.

Laurence, William L., "Bikini's King Gets Truman's Thanks," *New York Times*, July 17, 1946.

Laurence, William L., *Trinity draft press releases, May 14, 1945*, Memorandum, Nuclear Security Blog, Page 3.

Lawrence Livermore National Laboratory, *Caging the Dragon*, The containment of Underground Explosions, Defense Nuclear Agency, DNA-TR-95-74, June 1995

Lawrence Livermore National Laboratory, *Enewetak Atoll, Post Testing Era and Cleanup Activities*, LLNL-WEB-400363.

Lawrence Livermore National Laboratory, "Helping Bridge the Gap in Support of Rongelap Atoll Resettlement," *Marshall Islands Monitor*, Vol. 1, No. 4, December 2009.

Leahy, Paul Gregory, *Eisenhower's Dilemma, How to Talk about Nuclear Weapons*, University of Michigan Press, March 2009.

Lee, Capt. Fitzhugh, USN, *The Press at Operations Crossroads*, Army Information Digest, Vol. 2, No. 3, March 1947.

Lehman, Michael R., *Nuisance or Nemesis: Nuclear Fallout and Intelligence as Secrets, Problems and Limitations on the Arms Race 1940 to 1964*, University of Illinois at Urbana-Champaign, 2016.

Leonard, Graham, *The 1954 Shunkotsu Maru Expedition and American Atomic Secrecy*, Osaka School of International Public Policy, Osaka University, 2011.

Lessard, Edward T., *Letter to J. W. Thiessen, Department of Energy, enclosing "Protracted Exposure to Fallout: The Rongelap and Utirik Experience,"* September 13, 1983.

Lessard, Edward T., *Review of Marshall Islands Fallout Studies*, Brookhaven National Laboratory, 2007.

Letman, Jon, "Rising Seas give island nation a stark choice: relocate or elevate," *National Geographic*, November 19, 2018.

Light, Joe, *Why the Marshall Islands Is Trying to Launch a Cryptocurrency*, Bloomberg Businessweek, December 14, 2018.

Loeak, Christopher, *Future will be like living in a war zone*, Climate Home News, January 4, 2014.

Los Alamos Historical Document Retrieval and Assessment (LAHDRA) Project, *Final Report of CDC's LAHDRA Project*, November 2010.

Los Alamos National Laboratory, *Manhattan District History, Book VIII, Los Alamos Project (Y), Volume 3, Auxiliary Activities, Chapter 8, Operation Crossroads.*

Lum, Thomas et al., *Republic of the Marshall Islands Changed Circumstances Petition to Congress,* CRS Report for Congress, May 16, 2005.

McClurkin, Robert J.G., *Bikini Compensation,* Memorandum of Conversation by the Acting Director of the Office of Northeast Asian Affairs (McClurkin), Document No. 817, Foreign Relations of the United States, 1952–1954, China and Japan, Vol. XIV, Part 2, October 23, 1954.

McCraw, Tommy F., *Department of Energy (DOE) Involvement in the Evacuation of Rongelap Atoll,* Memo to Edward J. Vallario, July 22, 1985.

McDougal, Myres S. and Norbert A. Schlei, *The Hydrogen Bomb Tests in Perspective: Lawful Measures for Security,* The Yale Law Journal, Vol. 64, New Haven, Conn., 1955.

McElfish, Pearl Anna et al., "Effect of US Health Policies on Health Care Access for Marshallese Migrants," *American Journal of Public Health,* April 2015.

McElfish, Pearl Anna et al., Health Beliefs of Marshallese Regarding Type 2 Diabetes, *American Journal of Health Behavior* Vol. 40, No. 2, March 2016.

Maier, Thomas, "Cold War Fallout for Brookhaven National Lab," *Newsday,* August 21, 2009.

Manhattan Project File '42–'46, *Minutes of the second meeting of the Target Committee, Los Alamos, May 10–11, 1945.*

Marshall Islands Auditor General, *Audits of Local Governments,* Various Dates.

Marshall Islands Compendium, *A Compilation of Guidebook References and Cruising Reports.*

Marshall Islands Guide, "Rongelap Atoll, Rongelap: Big Hole."

Marshall Islands Journal, "Nitijela Backs RASAR Plan," April 2, 2020.

Marshall Islands Journal, "Utrik Gets Fish Base," August 30, 2018.

Marshallese Education Initiative, *Marshallese Community in Arkansas.*

Meade, Roger, *Discoveries and Collisions, The Atom, Los Alamos, and the Marshall Islands,* Arizona State University, May 2015.

Micronesia Support Committee, "Marshall Islands: A Chronology 1944–1981," Honolulu, Hawaii, 1982.

Miller, Jerry, *Stockpile,* Annapolis, MD, Naval Institute Press, 2010.

Mitchell-Eaton, Emily, *New Destinations of Empire: Imperial Migration from the Marshall Islands to Northwest Arkansas Islands to Northwest Arkansas,* Syracuse University, August 2016.

Moody, Walton S., *Building a Strategic Air Force,* Air Force History and Museum Program, 1995.

Murkowski, Chairman Lisa, *Opening Statement, Legislative Hearing on S. 2182 (Bikini) and S. 2325 (CNMI),* Senate Committee on Energy and Natural Resources, February 6, 2018.

Murphy, Robert, *Memorandum by the Acting Secretary of State to the President,* Document No. 763, Foreign Relations of the United States, 1952–1954, China and Japan, Vol. XIV, Part 2, May 29, 1954.

National Nuclear Security Administration, *Project Nutmeg,* Nevada National Security Site History, DOE/NV—767, August 2013.

National Research Council (US) Committee on Radiological Safety in the Marshall Islands, *Radiological Assessments for Resettlement of Rongelap in the Republic of the Marshall Islands,* Washington (DC): National Academies Press (US); 1994.

National Research Council (US) Committee on Radiological Safety in the Marshall Islands, *Radiological Assessments for Resettlement of Rongelap in the Republic of the Marshall Islands,* National Academies Press, Washington, DC, 1994, Executive Summary.

National Security Archive, *The Atomic Bomb and the End of World War II, A Collection of Primary Sources*, Edited by William Burr, Third update—August 7, 2017.

National Security Archive, *60th Anniversary of Castle BRAVO Nuclear Test, the Worst Nuclear Test in US History*, National Security Archive Electronic Briefing Book No. 459, February 28, 2014.

National Security Archive, Document 7: *US Embassy Tokyo Airgram 1482, "Public and Private Official Papers Relating to the Case of the Fukuryu Maru No. 5: Documentation March 17–April 23 1954,"* 30 April 1954, Secret, excerpts, Enclosure 14, Aide Memoire, Page 1 (Page 16 in online version).

National Security Archive, Document 8: *Admiral Arthur Radford, Chairman, Joint Chiefs of Staff, Memorandum to Secretary of Defense, "A Proposal for a Moratorium on Future Testing of Nuclear Weapons"*, April 30, 1954, Secret.

Naval History and Heritage Command, *Operation Crossroads: Fact Sheet*, The Navy Department Library, April 2015.

NavSource Online: Battleship Photo Archive, *BB-34 USS New York*.

Niedenthal, Jack, *A History of the Bikini People Following Nuclear Weapons Testing in the Marshall Islands: With Recollections and Views of Elders of Bikini Atoll*, Health Physics Society, March 1997.

Niedenthal, Jack, *For the Good of Mankind, a History of the People of Bikini and their Islands*, A Bravo Book, Republic of the Marshall Islands, 2001.

Nishiwaki, Yasushi, *Studies on the Radioactive Contamination Due to Nuclear Detonations I–VI*, Annual report of Nuclear Reactor Laboratory, Kinki University, Vol. 1, 1961.

Nitze, Paul H., *Paul H. Nitze Oral History Interviews, August 5 and August 6*, 1975, Harry S. Truman Library, Independence, Missouri.

Noble, John Wilford, "US Settles 75 on Pacific Atoll Evacuated for Bomb Tests in 40's," *New York Times*, April 11, 1977.

Norris, Robert Standish, and Thomas B. Cochran, *United States Nuclear Tests, July1945 to 31 December 1992*, Natural Resources Defense Council, Washington, DC, February 1, 1994.

Nuclear Weapon Archive, The, *Operation Castle 1954—Pacific Proving Ground*, May 2006.

Nuclear Weapon Archive, The, *Operation Greenhouse 1951*, August 2003.

Ogle, William E., *An Account of the Return to Nuclear Weapons Testing by the United States After the Moratorium 1958–1961*, US Department of Energy, October 1985.

Oishi, Matashichi, *The Day the Sun Rose in the West*, University of Hawaii Press, Honolulu, 2011.

Office of the Assistant to the Secretary of Defense (Atomic Energy), *History of the Custody and Deployment of Nuclear Weapons, July 1945 Through September 1977*, February 1978.

Office of the Director of National Intelligence, *Cold War Counterterrorism*, Washington, DC, 2007.

O'Keefe, Bernard J., *Nuclear Hostages*, Houghton Mifflin, Boston, 1983.

Olson, James C., *Stuart Symington: A Life*, University of Missouri Press, Columbia, MO, 2003.

Operation Crossroads—1946, sonicbomb.com, Page 1.

Palafox, Neal A., *Effects of US Nuclear Testing Program on the Marshall Islands*, Prepared Statement, Senate Committee on Energy and Natural Resources, S. Hrg. 109-178, July 19, 2005.

Palafox, Neal A. MD, *Marshall Islands Medical Program Fiscal Year 2001 Report*, Pacific Health Research Institute, Honolulu, Hawaii, July 31, 2002.

Parsons, Keith M., Robert A. Zaballa, *Bombing the Marshall Islands: A Cold War Tragedy*, Cambridge University Press, New York, 2017.

Pevec, Davor, *The Marshall Islands Nuclear Claims Tribunal: The Claims of the Enetwetak People*, Denver Journal of International Law and Policy, Vol. 35, No. 1, 2006.

Pevec, Davor, *The Marshall Islands Nuclear Claims Tribunal: The Claims of the Enewetak People*, Denver Journal of International Law and Policy, Vol. 35:1, March 10, 2008.

Pincus, Walter, "The Day the Ash Fell on Rongelap," *Washington Post*, June 16, 1977.

Pincus, Walter, "Bikinians Must Quit Island for at Least 30 Years, Hill Told," *Washington Post*, May 23, 1978.

Pincus, Walter, "Islanders Return to Atomic Test Area Amid Controversy," *Washington Post*, April 7, 1980.

Pincus, Walter, "Landowners Held in Sit-In On Kwajalein," *Washington Post*, June 22, 1982.

Podonsky, Glenn, *Statement Before the House Subcommittee on Asia, the Pacific and Global Environment*, May 20, 2010.

Public Law 95-348, *To authorize appropriations for certain insular areas of the United States, and for other purposes*, 95th Congress, August 18, 1978.

Pula, Nikolao, *Statement Before the House Subcommittee on Asia, the Pacific and Global Environment*, Director, Office of Insular Affairs, Department of the Interior, May 20, 2010.

Putnam, Frank W., *The Atomic Bomb Casualty Commission in retrospect*, Proceedings of the National Academy of Sciences, USA 95 (1998).

Rapoport, Roger, *The Great American Bomb Machine*, Dutton, New York, 1971.

Reardon, Steven L., *Council of War, A History of the Joint Chiefs of Staff 1942–1991*, Joint History Office, National Defense University Press, Washington, DC, 2012.

Reed, Thomas C., *At the Abyss: An Insider's History of the Cold War*, New York Presidio Press, 2005, Page 51.

Republic of the Marshall Islands, *State of Environment Report 2016*, Apia, Samoa, 2016.

Rhodes, Richard, *Dark Sun, The Making of the Hydrogen Bomb*, Simon & Schuster, New York,1995.

Rhodes, Richard, *Harold Agnew's Interview*, Voices of the Manhattan Project, National Museum of Nuclear Science & History, Albuquerque, New Mexico. 1994.

Robbins, Jack, William H. Adams, *Radiation Effects in the Marshall Islands*, Brookhaven National Laboratory, Upton, NY, 2017.

Robertson, Walter S., *Bikini Compensation*, Memorandum by the Assistant Secretary of State for Far Eastern Affairs (Robertson) to the Under Secretary of State (Hoover), Document No. 844, Foreign Relations of the United States, 1952-1954, China and Japan, Vol. XIV, Part 2, December 29, 1954.

Roche, Walter F., "Lease dispute threatens future of islands," *Baltimore Sun*, May 17, 2003.

Robbins, Jacob, William H. Adams, *Radiation Effects in the Marshall Islands*, National Institutes of Health, Bethesda, Maryland, 2017.

Rongelap Atoll Local Government, *Financial Statements and Independent Auditors' Report, September 30, 2015*.

Rust, Susanne, "They came here after the US irradiated their islands. Now they face an uncertain future," *Los Angeles Times*, December 31, 2019.

Satoru, Okuari, "How Japanese scientists confronted the US and Japanese governments to reveal the effects of Bikini H-bomb tests," *Asia-Pacific Journal,* Vol. 17, Issue 17, Number 2, September 1, 2019.

Schull, William J., *Effects of Atomic Radiation: A Half-Century of Studies from Hiroshima and Nagasaki,* United Kingdom: John Wiley & Sons Inc., October 1995.

Scowcroft, Brent, *Brent Scowcroft Oral History Part I, II*, Miller Center, University of Virginia, November 12–13, 1999.

Senate Armed Services Committee, *Military Construction Authorization Act FY 1976*, White House Records Office: Legislation Case Files at the Gerald R. Ford Presidential Library.

Senate Committee on Energy and Natural Resources, *Effects of US Nuclear Testing Program in the Marshall Islands*, S. Hrg. 109–178, July 19, 2005.

Senate Committee on Energy and Natural Resources, *Resettlement of Rongelap Atoll, Republic of the Marshall Islands*, S. Hrg. 102–316, September 19, 1991.

Sharp, Robert and William H. Chapman, *Exposure of Marshall Islanders and American Military Personnel to Fallout*, Operation Castle—Project 4.1 Addendum, Naval Medical Research Institute, Bethesda, Maryland, March 1957.

Shurcliff, William A., *Bombs at Bikini: The Official Report of Operation Crossroads*, W. H. Wise, New York, 1947.

Silk, Mary, *Interview of Lijon Eknilang*, Marshall Islands Story Project, March 5, 2008.

Simon, S. L., J. C. Graham, *Findings of the first comprehensive radiological monitoring program of the Republic of the Marshall Islands*, Health Physics, July 1997.

Simon, Steven L. et al., "Radiation doses and cancer risks in the Marshall Islands associated with exposure to radioactive fallout from Bikini and Enewetak nuclear weapons," *Health Physics*, August 2010.

Skelton, Lt. Col. John D., *The Forbidden Weapon—The Employment of Army Tactical Nuclear Weapons*, Army Command and General Staff College, Fort Leavenworth, Kansas, 1991.

Smith-Norris, Martha, "American Cold War Policies and the Enewetakese," *Journal of the Canadian Historical Association*, Vol. 22, Issue 2, 2011.

Smith-Norris, Martha, "Only as Dust in the Face of the Wind": An Analysis of the BRAVO Nuclear Incident in the Pacific," *The Journal of American-East Asian Relations*, Vol. 6, No. 1, Spring 1997.

Special Joint Committee Concerning Rongelap and Utirik Atolls, *Medical Aspects of the March 1, 1954 Incident: Injury, Examination, and Treatment*, Fifth Congress of Micronesia, First Regular Session, February 1973.

Special Joint Committee Concerning Rongelap and Utirik Atolls, *Compensation for the People of Rongelap and Utirik*, Fifth Congress of Micronesia, Second Regular Session, February 1974.

Smith, Allen E. and William E. Moore, *Report of the Radiological Clean-Up of Bikini Atoll*, Environmental Protection Agency, Washington DC, January 1972.

Smith, Mackenzie, *Remote Marshall Islands Atoll Plans to become the "Next Hong Kong,"* RNZ, September 21, 2018.

Special Committee Concerning Rongelap and Utirik Atolls, *Compensation for the People of Rongelap and Utirik*, Fifth Congress of Micronesia, February 1974.

Special Committee Concerning Rongelap and Utirik Atolls, *Medical Aspects of the March 1, 1954 Incident: Injury, Examination and Treatment*, Fifth Congress of Micronesia, February 1973.

Stassen, Harold E., *Memorandum by the Director of the Foreign Operations Administration (Stassen) to the President*, Document No. 774, June 29, 1954.

Strauss, Lewis, *Men and Decisions*, Doubleday, Garden City, NY, 1962.

Stewart, Esmond, *Defense Nuclear Agency 1947–1997*, Defense Threat Reduction Agency, Government Printing Office, Washington, DC, 2017.

Strope, Walmer E. (Jerry), *Autobiography of a Nerd*, Chapter 10, [unpublished online autobiography] 2003.

Sullivan, Walter, "Marshall Islander's Death Tied to Fallout," *New York Times*, November 21, 1972, Page 26.

Survey Research Center, *Public Reaction to the Atomic Bomb and World Affairs*, University of Michigan, Ann Arbor, Michigan, February 1947.

Szasz, Ferenc Morton, *The Day the Sun Rose Twice*, University of New Mexico Press, 1984, Page 63.

Target Committee, *Minutes of the second meeting*, Los Alamos, May 10–11, 1945, US National Archives, Record Group 77, Records of the Office of the Chief of Engineers, Manhattan Engineer District, TS Manhattan Project File '42–'46, folder 5D Selection of Targets, 2 Notes on Target Committee Meetings.

"This Little Pig Came Home." *Time Magazine*, April 11, 1969.

Tingley, Kim, "Secrets of the Wave Pilots," *New York Times Magazine*, March 17, 2016.

Tipton, W. J. and R. A. Meibaum, *An Aerial Radiological and Photographic Survey of Eleven Atolls and Two Islands within the Northern Marshall Islands*, 1978.

Tokyo Tech Museum and Archives, *Scientist Yasushi Nishiwaki in the Nuclear Age*.

United Nations Human Rights Council, *Mission to the Marshall Islands: Report of the Special Rapporteur on the implications for human rights of the environmentally sound management and disposal of hazardous substances and waste*s, A/HRC/21/48/Add.2, September 2012.

United Nations Security Council, Resolution of 2 April 1947, S/RES/21(1947), Pages 16–21 (Pages 1–6 in the online version).

United Press, "Bikini King Amazed at Changed Realm," *New York Times*, July 24, 1946.

US Department of State, "Trusteeship Agreement," appendix B, reprinted in Trust Territories of the Pacific Islands, 1993.

United States Marines, *Regional Study Oceania Study 3*, 2012.

US Naval Institute, *Oral History, Fitzhugh Lee, Vice Admiral (Retired)*, 1970.

United States Nuclear Regulatory Commission, *Radiation Basics*.

United States Strategic Bombing Survey, *Japan's Struggle to End the War*, US Government Printing Office, Washington, DC, 1946.

United States Strategic Bombing Survey, *Summary Report (Pacific War)*, US Government Printing Office, Washington, DC, 1946.

Utirik Atoll Local Government, *Financial Statements and Independent Auditors' Report, Year Ended September 30, 2014*, August 2020.

Vaswani, Ashok N. MD, *Medical Follow-Up in the Marshall Islands: An Overview of Sixty Years of Clinical Experience*, International Radiation and Thyroid Cancer Workshop, Tokyo, Japan, February 21–23, 2014.

Wadsworth, George, 350/5–354: Telegram, *The Deputy United States Representative at the United Nations (Wadsworth) to the Department of State*, Foreign Relations of the United States, 1952–1954, United Nations Affairs, Vol. III, Document 933.

Wall, Kim et al., -"The Poison and the Tomb, Part 6: The Big Scrape," *Mashable*, February 25, 2018.

Walsh, Julianne M., *Imagining the Marshalls: Chiefs, Tradition, and the State on the Fringes of US Empire*, University of Hawaii, 2003.

Warren, Col. Stafford M., *Occupancy of Target Vessels as Influenced by Intensity of Radiation of Various Types on Target Vessels, August 7, 1946*, Document 24, National Security Archives, George Washington University.

Warren, Dr. Stafford L., "Conclusions, Tests Proved Irresistible Spread of Radioactivity," *LIFE Magazine*, August 11, 1947.

Weisgall, Jonathan M., "The Able-Baker-Where's-Charlie Follies," *The Bulletin of the Atomic Scientists*, May/June 1994.

Weisgall, Jonathan M., *Operation Crossroads*, Annapolis, MD, Naval Institute Press, 1994.

Weisgall, Jonathan M., *Petition for Writ of Certiorari, People of Bikini v. United States of America*, Supreme Court, Washington, DC, 2009.

Weisgall, Jonathan M., *Statement by Legal Counsel to the People of Bikini*, House Natural Resources Committee, February 24, 1994.

Wellerstein, Alex, *What the NUKEMAP Taught Me About Fallout*, The Nuclear Secrecy Blog, August 2, 2013.

Wellerstein, Alex, *Why Nagasaki?*, The Nuclear Secrecy Blog, August 9, 2013.

Wellerstein, Alex, *Castle Bravo at 60*, The Nuclear Secrecy Blog, February 28, 2014.

Whaley, Barton, *Soviet Clandestine Communication Nets*, Center for International Studies, Massachusetts Institute of Technology, Sponsored by a Defense Department grant, Cambridge, Massachusetts, September 1969, Page 131.

Wittner, Lawrence S., *Resisting the Bomb, A History of the World Nuclear Disarmament Movement, 1954–1970*, Chapter One: The Gathering Storm, 1954–56, Stanford University Press, December 1997.

Woodward, K. T. et al., *The Determination of Internally Deposited Radioactive Isotopes in the Marshallese People by Excretion Analysis*, DASA-1180, Walter Reed Army Medical Center. Inst. of Research, Washington, DC, January 1, 1959.

Yehle, Emily, *Acquaculture Buoys Islands Devastated by Bomb Tests*, E&E News, April 23, 2015.

Yoshihara, H.K., *The dawn of radiochemistry in Japan*, Kusakidai, April 12, 2013, Iwaki City, Japan.

Zak, Dan, "A Ground Zero Forgotten," *Washington Post*, November 27, 2015.

NOTES and SOURCES

PROLOGUE

x. _The islands were originally settled_ (Susan Denyer, _Cultural Landscapes of the Pacific Islands_, ICOMOS, December 2007, Page 24. Kim Tingley, "Secrets of the Wave Pilots," _New York Times Magazine_, March 17, 2016.) **x. _Local chieftains controlled the atolls_** (Julianne M. Walsh, _Imagining the Marshalls: Chiefs, Tradition, and the State on the Fringes of US Empire_, University of Hawaii. 2003. Desmond N. Doulatram, _The Lolelaplap (Marshall Islands) in Us_, University of San Francisco, Master's Projects and Capstones, 2018.) **xi. _On March 1, 1954, eighty-two men, women, and children_** (National Research Council (US) Committee on Radiological Safety in the Marshall Islands. "Radiological Assessments for Resettlement of Rongelap in the Republic of the Marshall Islands," Washington (DC): National Academies Press (US); 1994.) **xii. _Life returned to normal until January 1954_** (Micronesia Support Committee, "Marshall Islands: A Chronology 1944–1981," Honolulu, Hawaii, 1982, Page 7.) **xiii. _Just after 3 a.m. March 1, 1954, some ninety miles west of Bikini_** (Matashichi Oishi, _The Day the Sun Rose in the West_, University of Hawaii Press, Honolulu, 2011. Ralph E, Lapp, _The Voyage of the Lucky Dragon_, Harper & Brothers, New York, 1958.) **xiii. _When the Lucky Dragon had left Japan on January 22_** (_The Japan Times_, "Lucky Dragon's Lethal Catch," March 18, 2012.)

Part I
CHAPTER 1

3. _The idea had first emerged from a May 17, 1941_ (Arthur H. Compton to Frank E, Jewett, May 17, 1941, _Report of National Academy of Sciences Committee on Atomic Fission._) **4. _An October 30, 1943, report_** (30 October 1943 Memorandum written to Brig. Gen. L. R. Groves, Manhattan Project from Drs. James B. Conant, Chairman, A. H. Compton, and H. C. Urey, comprising a Subcommittee of the S-1 Executive Committee.) **5. _A two-day meeting of what was called the Target Committee_** (Minutes of the second meeting of the Target Committee, Los Alamos, May 10–11, 1945; US National Archives, Record Group 77, Records of the Office of the Chief of Engineers, Manhattan Engineer District, TS Manhattan Project File '42–'46, folder 5D Selection of Targets, 2 Notes on Target Committee Meetings.) **6. _Nagasaki wasn't even on the original target list_** (Alex Wellerstein, _Why Nagasaki?_ The Nuclear Secrecy Blog, August 9, 2013.) **7. _On May 31, 1945, another key meeting was held_** (Richard G. Hewlett and Oscar E. Anderson Jr. _The New World, 1939–1946_, Pennsylvania University Press, University Park,

PA, 1962, pages 356–359. F. G. Gosling, *The Manhattan Project: Making the Atomic Bomb* (DOE/ MA-0001; Washington: History Division, Department of Energy, January 1999), 87–89. The Manhattan Project An Interactive History, *Debate Over How to Use the Bomb, Late Spring 1945*, US Department of Energy, Office of History and Heritage Resources. Vincent C. Jones, *Manhattan: The Army and the Atomic Bomb*, Center of Military History, United States Army, Washington, DC, 1985. 530–532.)

CHAPTER 2

11. The expected explosion, Hirschfelder later wrote (Hirschfelder, J. O. and J. Magee to K. Bainbridge, "Danger from Active Material Falling from Cloud Desirability of Bonding Soil Near Zero with Concrete and Oil," June 16, 1945, DOE/NV Nuclear Testing Archive, Las Vegas, Nevada.) **11. Their initial estimates, according to Trinity historian Ferenc Szasz** (Ferenc Morton Szasz, *The Day the Sun Rose Twice*, University of New Mexico Press, 1984, Page 63.) **11. "In spite of all this work, very few people believed** (Hirschfelder, J. O. and Magee, J., *Improbability of Danger from Active Material Falling from Cloud* (memorandum to K.T. Bainbridge dated July 6, 1945, from LANL Records Center Location M-7-1, SFSL-561), Project Y.) **12. A military detachment of 160 men** (K. T. Bainbridge, *Trinity*, LA-6300-H History Special Distribution, Los Alamos National Laboratory, Los Alamos, NM, Issued: May 1976, Page 37.) **12. The CIC members were armed with a cover story** (William L. Laurence, *Trinity draft press releases, May 14, 1945*, Memorandum, Nuclear Security Blog, Page 3.) **12. "You boys must have been up to something** (Joseph O. Hirschfelder, *The Scientific and Technological Miracle at Los Alamos* in *Reminiscences of Los Alamos*, ed. Lawrence Badash, Joseph O. Hirschfelder, and Herbert P. Broida (Boston: D. Reidel Publishing Company, 1980) Page 77.) **13. An elderly couple, the Raitliffs** (Barton C. Hacker, *The Dragon's Tale*, University of California Press, Berkley. California, 1987. Pages 104–105.) **13. Months after the test, The Eastman Kodak Company** (Joseph O. Hirschfelder, "The Scientific and Technological Miracle at Los Alamos," Page 75.) **13. The AEC's Merril Eisenbud, then-director** (Oral History of Merril Eisenbud, Conducted January 26, 1995, DOE/EH-0456, United States Department of Energy, Office of Human Radiation Experiments, May 1995.) **14. At the end of July, the War Department authorized Los Alamos medical personnel** (Los Alamos Historical Document Retrieval and Assessment (LAHDRA) Project, Final Report of CDC's *LAHDRA Project*, November 2010, Chapter 10, Page 278. "In a July 31, 1945 War Department memorandum to Dr. Louis Hempelmann (reproduced in Hempelmann 1947), Lt. Daniel Dailey of the Corps of Engineers refers to requests from Hempelmann that the health of persons in a certain house near Bingham, NM be discretely investigated. Over the two years following Trinity, at least seven visits were made to the Ratliff ranch . . . ") **14. Dr. Stafford Warren, then-chief of the Manhattan Project medical section** (Top Secret, July 21, 1945, To: Major Gen. Groves, SUBJECT: Report on Test II at Trinity, July 16, 1945.) **15. At 2 a.m. August 10, Prime Minister Admiral Baron Kantaro Suzuki pleaded** (The United States Strategic Bombing Survey, *Japan's Struggle to End the War*, US Government Printing Office, Washington, DC, 1946, Page 8. "Hoshina Memorandum" on the Emperor's "Sacred Decision [*go-seidan*]," 9–10 August, 1945. Zenshiro Hoshina, *Daitoa Senso Hishi: Hoshina Zenshiro Kaiso-roku* [Secret History of the Greater East Asia War: Memoir of Zenshiro Hoshina] (Tokyo, Japan: Hara-Shobo, 1975), excerpts from Section 5, "The Emperor made *go-seidan* [the sacred decision] – the decision to terminate the war," 139–149 [translation by Hikaru Tajima].) **16. *The United States Strategic Bombing Survey later reported*** (United States Strategic Bombing Survey, *Summary Report (Pacific War)*, US Government Printing Office, Washington, DC, 1946, Page 24.)

CHAPTER 3

17. *One day after Japan's Emperor Hirohito gave a recorded radio address* (Lewis Strauss, *Men and Decisions*, Doubleday, Garden City, NY, 1962. Page 208.) **19. *It was Blandy who gave the plan the name*** (Defense Threat Reduction Agency, Operation Crossroads, Fact Sheet. 2015, Page 1. William A. Shurcliff, *Bombs at Bikini: The Official Report of Operation Crossroads*, W. H. Wise, New York, 1947. Page 27.) **19. *An Army and Navy group agreed that a protected body of water*** (William A. Shurcliff, *Bombs at Bikini: The Official Report of Operation Crossroads*, W. H. Wise, New York, 1947. Page 16.) **20. *In January 1946, the American delegate*** (Kimi Hara, *Cold War Frontiers in the Asia-Pacific: Divided Territories in the San Francisco System*, Nissan Institute/Routledge Japanese Studies, Routledge, New York, 2007, Page 110.) **20. *During a February 21, 1946, evening speech*** (Jonathan M. Weisgall, *Operation Crossroads: The Atomic Tests at Bikini Atoll*, Naval Institute Press, Annapolis, Maryland, 1994, Page 64.) **21. *The use of Bikini Atoll required the relocation*** (Defense Nuclear Agency, *Operation Crossroads 1946*, DNA 6032F, Washington, DC, 1984, Page 19.) **22. *On March 7, 1946, with cisterns initially filled*** (Defense Nuclear Agency, *Operation Crossroads 1946*, DNA 6032F, Washington, DC, 1984, Page 19.)

CHAPTER 4

23. *At the time, syndicated columnists Joseph Alsop and Stewart Alsop* (Joseph Alsop and Stewart Alsop, "Atom Bomb Tests Must First Overcome Inter-Service Rivalry," *Washington Post*, Feb. 10, 1946.) **24. *The overall targets for the test would be a ninety-ship fleet*** (Naval History and Heritage Command, Operation Crossroads: Fact Sheet, The Navy Department Library, April 2015.) **25. *Six B-17, World War II bombers and four Navy F6F Hellcat fighters*** (Defense Nuclear Agency, *Operation Crossroads 1946*, DNA 6032F, Washington, DC, 1984, Page 86.) **26. *A Naval Medical Research Section was created*** (Director of Ship Material, Joint Task Force One, *Historical Report, Atomic Bomb Tests Able and Baker*, Washington, DC, Vol. 2 of 3, Page 29 (of document, Page 181 of original report).)

CHAPTER 5

29. *For the Crossroads public affairs job, Joint Task Force Commander Blandy* (William A. Shurcliff, *Bombs at Bikini: The Official Report of Operation Crossroads*, W. H. Wise, New York, 1947. Page 34. Joint Task Force One, *Public Information Estimate No 1*, 7 March 1946.) **30. *Among Lee's first problems was the number of media*** (Capt. Fitzhugh Lee USN, *The Press at Operations Crossroads*, Army Information Digest, Vol. 2, No. 3, March 1947, Page 25.) **31. *With a group that large, only a few people*** (William A. Shurcliff, *Bombs at Bikini: The Official Report of Operation Crossroads*, W. H. Wise, New York, 1947. Page 37–49.) **32. *Ten House members and four Senators*** (William A. Shurcliff, *Bombs at Bikini: The Official Report of Operation Crossroads*, W. H. Wise, New York, 1947. Page 185.) **33. *Penney also was among those who advised*** (Atomic Heritage Foundation, *William Penney, Mathematician and Physicist*, 2019.) **34. *It was no surprise that Moscow took up the US invitation*** (David Holloway, *Stalin and the Bomb*, Yale University Press, New

Haven, CT, 1994, Pages 163 and 227.) **34. *The Soviet nuclear weapons program had been accelerated*** (David Holloway, *Stalin and the Bomb*, Yale University Press, New Haven, CT, 1994, Page 129.) **34. *Among other VIP observers was Stuart Symington*** (James C. Olson, *Stuart Symington: A Life*, University of Missouri Press, Columbia, MO, 2003, Pages 81–83.) **35. *Another late arrival was then-Navy Secretary James V. Forrestal*** (James V. Forrestal (author), Walter Millis (editor), *The Forrestal Diaries*, Viking Press, New York, 1951, Page 170.)

CHAPTER 6

36. *The B-29 chosen for Test Able was named* Dave's Dream (*Operation Crossroads—1946*, sonicbomb.com, Page 1.) **37. Dave's Dream *lifted from the runway at 5:55 a.m.*** (Connie Goldsmith, *Bombs Over Bikini: The World's First Nuclear Disaster*, Twenty-First Century Books, April 2014, Pages 21–22.) **37. *On that same early morning, a metronome*** (Jonathan M. Weisgall, *Operation Crossroads: The Atomic Tests at Bikini Atoll*, Naval Institute Press 1994, Pages 1 and 184.) **38. *At 9 a.m., from twenty-eight thousand feet, Bombardier Wood called out*** (William A. Shurcliff, *Bombs at Bikini: The Official Report of Operation Crossroads*, W. H. Wise, New York, 1947, Page 105.) **39. *With a surface temperature of one hundred thousand degrees Fahrenheit*** (William A. Shurcliff, *Bombs at Bikini: The Official Report of Operation Crossroads*, W. H. Wise, New York, 1947, Page 117.) **40. *As Navy Captain Lee would later write*** (Capt. Fitzhugh Lee, USN, *The Press at Operations Crossroads*, Army Information Digest, Vol. 2, No. 3, March 1947, Page 25.) **40. *From the* USS Appalachian, *the* Washington Post's *Gerald G. Gross cabled*** (Gerald G. Gross, "A-Bomb Sinks 2 Ships and Damages 17; Capital Vessels Escape Extensive Harm; Destroyer Capsizes; Experiment 'Success,'" *Washington Post*, July 1, 1946, Page 1.) **41. *Symington, aboard a C-54 observer plane*** (Jonathan M. Weisgall, *Operation Crossroads: The Atomic Tests at Bikini Atoll*, Naval Institute Press 1994, Page 186.) **41. *An hour and a half after the bomb was exploded*** (Gerald G. Gross, "A-Bomb Sinks 2 Ships and Damages 17; Capital Vessels Escape Extensive Harm; Destroyer Capsizes; Experiment 'Success,'" *Washington Post*, July 1, 1946, Page 1.)

CHAPTER 7

42. *When the Able bomb detonated, its fireball* (Defense Nuclear Agency, *Operation Crossroads 1946*, DNA 6032F, Washington, DC, 1984, Page 452 (Page 457 in the online version).) **43. *In his post-shot broadcast, Joint Task Force Commander Admiral Blandy reported*** (Gerald G. Gross, "A-Bomb Sinks 2 Ships and Damages 17; Capital Vessels Escape Extensive Harm; Destroyer Capsizes; Experiment 'Success,'" *Washington Post*, July 1, 1946, Page 1.) **43. *After the clouds had cleared at 10 a.m., an observer aboard PBM Charlie*** (Commander, Joint Task Force One, *Report on Atomic Bomb Tests Able and Baker Operation Crossroads*, Vol. I, Defense Nuclear Agency, December 5, 1979, Page VI-(B)-9 and thereafter (Page 144 and thereafter in the online version).) **44. *Transmitters aboard the* USS Begor *guided the boats*** (The Office of the Historian, Joint Task Force One, *Operation Crossroads, The Official Pictorial Record*, Wm. H. Wise & Co. Inc., New York, 1946, Page 89 (Page 88 in the online version).) **44. *One of the radiological safety officers assigned*** (Robert A. Conard, *Fallout, The Experiences of a Medical Team in the Care of a Marshallese Population Accidentally Exposed to Fallout Radiation*, September 1992, Brookhaven National Laboratory, Upton, NY, September 1992, Page 6 (Page 7 in the online version).) **45. *Conard would later say, "We turned*** (Rear Admiral Robert Conard, *Oral History: Operation Crossroads, Nuclear Weapons Test at*

Bikini Atoll, 1946, Naval History and Heritage Command, November 9, 1993.) **46. *Seven hours after the detonation, they began retrieving*** (William A. Shurcliff, *Bombs at Bikini: The Official Report of Operation Crossroads*, W. H. Wise, New York, 1947, Page 140.) **46. *A fifty-pound, six-month-old, China Poland sow*** (*This Little Pig Came Home*, *Time* magazine, April 11, 1969.) **47. *On August 2, 1946, the Presidential Evaluation Board made a preliminary report*** (William A. Shurcliff, *Bombs at Bikini: The Official Report of Operation Crossroads*, W. H. Wise, New York, 1947, Page 195.) **47. *On July 2, as more areas of the lagoon were inspected*** (Combined American Press Dispatch, "Forrestal and Blandy Get Close-Up of Sakawa Sinking and Other Ruins," *New York Times*, July 3, 1946, Page 3.) **48. *The admiral and Forrestal then posed*** (NavSource Online: Battleship Photo Archive, *BB-34 USS New York.)* **48. *Speaking to the press after his lagoon tour, Forrestal described*** (Associated Press, "Forrestal After Survey Said Navies Will Go On," *New York Times*, July 2, 1946, Page 18.) **49. *A supporter of air power, Symington*** (James C. Olson, *Stuart Symington: A Life*, University of Missouri Press, Columbia, MO, 2003, Page 83.)

CHAPTER 8

50. *Two days after the test, Captain Lee shepherded a group* (US Naval Institute, *Oral History, Fitzhugh Lee, Vice Admiral (Retired),* 1970. Capt. Fitzhugh Lee USN, The Press at Operations Crossroads, Army Information Digest, Vol. 2, No. 3, March 1947, Page 25–31.) **51. *On July 11, the Joint Chiefs Evaluation Board released its preliminary statement*** (William A. Shurcliff, *Bombs at Bikini: The Official Report of Operation Crossroads*, W. H. Wise, New York, 1947, Pages 192–195.)

CHAPTER 9

53. *Back in October 1945, then-Navy Commodore Deak Parsons* (Roger Meade, *Discoveries and Collisions, The Atom, Los Alamos, and the Marshall Islands*, Arizona State University, May 2015, Page 89 (Page 101 in the online version).) **55. *To prepare for this first underwater nuclear detonation*** (Defense Nuclear Agency, *Operation Crossroads* 1946, DNA 6032F, Washington, DC, 1984, Page 60 (Page 65 in the online version).) **56. *During that same period, on July 16, there was a ceremony*** (William L. Laurence, "Bikini's King Gets Truman's Thanks," *New York Times*, July 17, 1946, Page 7.) **57. *Four days before the planned shot*** (William A. Shurcliff, *Bombs at Bikini: The Official Report of Operation Crossroads*, W. H. Wise, New York, 1947, Page 147.) **58. *King Juda's arrival at Bikini surprised*** (United Press, "Bikini King Amazed at Changed Realm," *New York Times*, July 24, 1946, Page 12.) **59. *At 5:30 a.m., bunting was seen flying*** (Defense Nuclear Agency, *Operation Crossroads 1946*, DNA 6032F, Washington, DC, 1984, Page 97 (Page 102 in the online version).)

CHAPTER 10

61. *"Things happened so fast in the next five seconds* (William A. Shurcliff, *Bombs at Bikini: The Official Report of Operation Crossroads*, W. H. Wise, New York, 1947, Page 151.) **63. *The Baker detonation dug a 2,000-foot-wide crater*** (James P. Delgado and Larry E. Murphy, *The Archeology of the Atomic Bomb, Chapter Two: Operation Crossroads*, National Park Service. Samuel Glasstone and Philip J. Dolan, *The Effects of Nuclear Weapons*, Third Edition, US Department of Defense, 1977, Page 251 (or Page 258 in the online version).) **63. *Much of the initial nuclear radiation***

had been absorbed (Samuel Glasstone and Philip J. Dolan, *The Effects of Nuclear Weapons*, Third Edition, US Department of Defense, 1977, Page 413 (page 420 in the online version).)

CHAPTER 11

65. *The drones, however, were so radioactive* (Defense Nuclear Agency, *Operation Crossroads 1946*, DNA 6032F, Washington, DC, 1984, Page 101 (Page 106 in the online version).) **67.** *Dr. Stafford Warren wrote his wife* (Jonathan M. Weisgall, *The Able-Baker-Where's-Charlie Follies*, The Bulletin of the Atomic Scientists, May/June 1994, Page 27.) **68.** *The Joint Task Force leadership then realized, according to a later Pentagon technical study* (Defense Threat Reduction Agency, Operation Crossroads, Fact Sheet, May 2015, Page 4.) **68.** *The Radiological Safety Section convened a meeting on July 27* (Defense Nuclear Agency, *Operation Crossroads 1946*, DNA 6032F, Washington, DC, 1984, Pages 110–120 (Pages 114–124 in the online version).)

CHAPTER 12

69. *On July 31, Admiral Solberg issued a memorandum entitled "Decontamination Procedures on Target Vessels,"* (*Radiological Decontamination of Target and Non-Target Vessels*, Vol. 3, Operation Crossroads Joint Task Force One, AD473908, 1946, Page 4 (Page 7 of the online version).) **70.** *"There is very little leeway in this"* (Col. Stafford M. Warren, *Occupancy of Target Vessels as Influenced by Intensity of Radiation of Various Types on Target Vessels, August 7, 1946*, Document 24, National Security Archives, George Washington University.) **71.** *On August 4, Admiral Solberg, as director of ship material* (Director of Ship Material Technical Inspection Report, *Radiological Decontamination of Target and Non-Target Vessels*, Vol. 3, Operation Crossroads Joint Task Force One, AD473908, Page 8 (Page 11 of the online version).) **72.** *The support ships that had entered Bikini Lagoon* (William A. Shurcliff, *Bombs at Bikini: The Official Report of Operation Crossroads*, W. H. Wise, New York, 1947, Pages 170–171.) **72.** *During a later oral history, Conard recalled* (*Oral History: Operation Crossroads, Nuclear Weapons Test at Bikini Atoll*, 1946, Naval History and Heritage Command, November 9, 1993.) **73.** *By August 8, however, Blandy recognized some ships* (Defense Nuclear Agency, *Operation Crossroads 1946*, DNA 6032F, Washington, DC, 1984, Page 115 (Page 119 in the online version).) **73.** *On August 9, an inspection of decontamination progress took place* (Defense Nuclear Agency, *Operation Crossroads 1946*, DNA 6032F, Washington, DC, 1984, Pages 116–118 (Pages 120-122 in the online version).) **73.** *On August 10, Blandy called a conference* (Jonathan M. Weisgall, *The Able-Baker-Where's-Charlie Follies*, The Bulletin of the Atomic Scientists, May/June 1994, Page 32.) **74.** *On August 19, the Navy began towing target ships* (Defense Nuclear Agency, *Operation Crossroads 1946*, DNA 6032F, Washington, DC, 1984, Page 108 (Page 112 in the online version).) **74.** *The chief of naval operations later declared Bikini Lagoon* (Commander, Joint Task Force One, *Report on Atomic Bomb Tests Able and Baker (Operation Crossroads*, Vol. I, Defense Nuclear Agency, December 5, 1979, Pages V-(D)–5 and V–(D)–6 (Pages 126–127 in the online version).)

CHAPTER 13

77. *A University of Michigan Survey Research Center poll* (Survey Research Center, *Public Reaction to the Atomic Bomb and World Affairs*, University of Michigan, Ann Arbor, Michigan,

February 1947, Page 15 (Page 26 of the online version).) **77. Captain Lee later wrote, "If I were doing this job** (Capt. Fitzhugh Lee USN, *The Press at Operations Crossroads*, Army Information Digest, Vol. 2, No. 3, March 1947, Page 28.) **78. Betts recommended that for future tests** (Commander, Joint Task Force One, *Report on Atomic Bomb Tests Able and Baker Operation Crossroads*, Vol. I, Defense Nuclear Agency, December 5, 1979, Page III-B-2 (Page 61 in the online version).) **78. For example, Los Alamos scientists had a different view** (Los Alamos National Laboratory, *Manhattan District History, Book VIII, Los Alamos Project (Y), Volume 3, Auxiliary Activities, Chapter 8, Operation Crossroads*, Page 8.1 (Page 5 in the online version).)

CHAPTER 14

81. Nonetheless, the Navy examined tens of thousands of specimens (Defense Nuclear Agency, *Operation Crossroads 1946*, DNA 6032F, Washington, DC, 1984, Page 158 (Page 163 in the online version).) **82. In May 1947, having had food shortages** (Jack Niedenthal, A Short History of the People of Bikini Atoll, from his book, *For the Good of Mankind.*) **82. On July 18, 1947, the United Nations Security Council** (United Nations Security Council, *Resolution of 2 April 1947*, S/RES/21(1947), Pages 16–21 (Pages 1–6 in the online version).)

CHAPTER 15

84. At Kwajalein, a 1,500-person Kwajalein Maintenance Force (KMF) (Defense Nuclear Agency, *Operation Crossroads 1946*, DNA 6032F, Washington, DC, May 1984, Page 121.) **86. Then-Lieutenant Commander Conard recalled transferring** (US Naval Institute Oral Histories, Operation Crossroads, Nuclear Weapons Test at Bikini Atoll, 1946, *Reminiscences of Rear Admiral Robert Conard, MC, USNR (Ret.)*, Nov. 9, 1993.) **87. In early September 1946, the destroyer** USS **Laffey** (Director of Ship Material Technical Report, *Radiological Decontamination of Target and Non-Target Vessels*, Vol. 2 of 3, Defense Nuclear Agency, Washington, DC, Page 5.) **88. After Admiral Solberg conducted an inspection of the Laffey** (Director of Ship Material Technical Report, *Radiological Decontamination of Target and Non-Target Vessels*, Vol. 2 of 3, Defense Nuclear Agency, Washington, DC, Page 27.)

CHAPTER 16

90. In his final classified report as commander of Joint Task Force One, (Commander Joint Task Force One, *Historical Report, Atomic Bomb Tests Able and Baker*, Washington, DC, Vol. 1, 1946, Page II-(A)-6 in actual document (Page 33 of online version).) **91. The President's Evaluation Committee released** (William A. Shurcliff, *Bombs at Bikini: The Official Report of Operation Crossroads*, W. H. Wise, New York, 1947, Page 204.) **91. Army Air Force Major General W. W. Kepner** (Commander Joint Task Force One, *Historical Report, Atomic Bomb Tests Able and Baker*, Washington, DC, Vol. 1, 1946, Page VII-(E)-230 (Page 682 in online version).) **92. While Blandy and Kepner's contributions to the Joint Task Force final report** (Commander Joint Task Force One, *Historical Report, Atomic Bomb Tests Able and Baker*, Washington, DC, Vol. 1, 1946, Page II-(C)-3 (Page 47 in online version).) **93. Major General Leslie Groves and other senior Los Alamos officials** (Joint Chiefs of Staff, *Joint Chiefs of Staff Decision Amending J C S 1552 74 Test C Operation Crossroads*, JCS 1901 10, 24 August 1946, Top Secret, Appendix C, Page 213 (National Security Archive Document 30, Page 7).) **95. "Everyone knew that I was in a caretaker's position** (General

Leslie R. Groves, *Now It Can Be Told: The Story Of The Manhattan Project*, (paperback) Da Capo Press Inc., 1983, Page 395.)

CHAPTER 17

98. *The next decision to be made by the AEC* (Department of Energy, *Enewetak Radiological Support Project, Final Report*, NVO-213, Las Vegas, Nevada, September 1982, Page 4 (Page 24 online document).) **99. *In late November 1947, twenty Navy Seabees*** (Micronesian Support Committee, *Marshall Islands, A Chronology: 1944–1981*, Honolulu, Hawaii, 1981, Page 10.) **100. *As for the 145 Enewetak Marshallese, their new homes*** (Department of Energy, *Enewetak Radiological Support Project, Final Report*, NVO-213, Las Vegas, Nevada, September 1982, Page 4–5 (Page 24–25 online document).) **100. *Marshall Islanders in general, and Enewetak people*** (Davor Pevec, *The Marshall Islands Nuclear Claims Tribunal: The Claims of the Enewetak People*, Denver Journal of International Law and Policy, Vol. 35:1, March 10, 2008, Page 223 (Page 3 online version).) **100. *The day after Truman's announcement, December 3, 1947*** (Davor Pevec, *The Marshall Islands Nuclear Claims Tribunal: The Claims of the Enewetak People*, Denver Journal of International Law and Policy, Vol. 35:1, March 10, 2008, Page 227 (Page 7 online version).) **101. *Meanwhile, the Bikinians still on Rongerik*** (Jack Niedenthal, *For the Good of Mankind, a History of the People of Bikini and their Islands*, A Bravo Book, Republic of the Marshall Islands, 2001, Page 4.)

CHAPTER 18

103. *AEC Commissioner Lewis Strauss even warned* (Esmond Stewart, *Defense Nuclear Agency 1947–1997*, Defense Threat Reduction Agency, Government Printing Office, Washington, DC, 2017, Page 51.) **104. *The yield of the first test, X-Ray*** (Defense Nuclear Agency, *Operation Sandstone:1948*, DNA 6033F, December 1983, Pages 1 and 105 (Pages 5 and 109 in online document).) **105. *In September 1948, as tensions over Berlin increased*** (James V. Forrestal (author), Walter Millis (editor), *The Forrestal Diaries*, Viking Press, New York, 1951, Pages 487–488.) **106. *The site had originally been chosen by Soviet Secret Police Chief Beria*** (David Holloway, *Stalin and the Bomb*, Yale University Press, New Haven, CT, 1994, Page 213.) **108. *The Armed Forces Special Weapons Project, at the direction of the Joint Chiefs*** (Terrence R. Fehner and F. G. Gosling, *Origins of the Nevada Test Site*, History Division, Department of Energy, DOE/MA-0518, Washington, DC, December 2000, Page 40 (Page 48 in the online version).) **109. *The expert panel determined there was*** (Terrence R. Fehner and F. G. Gosling, *Origins of the Nevada Test Site*, History Division, Department of Energy, DOE/MA-0518, Washington, DC, December 2000, Page 46 (Page 54 in the online version).) **110. *The NSC went back to Dean*** (Terrence R. Fehner and F. G. Gosling, *Origins of the Nevada Test Site*, History Division, Department of Energy, DOE/MA-0518, Washington, DC, December 2000, Page 48 (Page 56 in the online version).) **110. *The handbill warned that "NO PUBLIC ANNOUNCEMENT*** (Terrence R. Fehner and F. G. Gosling, *Origins of the Nevada Test Site*, History Division, Department of Energy, DOE/MA-0518, Washington, DC, December 2000, Page 56 (Page 46 in the online version).)

CHAPTER 19

111. *Greenhouse's George test at Enewetak* (The Nuclear Weapon Archive, *Operation Greenhouse 1951*, August 2003.) **114. *That was three days before the US presidential election*** (Richard G.

Hewlett and Francis Duncan, *Atomic Shield, 1947–1952*, Pennsylvania State University Press, 1969, Pages 647–649 (Page 665–667 in online version).) **114. *At about the same time in the Pacific, a British cargo ship*** (Defense Nuclear Agency, *Operation Ivy 1952*, DNA 6036F, Pages 180–181, (Pages 187–188 in online version), December 1982.) **115. *A month later, Dr. Lauren R. Donaldson wrote in a report*** (Lauren R. Donaldson, *Radiobiological Studies at Enewetak Following the Mike Shot of November 1952*, Page 1 (Page 11 in online version), University of Washington, Seattle, Washington, June 1953.) **115. *He did, however, make arrangements*** (Richard Rhodes, *Dark Sun, The Making of the Hydrogen Bomb*, Simon & Schuster, New York,1995, Pages 510–511.) **116. *When later reporting the Mike results*** (Joanne Lamm, *The Island is Missing!*, US Army Heritage and Education Center, October 28, 2010.)

CHAPTER 20

120. *In June 1952, the Joint Chiefs of Staff* (Omar N. Bradley, *A General's Life: An Autobiography by General of the Army Omar N. Bradley*, Simon & Schuster, New York, 1983, Page 515.) **120. *The high commissioner for the Trust Territories, Elbert D. Thomas*** (House Subcommittee on Oversight and Investigations, *Radiation Exposure From Nuclear Tests in the Pacific*, Hearing February 24, 1994, Washington, DC, Committee on Natural Resources, Serial No. 103—68, Page 26.) **121. *As a result of this objection, the danger area*** (House Subcommittee on Oversight and Investigations, *Radiation Exposure From Nuclear Tests in the Pacific*, Hearing February 24, 1994, Washington, DC, Committee on Natural Resources, Serial No. 103—68, Page 7.) **122. *One month later, a notice to mariners*** (Myres S. McDougal and Norbert A. Schlei, *The Hydrogen Bomb Tests in Perspective: Lawful Measures for Security*, The Yale Law Journal, Vol. 64, New Haven, Conn., 1955, Page 651 (Page 8 in the online version).) **122. *An October 31, 1953, letter*** (Technical Branch, Joint Task Force Seven, *Operation Castle: Radiological Safety*, Final Report, Vol. 1, (Extract) Defense Nuclear Agency, Washington, DC, September 1985, Page C-3 (Page 153 in the online version).) **123. *To assure that adequate weather information*** (Defense Nuclear Agency, *Castle Series 1954*, DNA 6035F, Washington, DC, 1982, Page 113 (Page 118 in online version).) **124. *They proposed floating one hundred buoys*** (Defense Nuclear Agency, *Castle Series 1954*, DNA 6035F, Washington, DC, 1982, Pages 52, 65, 174, 179.) **126. *The information available to the Bravo planners*** (Defense Nuclear Agency, *Castle Series 1954*, DNA 6035F, Washington, DC, 1982, Page 41.) **127. *Two days after the commission's announcement*** (Report of Task Group Commander Group 7.1, *Operation Castle Pacific Proving Grounds, March–May 1954*, Los Alamos Scientific Laboratory, Los Alamos, New Mexico, June 1954. Pages 39–40 (Pages 41–42 in online version).) **127. *Meanwhile, at a point on the northern part of the Bikini reef*** (The Nuclear Weapons Archive, *Operation Castle 1954—Pacific Proving Ground*, May 2006.) **128. *Nevertheless, at 11 a.m. that morning*** (House Subcommittee on Oversight and Investigations, *Radiation Exposure From Nuclear Tests in the Pacific*, Hearing February 24, 1994, Washington, DC, Committee on Natural Resources, Serial No. 103—68, Page 8.) **129. *At the 6 p.m. weather briefing*** (Defense Nuclear Agency, *Castle Series 1954*, DNA 6035F, Washington, DC, 1982, Page 202 (Page 209 in online version).) **129. *At dusk on February 28, John C. (Joe) Clark*** (Keith M. Parsons, Robert A. Zaballa, *Bombing the Marshall Islands: A Cold War Tragedy*, Cambridge University Press, New York, 2017, Pages 57–59.)

CHAPTER 21

132. *From Building 70, the announcement was made* (Dr. John C. Clark, *We Were Trapped by Radioactive Fallout*, Saturday Evening Post, July 20, 1957.) **133.** *Instead, it turned out to be a one-hundred-mile-wide cloud* (Thomas Kunkle, Bryan Ristvet, *Castle Bravo: Fifty Years of Legend and Lore*, Defense Threat Reduction Agency, Fort Belvoir, Va., January 2013, Page 52 (Page 65 in online version).) **135.** *Fallout, described as "pinhead-sized, white and gritty snow"* (Defense Nuclear Agency, *Castle Series 1954*, DNA 6035F, Washington, DC, 1982, Page 210 (Page 217 in the online version).) **135.** *A chemical decomposition accompanied* (House Subcommittee on Oversight and Investigations, *Radiation Exposure From Nuclear Tests in the Pacific*, Hearing February 24, 1994, Washington, DC, Committee on Natural Resources, Serial No. 103—68, Page 130 (Page 135 in the online version).)

CHAPTER 22

137. *At 9:49 a.m., according to a Defense Nuclear Agency report* (Defense Nuclear Agency, *Castle Series 1954*, DNA 6035F, Washington, DC, 1982, Pages 212–213 (Pages 219–220 in online version).) **138.** *Later on the morning of the explosion, John Anjain* (John Anjain, *The Voices of Rongelap: Cautionary Tales From a Nuclear War Zone*, Barbara Rose Johnston and Holly M. Barker, posted November 22, 2008.) **140.** *The primary radiation hazard to the human body* (United States Nuclear Regulatory Commission, *Radiation Basics.*) **142.** *On Rongerik, hours after the detonation* (Thomas Kunkle, Bryan Ristvet, Castle Bravo: *Fifty Years of Legend and Lore*, Defense Threat Reduction Agency, Fort Belvoir, VA, January 2013, Page 106 (Page 118 in the online version).) **143.** *At Bikini Atoll just before sunset* (Dr. John C. Clark, *We Were Trapped by Radioactive Fallout*, Saturday Evening Post, July 20, 1957.) **143.** *In the Army's van on Rongerik* (Robert Sharp and William H. Chapman, *Exposure of Marshall Islanders and American Military Personnel to Fallout*, Operation Castle—Project 4.1 Addendum, Naval Medical Research Institute, Bethesda, Maryland, March 1957, Page 40.) **144.** *An error in communications* (Defense Nuclear Agency, *Castle Series 1954*, DNA 6035F, Washington, DC, 1982, Pages 220–221 (Pages 227–228 in online version).)

CHAPTER 23

146. *At 9:45 a.m. on March 2* (Capt. Louis B. Chrestensen, *SUBJECT: Island Evacuation, To Whom It May Concern*, Pages 11–15 of online document. Plus attached documents.) **147.** *After the lunch on March 2* (Thomas Kunkle, Bryan Ristvet, Castle Bravo: *Fifty Years of Legend and Lore*, Defense Threat Reduction Agency, Fort Belvoir, Va., January 2013, Page 114 (Page 126 in the online version).) **149.** *A decision was made in Washington* (Major General P. W. Clarkson, *History of Operation Castle*, Defense Nuclear Agency, Washington, DC, June 1983, Page 110 (Page 127 in the online version).) **150.** *It was 7:30 a.m., March 3, when the* **USS Philip** *dropped anchor* (Thomas Kunkle, Bryan Ristvet, *Castle Bravo: Fifty Years of Legend and Lore*, Defense Threat Reduction Agency, Fort Belvoir, VA, January 2013, Page 115 (Page 127 in the online version).) **156.** *Significant exposure to radiation depresses* (Samuel Glassman and Philip J. Dolan, *The Effects of Nuclear Weapons*, Third Edition, US Departments of Defense and Energy, Washington, DC, 1977, Pages 575–587.) **157.** *Dr. Conard would late write* (Robert A. Conard, *Fallout: The Experiences of a Medical Team in the Care of a Marshallese Population Accidentally Exposed* *to Fallout Radiation*, Brookhaven National Laboratory, Upton, NY, September 1992, Page 6 (Page 14 in the online version).)

CHAPTER 24
159. *On March 4, Dr. Alvin C. Graves* (House Subcommittee on Oversight and Investigations, *Radiation Exposure From Nuclear Tests in the Pacific*, Hearing February 24, 1994, Washington, DC, Committee on Natural Resources, Serial No. 103—68, Page 22 (Page 37 in online version).) **160.** *Original planning for all scientific investigations* (Thomas Kunkle, Bryan Ristvet, *Castle Bravo: Fifty Years of Legend and Lore*, Defense Threat Reduction Agency, Fort Belvoir, VA, January 2013, Page 130 (Page 142 in the online version).) **162.** *On March 8, Cronkite and his medical team* (ACHRE Report, *The Marshallese*, Part II, Chapter 12, Advisory Committee on Human Radiation Experiments, US Government Printing Office, Washington, DC, 1995.) **163.** *On March 8, Herbert J.* (Pete) Scoville Jr. (Thomas Kunkle, Bryan Ristvet, *Castle Bravo: Fifty Years of Legend and Lore*, Defense Threat Reduction Agency, Fort Belvoir, VA, January 2013, Page 93 (Page 106 in online version).) **163.** *Walmer Strope, a radiation protection researcher* (Walmer E. (Jerry) Strope, *Autobiography of a Nerd*, Chapter 10, (unpublished online autobiography) 2003, Pages 123 and beyond.)

Part II: Long-Term Problems
CHAPTER 25
171. *It began March 3, two days after the detonation, with Marine Corporal Don Whitaker* (Congressional Record—House, *The Three Surprises*, March 17, 2004, Page H1139.)

CHAPTER 26
176. *On March 12, as they neared home, the crewmen* (Ralph Lapp, *The Voyage of the Lucky Dragon*, Harper & Brothers, New York 1957, Page 52.) **178.** *Dr. Tsuzuki had just been put in charge* (Aya Homei, *The contentious death of Mr Kuboyama: science as politics in the 1954 Lucky Dragon incident*, Japan Forum, 25:2, December 2012, Pages 212–232.) **181.** *Professor Takanobu Shiokawa, who taught chemistry* (Ralph Lapp, *The Voyage of the Lucky Dragon*, Harper & Brothers, New York 1957, Pages 80–84.) **182.** *That same morning, two hundred miles west of Yaizu* (Ralph Lapp, *The Voyage of the Lucky Dragon*, Harper & Brothers, New York 1957, Pages 92–98.) **184.** *Nishiwaki decided to write a letter* (Tokyo Tech Museum and Archives, *Scientist Yasushi Nishiwaki in the Nuclear Age*.)

CHAPTER 27
188. *A cable had come in from Ambassador Allison* (American Embassy Tokyo to Department of State, Embassy Dispatch 1428, Secret, April 16, 1954, *Public and Private Official Papers Relating to the case of "Fukuryu Maru No. 5" [Lucky Dragon]*, Documentation March 17 to April 23, 1954. The Ambassador in Japan (Allison) to the Department of State, Top Secret, Tokyo, May 20, 1954—2 p.m., Telegram *2853. Subject:* Fukuryu Maru, Foreign Relations of the United States, 1952–1954, China and Japan, Vol. XIV, Part 2.) **189.** *Marshall Islanders, Dunham said at a 1967 meeting* (United States Defense Atomic Support Agency, *Proceedings of the Second Interdisciplinary Conference on Selected Effects of a General War*, Vol. II, DASA 2019–2, July 1969, Page 37 (Page 53 in the online

version).) **190. *Dr. Tsuzuki held the rank of rear admiral*** (Frank W. Putnam, *The Atomic Bomb Casualty Commission in retrospect*, Proceedings of the National Academy of Sciences, USA 95 (1998) Page 5427 (Page 2 in online version).) **190. *Professor Kenjiro Kimura—a distinguished, internationally respected*** (H. K. Yoshihara, *The dawn of radiochemistry in Japan*, Kusakidai April 12, 2013, Iwaki City, Japan, Pages 525–526 (Pages 3–4 in the online version).) **192. *He said he found the two* Lucky Dragon *seamen*** (Roger Rapoport, *The Great American Bomb Machine*, Dutton, New York, 1971, Page 58.) **192. *Unknown to the public, and a deep secret*** (Office of the Director of National Intelligence, *Cold War Counterterrorism*, Washington, DC, 2007, Page 66.) **193. *Rastvorov provided his CIA interrogators*** (Barton Whaley, *Soviet Clandestine Communication Nets*, Center for International Studies, Massachusetts Institute of Technology, Sponsored by a Defense Department grant, Cambridge, Massachusetts, September 1969, Page 131 (Page 139 in online version). Serge F. Kovaleski, "The Most Dangerous Game," *Washington Post Magazine*, January 15, 2006.) **193. *On March 22, 1954, days after the* Lucky Dragon *had surfaced*** (Matthew Jones, *After Hiroshima: The United States, Race and Nuclear Weapons in Asia, 1945–1965*, Cambridge University Press, New York, 2010, Page 188.) **194. *The project engineer on the test of the Romeo device*** (Richard Rhodes, *Harold Agnew's Interview*, Voices of the Manhattan Project, National Museum of Nuclear Science & History, Albuquerque, New Mexico. 1994.) **196. *Russian scientists working on the H-bomb project*** (David Holloway, *Stalin and the Bomb: The Soviet Union and Atomic Energy, 1939–1956*, Yale University Press, New Haven, CT, 1994, Page 337.) **197. *In the wake of Bravo problems, Dr. Merril Eisenbud (House Subcommittee on Oversight and Investigations*** , *Radiation Exposure From Nuclear Tests in the Pacific*, Hearing February 24, 1994, Washington, DC, Committee on Natural Resources, Serial No. 103—68, Pages 95–111 (Pages 100–121 in the online version).)

CHAPTER 27

199. *On March 24, the White House Operations Coordinating Board* (Records of the White House Operations Coordinating Board which I photographed at the Eisenhower Library, Abilene, Kansas, along with notes taken at National Security Council meetings are the sources for material in the early portions of this chapter. A reference to the above session is found in the following: Peter Kuznick, *Japan's nuclear history in perspective: Eisenhower and atoms for war and peace*, Bulletin of the Atomic Scientists, April 13, 2011.) **202. *Asked at the meeting's end to come up with a plan*** (Thomas C. Reed, *At the Abyss: An Insider's History of the Cold War* (New York: Presidio Press, 2005), Page 51.) **203. *The seventy-nine-year-old prime minister, shaken*** (Hansard 1803–2995, *Summary Information for Mr. Winston Churchill, 1954*.)

CHAPTER 28

206. *The Tokyo government four days earlier* (National Security Archive, Document 7: *US Embassy Tokyo Airgram 1482, "Public and Private Official Papers Relating to the Case of the Fukuryu Maru No. 5: Documentation March 17–April 23 1954," 30 April 1954, Secret, excerpts*, Enclosure 14, Aide Memoire, Page 1 (Page 16 in online version).) **210. *Three weeks later, Joint Chiefs of Staff Chairman Adm. Arthur Radford*** (National Security Archive, Document 8: *Admiral Arthur Radford, Chairman, Joint Chiefs of Staff, Memorandum to Secretary of Defense, "A Proposal for a Moratorium on Future Testing of Nuclear Weapons," 30 April 1954, Secret*.) **211. *As historians Richard***

Hewlett and Jack Holl described it (Richard G, Hewlett and Jack M. Holl, *Atoms for Peace and War 1953–1961*, University of California Press, Berkley, CA, 1969, Page 181 (Page 209 in the online version).) **211. *Cronkite later said that after the March 31 press conference*** (United States Advisory Committee on Human Radiation Experiments, *Final Report*, US Government Printing Office, Washington, DC, October 1995, Page 587.) **212. *Eisenbud had also been angered over Strauss's remark*** (*Human Radiation Studies: Remembering the Early Years*, Oral History of Merril Eisenbud, Department of Energy, Office of Human Radiation Experiments, May 1995.)

CHAPTER 29

213. *The day after Strauss's press conference, Dr. Masao Tsuzuki* (Defense Atomic Support Agency, *Proceedings of the Second Interdisciplinary Conference on the Selected Effects of a General War*, DASIAC Special report 95, Vol. II, July 1969, Page 77 (Page 93 in online version).) **214. *On April 20, 1954, a committee of the Marshall Islands Congress*** (350/5–354: Telegram, The Deputy United States Representative at the United Nations (Wadsworth) to the Department of State, Foreign Relations of the United States, 1952–1954, United Nations Affairs, Vol. III, Document 933.) **216. *The issue of the* Lucky Dragon *being a spy ship*** (House Subcommittee on Oversight and Investigations, *Radiation Exposure From Nuclear Tests in the Pacific*, Hearing February 24, 1994, Washington, DC, Committee on Natural Resources, Serial No. 103—68, Page 31 (Page 36 in online version). Matthew Jones, *After Hiroshima: The United States, Race and Nuclear Weapons in Asia*, 1945–1965, Cambridge University Press, New York, 2010, Page 191. National Security Archive, Document 7: *US Embassy Tokyo Airgram 1482, "Public and Private Official Papers Relating to the Case of the Fukuryu Maru No. 5: Documentation March 17–April 23 1954," 30 April 1954, Secret, excerpts*, Enclosure 26, Aide Memoire, Page 1 (Page 16 in online version).) **218. *A Defense Department report on Castle's military effects*** (Defense Nuclear Agency, Castle Series 1954, DNA 6035F, Washington, DC, 1982, Page 186 (Page 192 in the online version).) **220. *John Anjain, who still served as magistrate*** (The Special Committee Concerning Rongelap and Utirik Atolls, *Compensation for the People of Rongelap and Utirik*, Fifth Congress of Micronesia, February 1974, Page 11 (Page 27 in the online version).)

CHAPTER 30

222. *In response, Robert Murphy, the acting secretary of state* (Robert Murphy, *Memorandum by the Acting Secretary of State to the President*, Document No. 763, Foreign Relations of the United States, 1952–1954, China and Japan, Vol. XIV, Part 2, May 29, 1954.) **222. *Gov. Harold Stassen, then-White House director of foreign operations*** (Harold E. Stassen, *Memorandum by the Director of the Foreign Operations Administration (Stassen) to the President*, Document No. 774, June 29, 1954.) **223. *On August 5, 1954, in connection with the Hiroshima anniversary*** (Ralph Lapp, *The Voyage of the Lucky Dragon*, Harper & Brothers, New York 1957, Page 165.) **225. *Smith's finding led to the commission's establishment of Project Gabriel*** (Minutes, *Thirty-sixth Meeting of the General Advisory Committee to the US Atomic Energy Commission*, Washington, DC, August 1953, Page 12 (Page 4 in the online version).) **227. *Work on strontium-90, through the Project Gabriel*** (Memorandum, Advisory Staff Committee, *Documentary Update on Project Sunshine "Body Snatching"*, Advisory Committee on Human Radiation Experiments, Washington, DC, June 9, 1995, Page 2.) **228. *The Japanese government tried to respond*** (Graham Leonard, *The 1954 Shunkotsu Maru Expedition* *and American Atomic Secrecy*, Osaka School of International Public Policy, Osaka University, 2011.) **229. *In a later report to Congress, the AEC said*** (US Atomic Energy Commission, *Eighteenth Semiannual Report of the Atomic Energy Commission*, July 1955, Page 93 (Page 97 in the online version).) **229. *In the summer of 1954, the Japanese data*** (Gordon M. Dunning (editor), *Radioactive Contamination of Certain Areas of the Pacific Ocean From Nuclear Tests*, US Atomic Energy Agency, Washington, DC, August 1957, Page 39 (Page 64 in the online version).) **230. *In a report filed with the AEC in August 1955*** (Lauren R. Donaldson, *A Radiological Study of Rongelap Atoll, Marshall Islands 1954–1955*, UWFL-42, University of Washington, Seattle, Washington, August 1955.) **231. *During an October 23, 1954, meeting*** (Memorandum of Conversation by the Acting Director of the Office of Northeast Asian Affairs (McClurkin), *Bikini Compensation*, Document No. 817, Foreign Relations of the United States, 1952–1954, China and Japan, Vol. XIV, Part 2, October 23, 1954.) **232. *In a January 5, 1955, settlement note*** (Memorandum by the Assistant Secretary of State for Far Eastern Affairs (Robertson) to the Under Secretary of State (Hoover), *Bikini Compensation*, Document No. 844, Foreign Relations of the United States, 1952–1954, China and Japan, Vol. XIV, Part 2, December 29, 1954.) **233. *Late in January 1955, the Utirik people*** (The Special Committee Concerning Rongelap and Utirik Atolls, *Compensation for the People of Rongelap and Utirik*, Fifth Congress of Micronesia, February 1974, Page 9 (Page 25 in the online version).)

CHAPTER 31

235. *On December 23, 1954, distinguished Manhattan Project physicist Hans Bethe* (Hans Bethe, *Letter to Dr. W. F. Libby*, December 17, 1954.) **237. *At a February 3, 1955, National Security Council session*** (S. Everett Gleason, *Memorandum of Discussion at the 235th Meeting of the National Security Council*, Washington, February 3, 1955, Document 5, Foreign Relations of the United States, 1955–1957, Regulation of Armaments; Atomic Energy, Vol. XX.) **237. *Strauss opened the following week's February 10, 1955, NSC meeting*** (S. Everett Gleason, *Memorandum of Discussion at the 236th Meeting of the National Security Council*, Washington, February 10, 1955, Document 7, Foreign Relations of the United States, 1955–1957, Regulation of Armaments; Atomic Energy, Vol. XX.) **238. *Democrats made "an argument*** (James Hagerty, *Diary Entry by the President's Press Secretary (Hagerty)*, Washington, January 4, 1955, Document 3, Foreign Relations of the United States, 1955–1957, National Security Policy, Vol. XIX.) **239. *The next day, on Tuesday, February 15, 1955*** (US Atomic Energy Commission, *Eighteenth Semiannual Report of the Atomic Energy Commission*, July 1955. Page 147.) **xxx. *The next series of American nuclear tests in 1955*** (Defense Nuclear Agency, *Operation Teapot. 1955*, DNA 6009F, Washington, DC, November 1981, Page 1 (Page 5 in the online version).) **243. *On February 21, 1955, the day before a postponed test shot*** (Richard G, Hewlett and Jack M. Holl, *Atoms for Peace and War 1953–1961*, University of California Press, Berkley, Calif., 1969, Page 289 (Page 316 in the online version).) **244. *Libby, who had been concerned*** (Terrence R. Fehner & F. G. Gosling, *Battlefield of the Cold War: The Nevada Test Site, Atmospheric Nuclear Weapons Testing1951–1963, Atmospheric Nuclear Weapons Testing 1951–1963*, Vol. I, National Nuclear Security Administration, Washington, Page 132 (Page 142 in the online version).)

CHAPTER 32

245. *Conard's second medical resurvey* (*Return of the Rongelapese to their Home Island*, Note by the Secretary, February 6, 1957, AEC 125/30, Page 24 (Page 26 of the online version).) **246.** *Two weeks later, Conard spelled out* (March 28, 1956 letter from Robert A. Conard MD to Dr. Charles Dunham, Director of Biology and Medicine, Atomic Energy Commission. First found in Laura J. Harkewicz, *The Ghost of the Bomb: The Bravo Medical Program, Scientific Uncertainty, and the Legacy of US Cold War Science, 1954–2005*, University of California, San Diego, California, 2010, Page 63 (Page 75 in the online version).) **247.** *An AEC memo, months later, showed* (Gordon M. Dunning (editor), *Radioactive Contamination of Certain Areas of the Pacific Ocean From Nuclear Tests*, US Atomic Energy Agency, Washington, DC, August 1957, Page 48 (Page 52 in the online version).) **248.** *On June 16, 1956, Holmes & Narver Co* (Holmes & Narver Inc., *Report of Repatriation of the Rongelap People*, November 1957, Page 1–3 (Page 7 of the online version).) **251.** *Before disembarking, all the Rongelapese* (Holmes & Narver Inc., *Report of Repatriation of the Rongelap People*, November 1957, Page 1–33 (Page 37 in the online version).) **251.** *In mid-March 1957, Conard had taken four of the exposed Rongelapese* (Robert A. Conard, *Fallout; The Experiences of a Medical Team in the Care of a Marshallese Population Accidentally Exposed to Fallout Radiation*, BNL—4644, Department of Energy, 1992, Page 51 (Page 59 in the online version).) **253.** *Since the whole-body counter was new* (Robert A. Conard MD et al., *Medical Survey of the Rongelap People, March 1958, Four Years After Exposure to Fallout*, Brookhaven National Laboratory, Upton, NY, May 1959, Pages 6–32 (Pages 9–35 in the online version).) **254.** *In his report that year to the AEC* (Robert A. Conard MD et al., *Medical Survey of the Rongelap People, March 1958, Four Years After Exposure to Fallout*, Brookhaven National Laboratory, Upton, NY, May 1959, Page 32.) **255.** *In one, dated June 3, 1959, he said of the Rongelap people* (Joint Committee on Atomic Energy, The Nature of Radioactive Fallout and Its Effects on Man, Special Subcommittee on Radiation, Part I, Washington, DC, May 27, 28, 29, June 3, 1957.) **256.** *He wrote to one of his AEC bosses* (Emory Jerry Jessee, *Radiation Ecologies: Bombs, Bodies, and Environment During the Atmospheric Nuclear Weapons Testing Period, 1942–1965*, Montana State University, Bozeman, Montana, January 2013, Page 256 (Page 276 in the online version).) **256.** *American government concern and interest* (Robert A. Conard, MD et al., *Medical Survey of Rongelap People Five and Six Years After Exposure to Fallout*, Brookhaven National Laboratory, Upton, NY, September 1960, Page 6 (Page 14 in online version).)

CHAPTER 33

258. *Over the years, Dr. Robert A. Conard and pediatricians* (Robert A. Conard MD et al., *Medical Survey of the People of Rongelap and Utirik Islands Nine and Ten years after Exposure to Fallout Radiation* (March 1963 and March 1964), Brookhaven National Laboratory, Upton, NY, September 1960, Pages 20 and 23 (Pages 25 and 28 in the online version).) **259.** *Early in 1963, Rep. Wayne Aspinall* (D-CO) (The Special Committee Concerning Rongelap and Utirik Atolls, *Compensation for the People of Rongelap and Utirik*, Fifth Congress of Micronesia, February 1974, Page 14 and thereafter (page 31 and thereafter in the online version).) **261.** *Conard decided surgery would be* (Robert A. Conard MD et al., *Medical Survey of the People of Rongelap and Utirik Islands Nine and Ten years after Exposure to Fallout Radiation (March 1963 and March 1964)*, Brookhaven National Laboratory, Upton, NY, September 1960, Page 20 (Page 25 in the online version).) **261.** *Unaware of the new medical discoveries* (The Special Committee Concerning Rongelap and Utirik Atolls, *Compensation for the People of Rongelap and Utirik*, Fifth Congress of Micronesia, February 1974, Page

18 and thereafter (Page 34 and thereafter in the online version).) **263. *The 1965 medical survey substantiated*** (Robert A. Conard MD et al., *Medical Survey of the People of Rongelap and Utirik Islands Eleven and Twelve Years after Exposure to Fallout Radiation (March 1965 and March 1966)*, Brookhaven National Laboratory, Upton, NY, April 1967, Page 31 (Page 38 in the online version).) **264. *(fn) The negotiations over Kwajalein took more*** (Congressional Record, June 16, 1966, Senate, Page 13501(Page 35 in the online version).) **266. *Conard's 1967 visit also showed*** (Robert A. Conard, *The 1968 Annual Medical Survey of the Rongelap People Exposed to Fallout in 1954: interim Report*, Brookhaven National Laboratory, Upton, NY, Page 8.) **267. *In 1968, however, two new thyroid cases*** (Robert A. Conard MD et al., *Medical Survey of the People of Rongelap and Utirik Islands Thirteen, Fourteen, and Fifteen Years After Exposure to Fallout Radiation (March 1967, March 1968, March 1969)*, Brookhaven National Laboratory, Upton, NY, June 1970, Page 21 and many thereafter (Page 28 and many thereafter in the online version).)

CHAPTER 34
270. *In early October 1967, a three-day meeting* (DASA Information and Analysis Center, *Second Interdisciplinary Conference on Selected Effects of a General War*, Vol. II, DASIAC Special Report, DASA 2019–2, July 1969, Pages 95–171 contains all the material in this chapter.)

CHAPTER 35
280. *Upon his return to the Marshall Islands, Balos* (The Special Joint Committee Concerning Rongelap and Utirik Atolls, *Medical Aspects of the March 1,1954 Incident: Injury, Examination, and Treatment*, Fifth Congress of Micronesia, First Regular Session, February 1973, Pages 55 and 122 (Pages 79 and 147 in the online version).) **281. *The Balos resolution established*** (The Special Joint Committee Concerning Rongelap and Utirik Atolls, *Medical Aspects of the March 1, 1954 Incident: Injury, Examination, and Treatment*, Fifth Congress of Micronesia, First Regular Session, February 1973, Page marked 5010253 (Page 11 in the online version).) **281. *Then, on the eve of the Japanese doctors' departure*** (The Special Joint Committee Concerning Rongelap and Utirik Atolls, *Medical Aspects of the March 1, 1954 Incident: Injury, Examination, and Treatment*, Fifth Congress of Micronesia, First Regular Session, February 1973, Pages 122–132 (Pages 147–157 in the online version).) **282. *Conard later wrote in his 1992 memory*** (Robert A. Conard, *Fallout; The Experiences of a Medical Team in the Care of a Marshallese Population Accidentally Exposed to Fallout Radiation*, BNL—4644, Department of Energy, 1992, Page 71 (Page 79 in the online version).) **283. *Senator Olympio T. Borja, the Northern Mariannas legislator*** (The Special Joint Committee Concerning Rongelap and Utirik Atolls, *Medical Aspects of the March 1, 1954 Incident: Injury, Examination, and Treatment*, Fifth Congress of Micronesia, First Regular Session, February 1973, Pages 122–132 (Pages 147–157 in the online version).) **284. *However, the most important discovery*** (The Special Joint Committee Concerning Rongelap and Utirik Atolls, *Medical Aspects of the March 1, 1954 Incident: Injury, Examination, and Treatment*, Fifth Congress of Micronesia, First Regular Session, February 1973, Pages 233–234 (Pages 261–262 in the online version).) **285. *On October 2, Lekoj's temperature returned*** (The Special Joint Committee Concerning Rongelap and Utirik Atolls, *Medical Aspects of the March 1, 1954 Incident: Injury, Examination, and Treatment*, Fifth Congress of Micronesia, First Regular Session, February 1973, Page 235 (Page 263 in the online version).) **285. *Then occurred a great coincidence*** (Stewart

Alsop, *Lekoj and the Unusable Weapon*, Newsweek, October 30, 1972. (Stewart Alsop, *Stay of Execution, a Sort of a Memoir*, Lippincott, New York, 1973, Pages 278–281 (Excerpted in *Micronesian Independent*, February 22, 1974, see below).) **286. His twenty-one-year-old brother George** (Robert A. Conard, *Letter to Mr. N. J. Emerson*, October 27, 1972.) **287. He died on November 15, 1972** (Walter Sullivan, "Marshall Islander's Death Tied to Fallout," *New York Times*, November 21, 1972, Page 26.) **287. But troubles were far from over** (Memo: Patient's Death, *1953–1972, Lekoj Anjain*, Page 5051856.) **288. On November 22, Dr. Conard wrote Trust Territory High Commissioner** (Robert A. Conard, *Letter to The Honorable Edward Johnston*, November 22, 1972.)

CHAPTER 36

290. The death of Lekoj stirred anger (Laura J. Harkewicz, *The Ghost of the Bomb: The Bravo Medical Program, Scientific Uncertainty, and the Legacy of US Cold War Science, 1954–2005*, University of California, San Diego, 2010, Page 151 (page 163 in the online version).) **290. Senator Borja's Special Joint Committee of the Micronesian Congress** (The Special Joint Committee Concerning Rongelap and Utirik Atolls, *Medical Aspects of the March 1,1954 Incident: Injury, Examination, and Treatment*, Fifth Congress of Micronesia, First Regular Session, February 1973, Dedication Page 5010263 (Page 21 in the online version).) **291. The panel's report read, "By the** (The Special Joint Committee Concerning Rongelap and Utirik Atolls, *Medical Aspects of the March 1,1954 Incident: Injury, Examination, and Treatment*, Fifth Congress of Micronesia, First Regular Session, February 1973, Page 152 (Page 178 in the online version).) **291. On February 28, 1974, Borja delivered** (The Special Committee Concerning Rongelap and Utirik Atolls, *Compensation for the People of Rongelap and Utirik*, Fifth Congress of Micronesia, February 1974, (Page 11 and thereafter in the online version).) **292. For example, on April 9, 1975, Nelson Anjain** (Nelson Anjain, *Letter to Dr. Robert Conard*, Rongelap Island, April 9, 1975.) **293. In an October 25, 1976, statement** (Robert A. Conard Draft Statement, *To The Chiefs and All The People in Utirik Atoll*, 401326, 10/25/76, Page 3 (Page 3 in the online version).) **293. On October 15, 1977, President Carter signed** (Ann C. Deines et al., *Marshall Islands Chronology 1944 to 1990*, History Associates Incorporated, Rockville, Maryland, 1991, Page 53 (Page 59 in the online version). Public Law 95–134, October 15, 1977, Title I, Section 104.) **293. In December 1977, Brookhaven released** (W. J. Tipton and R. A. Meibaum, *An Aerial Radiological and Photographic Survey of Eleven Atolls and Two Islands within the Northern Marshall Islands*, 1978, Pages 20–25 (Pages 22–27 in the online version).) **294. Dr. Conard retired in 1979, and a year later** (Robert A. Conard, *The 1954 Bikini Atoll Incident: An Update of the Findings in the Marshallese People*, published in The Medical Basis for Radiation Accident Preparedness, Elsiver North Holland, Inc., 1980, Page 57 (Page 3 in the online version).) **295. Relaxation of restrictions** (Tommy F. McCraw, *Department of Energy (DOE) Involvement in the Evacuation of Rongelap Atoll*, Memo to Edward J. Vallario, July 22, 1985, Pages 2–5 (Pages 2–6 in the online version).) **296. In 1984, after the US took no action** (Senator Jeton Anjain, *Statement on Behalf of the Rongelap Local Government and the Rongelap People Living in Exile at Mejato*, House Subcommittee on Insular and International Affairs, Washington, DC, November 16, 1989, Pages 14–15 (Pages 15–16 in online version).)

CHAPTER 37

298. *In 1956, the US drew up an agreement* (Weisgall, Jonathan M., *Petition for Writ of Certiorari, People of Bikini v. United States of America*, Supreme Court, Washington, DC, 2009, Page 19a (Page 51 of the online version).) **299. *The US signed a similar deal in 1956*** (Government Accountability Office, *Enewetak Atoll—Cleaning Up Nuclear Contamination*, PSAD-79-54, Washington, DC, May 8, 1979, Page 12 (Page 20 in the online version).) **299. *However, it was not until August 12, 1968*** (Special to the *New York Times*, "US to Let Bikinians Back on A-Test Isle," *New York Times*, August 13, 1968, Pages 1–2.) **299. *Plans were prepared, but it was understood*** (Allen E. Smith and William E. Moore, *Report of the Radiological Clean-Up of Bikini Atoll*, Environmental Protection Agency, Washington DC, January 1972, Pages 8–9.) **299. *Energy Department survey showed "that food plants*** (P. H. Gudiksen et al., *External Dose Estimates for Future Bikini Atoll Inhabitants*, Lawrence Livermore Laboratory UCRL-51879, Livermore, California, March 1976, Page 21 (Page 24 in the online version).) **299. *Adding to the data was an AEC 1975 urine analyses*** (Jonathan M. Weisgall, *Statement by Legal Counsel to the People of Bikini*, House Natural Resources Committee, February 24, 1994, Page 24 (Page 24 in the online version).) **300. *Months later, following a new radiation study*** (Walter Pincus, "Bikinians Must Quit Island for at Least 30 Years, Hill Told," *Washington Post*, May 23, 1978.) **301. *A Trust Territory official visited in August 1967*** (Martha Smith-Norris, American Cold War Policies and the Enewetakese, Journal of the Canadian Historical Association, Vol. 22, Issue 2, 2011, Paragraph 29 on the website.) **301. *"For Marshall Islanders in general*** (Davor Pevec, *The Marshall Islands Nuclear Claims Tribunal: The Claims of the Enetwetak People*, Denver Journal of International Law and Policy, Vol. 35, No. 1, 2006, Pages 221–222 (Page 3 in the online version).) **301. *Initial radiological studies, finished in April 1973*** (Defense Nuclear Agency, *Cleanup, Rehabilitation, Resettlement of Enewetak Atoll—Marshall Islands*, Washington, DC, April 1975, various pages such as Pages 37 and 109 in the online version.) **302. *By August 1977, it was agreed*** (Government Accountability Agency, *Enewetak Atoll—Cleaning Up Nuclear Contamination*, PSAD-79-54, Washington, DC, May 1979, Page 8 (Page 16 in the online version).) **302. *By 1980, 19,600 new coconut trees*** (Department of Energy, *Enewetak Radiological Support Project Final Report*, NVO-213, Las Vegas, Nevada, September 1982, Page 71 (Page 91 in the online version).) **303. *The 1980 DNA fact sheet claimed*** (Defense Nuclear Agency, *Fact Sheet—Enewetak Operation*, Washington, DC, April 1980, Page 2.) **303. *For a hint, Chernobyl, so far*** (The Chernobyl Forum, *Chernobyl's Legacy: Health, Environmental, and Socio-Economic Impacts*, IAEA Division of Public Information, Vienna, Austria, April 2006, Page 33 footnote: 6 (Page 32 in the online version).) **304. *Beginning in early April 1980, 450 Enewetak*** (John Wilford Noble, "US Settles 75 on Pacific Atoll Evacuated for Bomb Tests in 40's," *New York Times*, April 11, 1977, Pages 1, 8.)

CHAPTER 38

305. *Called the Compact of Free Association* (US Public Law 99-239, *Compact of Free Association Act of 1985*, January 14, 1986.) **306. *It was approved by the Marshallese*** (United States Marines, *Regional Study Oceania Study 3*, 2012, Page 310 (Page 50 in the online version).) **306. *Opponents, led by the Kwajalein Landowners Association*** (Walter Pincus, "Landowners Held in Sit-In On Kwajalein," *Washington Post*, June 22, 1982.) **307. *On October 30, 1986, the new Marshall Islands government*** (Government Accountability Office, *Marshall Islands Status of the Nuclear Claims Trust Fund*, GAO/NSIAD-92-229, Washington, DC, September 1992, Pages 1 and 13 (Pages 3 and

15 in the online version).) **307. *By the year 2000, the Marshall Islands government*** (Thomas Lum et al., *Republic of the Marshall Islands Changed Circumstances Petition to Congress*, CRS Report for Congress, May 16, 2005, Page CRS-1 (Page 5 in the online version).) **308. *In February 2014, Dr. Ashok N. Vaswani*** (Ashok N. Vaswani MD, *Medical Follow-Up in the Marshall Islands: An Overview of Sixty Years of Clinical Experience*, International Radiation and Thyroid Cancer Workshop, Tokyo, Japan, February 21st –23rd, 2014, Slides 20, 23 (Pages 565 and 568 in the online version).) **308. *A 2017 article on "Radiation Effects in the Marshall Islands,"*** (Jacob Robbins, William H. Adams, *Radiation Effects in the Marshall Islands*, Brookhaven National Laboratory, Upton, NY, Page 22 (Page 13 in the online version).) **309. *While they paid US income taxes*** (Pearl Anna McElfish et al., *Effect of US Health Policies on Health Care Access for Marshallese Migrants*, American Journal of Public Health, April 2015.) **309. *As outlined in a 2005 hearing before Congress*** (Prepared Statement of Neal A. Palafox, *Effects of US Nuclear Testing Program on the Marshall Islands*, Senate Committee on Energy and Natural Resources, S. Hrg. 109–178, July 19, 2005.) **309. *A study by the director of the Office*** (Pearl Anna McElfish et al., *Health Beliefs of Marshallese Regarding Type 2 Diabetes*, American Journal of Health Behavior, Vol. 40, Number 2, March 2016, Pages 248–257.) **310. *A draft National Climate Assessment, released in January 2013*** (Ingrid Ahlgreen MS, Seiji Yamada, MD, and Allen Wong, MD, Rising Oceans, *Climate Change, Food Aid, and Human Rights in the Marshall Islands*, Health and Human Rights Journal, Vol. 16, Number 1, June 2014, Pages 69–81.) **310. *"Nowhere else in the world*** (Christopher Loeak, *Future will be like living in a war zone*, Climate Home News, January 4, 2014.) **311. *However, the Kwajalein base rental money*** (Department of Interior, *Second Five-Year Review of the Compact of Free Association, as Amended, Between the Government of the United States and the Republic of the Marshall Islands*, Washington, DC, Page 30 (Page 32 in the online version).) **311. *Family members of three current*** (Julianne M. Walsh, *Imagining the Marshalls: Chiefs, Tradition, and the State on the Fringes of US Empire*, August 2003, Page 221 (Page 240 in the online version).)

CHAPTER 39

312. *The Bikini Atoll Rehabilitation Committee had been placed* (Bikini Atoll Rehabilitation Committee, *Report No. 4: Status March 31, 1986*, Department of Interior, Washington, DC, No. TT-158X08, Page 7 (Page 9 in the online version).) **312. *By 1996 there was an airport system*** (Nicole Baker, *Bikini Atoll Nomination by the Republic of the Marshall Islands for Inscription on The World Heritage List 2010*, January 2009, Page 23 (Page 21 in the online version).) **313. *On April 11, 2006, the people of Bikini*** (Bikini Atoll, US Reparations for Damages: People of Bikini vs. US Lawsuit, Court Filings and Updates 2006–2010.) **313. *The Bikinians, during the Obama administration*** (Department of Interior, *Interior Authorizes Full Decision-Making Power to Bikini Leaders over Annual Budget of the Bikini Resettlement Trust Fund*, Press release, November 28, 2017.) **314. *In August 2017, after President Trump's election*** (US Department of the Interior Assistant Secretary for Insular Areas Doug Domenech, *Interior Authorizes Full Decision-Making Power to Bikini Leaders over Annual Budget of the Bikini Resettlement Trust Fund,* Statement November 28, 2017.) **314. *At a February 2018 hearing before her committee*** (Chairman Lisa Murkowski, *Opening Statement, Legislative Hearing on S. 2182 (Bikini) and S. 2325 (CNMI)*, Senate Committee on Energy and Natural Resources, February 6, 2018.) **315. *Overall, as of 2021, the population of Bikini Atoll descendants*** (Bikini Atoll, *Bikinian Demographics*.) **315. *In 2000, the Enewetak/Ujelang local government signed*** (Lawrence Livermore National Laboratory, *Enewetak Atoll, Post Testing Era* *and Cleanup Activities*, LLNL-WEB-400363.) **316. In a 2008 article, Montana State University professor of anthropology** (Davor Pevec, *The Marshall Islands Nuclear Claims Tribunal: The Claims of the Enewetak People*, Denver Journal of International Law & Policy, March 2008, Page 236 (Page 16 in the online version).) **316. In a 2010 report, the Livermore team** (Department of Energy, *Report on the Status of the Runit Dome in the Marshall Islands*, Report to Congress, June 2020, Page 5 (Page 11 in the online version).) **317. Assisted by the University of South Florida** (Department of the Interior, *Budget Justifications and Performance Information Fiscal Year 2020: Office of Insular Affairs*, Pages 65–66 (Pages 69–70 in the online version).) **317. Many residents use their share** (Kim Wall et al., *The Poison and the Tomb, Part 6: The Big Scrape*, Mashable, February 25, 2018.) **317. For those living on Utirik, a 1999 radioactivity study** (Maveric K. I. L. Abella et al., *Background Gamma Radiation and Soil Activity Measurements in the Northern Marshall Islands*, PNAS, July 30, 2019, Vol. 116, No. 31, Pages 15426–15427 (Pages 2–3 in the online version).) **318. Utirik's Claims Trust Fund stood** (Utirik Atoll Local Government, *Financial Statements and Independent Auditors' Report, Year Ended September 30, 2014*, August 2020, Page 4 (Page 6 of the online version).) **318. In August 2018, Charge d'Affaires of the Japan embassy** (*Utirik Gets Fish Base*, Marshall Islands Journal, August 30, 2018.)

CHAPTER 40

319. The Department of Energy, under the 1986 Compact (Dr. Henry I. Kohn, *Report [of the] Rongelap Reassessment Project, Corrected Edition*, Berkeley, CA, March 1, 1989, Page 3 (Page 6 of the online version).) **319. At a 1991 Senate hearing, Peter Oliver** (Senate Committee on Energy and Natural Resources, *Resettlement of Rongelap Atoll, Republic of the Marshall Islands*, S. Hrg. 102–316, September 19, 1991, Page 22 (Page 26 in the online version).) **320. The National Academy of Sciences report, released in 1994** (National Research Council, (US) Committee on Radiological Safety in the Marshall Islands, *Radiological Assessments for Resettlement of Rongelap in the Republic of the Marshall Islands*, National Academies Press, Washington, DC, 1994, Executive Summary.) **320. On September 19, 1996, an agreement** (Office of Insular Affairs, *A Report on the State of the Islands 1999*, US Department of the Interior, Washington, DC, Page 8 (Page 13 of the online version).) **321. In 1998, the Rongelap government contracted** (Marshall Islands Monitor, *Helping Bridge the Gap in Support of Rongelap Atoll Resettlement*, Lawrence Livermore National Laboratory, Vol. 1, No. 4, December 2009, Pages 2 and 5.) **321. The general aim of the Rongelap garden projects** (Glenn Podonsky, *Statement Before the House Subcommittee on Asia, the Pacific and Global Environment*, May 20, 2010, Page 6 (Same in the online version).) **322. In response, the Rongelap local government council** (Nikolao Pula, Statement before Director, Office of Insular Affairs, US Department of the Interior, Statement Before the House Subcommittee on Asia, the Pacific and Global *Environment*, May 20, 2010.) **322. Late in 2011, a number of US Interior Department officials** (Marshall Islands Guide, *Rongelap Atoll, Rongelap: Big Hole*.) **323. In 2016, the Marshall Islands government released** (The Government of the Marshall Islands, State of the Environment Report 2016, SPREP 2016, Apia, Samoa, Page 136 (Page 137 in the online version).) **323. Deloitte & Touche LLC audited** (Rongelap Atoll Local Government, *Financial Statements and Independent Auditors' Report, September 30, 2015*, Pages 6, 19,20 (Pages 8, 21, 22 in the online version).) **323. The Rongelap Local Government invested $300,000** (Emily Yehle, Aquaculture Buoys Islands Devastated by Bomb Tests, E&E News, April 23, 2015.) **324. In April 2018, at the Asia World Expo** (Mackenzie Smith, Remote Marshall Islands Atoll Plans to become the "Next Hong Kong," RNZ, September 21, 2018.) **324. *Matayoshi's plan had not gotten support*** *(Nitijela Backs RASAR Plan, Marshall Islands Journal,* April 2, 2020.) **324. *In January 2013, Pete and Daria Friday sailed*** (When I first researched and wrote this section, the Fridays had written about their time on Rongelap on their own website. It was at: https://downtimecat .blogspot.com/2013/02/rongelap-atoll.html. That no longer exists. But the Fridays wrote another version with some of the same material on a different website: *The Marshall Islands Compendium, A Compilation of Guidebook References and Cruising Reports,* Page with 6.8 Rongelap (Page 111 in the online version).)

INDEX

ABOUT THE AUTHOR

▪ **WALTER PINCUS** reported on intelligence, defense, and foreign policy for the *Washington Post* from 1966 through 2015. He was among *Post* reporters awarded the 2002 Pulitzer Prize for national reporting. Among many other honors were the 1977 George Polk Award for articles exposing the neutron warhead, a 1981 Emmy for writing a CBS documentary on strategic nuclear weapons, and the 2010 Arthur Ross Award from the American Academy for Diplomacy for columns on foreign policy. Currently a contributing senior national security columnist for the *Cipher Brief*, he lives in Washington, DC.